Spacecraft Formation Flying: Dynamics, control and navigation

Butterworth-Heinemann is an imprint of Elsevier
Boulevard, Langford Lane, Kidlington, Oxford OX5 1GB, UK
Corporate Drive, Suite 400, Burlington, MA 01803, USA

First edition 2010

British Library Cataloguing in Publication Data
A catalogue record for this book is available from the British Library

ISBN: 978-0-75-068533-7

For information on all Butterworth-Heinemann publications
visit our web site at books.elsevier.com

Transferred to Digital Printing 2010

Spacecraft Forma
Flying: Dynamics, (
and navigation

Kyle T. Alfriend

Srinivas R. Vadali

Pini Gurfil

Jonathan P. How

Louis S. Breger

Astrodynamics Series Editor: Pini Gurfil

ELSEVIER

AMSTERDAM • BOSTON • HEIDELBERG • LONDON
NEW YORK • OXFORD • PARIS • SAN DIEGO
SAN FRANCISCO • SINGAPORE • SYDNEY • TOKYO
Butterworth-Heinemann is an imprint of Elsevier

Dedication

I want to dedicate this book to my wife and best friend of many years, Bonnie Alfriend. Without her support and encouragement, my research that led to my portion of this book would not have been possible. I also want to dedicate it to my two children, Kyle and Kim, and my four grandsons, Brandon, Travis, Tyler and Erik; they enrich my life.

Terry Alfriend

To my wife Sharada, for her ever-present love; to my parents Venkat Rao and Sita for their guidance; to Vivek and Meera for the motivation.

Srinivas Vadali

To my wife, Michal, for the road; to my parents, Arie and Sara, for the wisdom; to my sons, Eytam and Oshri-Yahel, for the meaning.

Pini Gurfil

To my wife Carolyn and daughter Amelia.

Jonathan How

To my family, friends, and colleagues – your support, advice, and patience made this work a pleasure.

Louis Breger

About the authors

Dr. Terry Alfriend is currently the TEES Distinguished Research Chair Professor of Aerospace Engineering at Texas A&M University, where he was the Department Head for 1997–2001. He is also a Visiting Professor at the Naval Postgraduate School. He is a member of the NAE, a Fellow of the AAS and AIAA and a member of the International Academy of Astronautics. He has served as an Associate Editor and Editor-in-Chief of both the AAS Journal of the Astronautical Sciences and the AIAA Journal of Guidance, Control and Dynamics. He is also the recipient of the AAS Dirk Brouwer Award and the AIAA Mechanics and Control of Flight Award. He is a member of the Air University Board of Visitors. In 2007 he was selected to give the Von Karman lecture at the 2008 Israel Aerospace Sciences Conference. In 2005 he, along with six others from the US and Russia, received the AAAS International Scientific Cooperation Award. His BS and PhD are from Virginia Tech in Engineering Mechanics and he received his MS in Applied Mechanics from Stanford University. His research interests are in space surveillance, astrodynamics, satellite attitude dynamics and control and spacecraft design.

Dr. Srinivas R. Vadali has been a member of the faculty at Texas A&M University since 1986, where he is currently the Stewart & Stevenson-I Endowed Professor of Aerospace Engineering. He served on the faculty of Iowa State University from 1983 till 1985. He received his B.Sc. (Hons) in Mechanical Engineering from Sambalpur University, India, his M.E. (Distinction) in Aeronautical Engineering from the Indian Institute of Science, and his Ph.D. in Engineering Mechanics from Virginia Tech. He is a Fellow of the AAS and an AIAA Associate Fellow. He has served as an associate editor of the AIAA Journal of Guidance, Control, and Dynamics.

Dr. Pini Gurfil is an Associate Professor of Aerospace Engineering at the Technion - Israel Institute of Technology. He received his Ph.D. in Aerospace Engineering from the Technion in 2000. From 2000 to 2003, he was with Princeton University's Department of Mechanical and Aerospace Engineering. In September 2003, Dr. Gurfil joined the Faculty of Aerospace Engineering at the Technion, where he founded the Distributed Space Systems Laboratory. Dr. Gurfil has been conducting research in Astrodynamics, Distributed Space Systems, Vision-Based Navigation and Optimization. He has published over 140 journal and conference articles in these areas. Dr. Gurfil is the Editor of the Elsevier Astrodynamics Book Series, Editorial Board Member of four aerospace engineering journals, including the AIAA Journal of Guidance, Control and Dynamics, Member of the AIAA GN&C and AAS Spaceflight Mechanics Technical Committees, Associate Fellow of AIAA, Member of

AAS, Member of Sigma Xi, and an Affiliate Member of the Division on Dynamical Astronomy.

Dr. Jonathan P. How is a Professor in the Department of Aeronautics and Astronautics at the Massachusetts Institute of Technology (tenured in 2003, promoted to Full Professor in 2007). He received a B.A.Sc. from the University of Toronto in 1987 and his S.M. and Ph.D. in Aeronautics and Astronautics from MIT in 1990 and 1993, respectively. He then studied for two years at MIT as a postdoctoral associate for the Middeck Active Control Experiment (MACE) that flew on-board the Space Shuttle Endeavour in March 1995. Prior to joining MIT in 2000, he was an Assistant Professor in the Department of Aeronautics and Astronautics at Stanford University. He has graduated a total of 28 Ph.D. students while at MIT and Stanford University on topics related to GPS navigation, multi-vehicle planning, and robust/hybrid control. He has published more than 200 articles in technical proceedings, and 59 papers in technical journals. Current research interests include: (1) Design and implementation of distributed robust planning algorithms to coordinate multiple autonomous vehicles in dynamic uncertain environments; (2) Spacecraft navigation, control, and autonomy, including GPS sensing for formation-flying vehicles; and (3) Adaptive flight control to enable autonomous agile flight and aerobatics. Professor How was the planning and control lead for the MIT DARPA Urban Challenge team that placed fourth in the recent race at Victorville, CA. He was the recipient of the 2002 Institute of Navigation Burka Award, a recipient of a Boeing Special Invention award in 2008, is the Raymond L. Bisplinghoff Fellow for MIT Aeronautics/Astronautics Department, an Associate Fellow of AIAA, and a senior member of IEEE.

Dr. Louis S. Breger is a spacecraft control systems and guidance engineer, specializing in formation flying and rendezvous operations. He obtained his S.B., S.M., and Ph.D. degrees at the Massachusetts Institute of Technology. Since co-authoring *Spacecraft Formation Flying: Dynamics, Control and Navigation*, he has begun work at the Charles Stark Draper Laboratory in Cambridge, Massachusetts.

Table of Contents

ix

Foreword

Spacecraft Formation Flying: Dynamics, Control and Navigation provides a comprehensive treatment of the subject in a widely accessible fashion. The text fills an important void and the authors have utilized the advantage they enjoy, being among the leaders in researching this subject, to write an authoritative exposition. The result is an up-to-date text that will be broadly useful to academics and professionals in governmental and industrial research. The focus is on spacecraft formations in low Earth orbit, with a detailed discussion of modeling, control, and navigation; with consideration of both translation and attitude relative motions. While this is an integrative sub-field in astrodynamics, and guidance, navigation and control, the number of unique challenges and contributions addressed make it clear that a text dealing with formation flying is needed. This book fulfils this need. Chapter 1 provides an introduction and defines the several kinds of formations: orbit tracking, leader/follower, virtual structures, and swarming. Also considered is an introduction to the drivers that dictate fuel consumption such as atmospheric drag and station-keeping. Chapter 2 provides a concise introduction to astrodynamics, considering both the unperturbed (Keplerian) motion as well as the method of variation of parameters. The perturbations of main focus are the zonal harmonics due to Earth oblateness. Chapter 3 presents fundamental ideas from mechanics, optimization, control, and estimation theory. This chapter develops the Lagrangian and Hamiltonian generalized formalisms for establishing equations of motion, and also discusses Lyapunov stability theory and its use in constructing stable control laws. Also presented are recent contributions to Kalman Filter theory with emphasis on the several alternative forms for the Unscented Kalman Filter that addresses the critical issue of maintaining covariance estimates that are reflective of nonlinear dynamics in the error propagation between measurement updates.

Chapters 4 and 5 develop the nonlinear and linearized dynamics of relative spacecraft motion, with emphasis on judicious coordinate choices, accounting for oblateness perturbations, and of central importance in Chapter 5, developing the several forms of the relative motion state transition matrix central to estimation and control. Chapters 6–8 focus on dynamical modeling of relative motion using orbit elements and the mitigation of the effects of perturbations. Chapter 6 considers short time motion with time series approximations whereas Chapter 7 utilizes generalizations of classical perturbation methods to examine **xiii** solutions for the J_2-perturbed relative motion. Chapter 8 introduces additional methods for analyzing J_2-perturbed relative motion, including selection of judicious initial conditions and techniques for the minimization of fuel consumption and achieving balanced fuel consumption. Chapter 9 addresses the problems that arise in rotation-translational coupling. Chapter 10 is devoted to several aspects of formation control, including control based on averaging, impulsive control based on variation of parameters, and various special cases that provide useful insights. Chapter 11 provides practical insight into how to implement impulsive maneuver commands. Chapter 12 is devoted to measurement models and Kalman Filter implementations for various navigation algorithms. Several examples are included. Chapter 13 is focused on approaches to achieve high fidelity formation flight simulations, and example results are presented. Finally Chapter 14 provides a summary and perspective looking to future developments in this field.

The overall treatment is at a level such that any individual with a senior undergraduate level background in orbital mechanics and control systems can access the majority of the material, and apply it efficiently in mission analysis or design. The level of abstraction is at an intermediate level with frequent appeal to physical and geometric reasoning to motivate the derivations and interpret the results obtained. For specialized aspects of text such as orbit perturbation theory, Kalman Filter and control theory, and the methods of classical mechanics, of course, additional reading of classical works and recent literature will be needed to supplement the treatments presented. The central contribution of this book is to have an authoritative, integrated presentation of spacecraft formation dynamics, navigation and control by these leaders in the field. I am delighted to recommend this book with conviction.

John L. Junkins

Texas A&M University

Preface

This book constitutes the second volume in the *Elsevier Astrodynamics Book Series*, and is devoted to spacecraft formation flying. As opposed to Volume I in the series, titled *Modern Astrodynamics*, the current volume is a *textbook* rather than an edited volume.

Spacecraft formation flying is a vast subject, which can be studied from many different viewpoints. This book develops the theory from an astrodynamical viewpoint, emphasizing modeling, control and navigation of formation flying satellites on Earth orbits – mostly low Earth orbit (LEO) missions. Spacecraft formation flying in deep-space and on libration-point orbits is not covered in the current text; this subject deserves a stand-alone book.

We have attempted to create a coherent exposition of spacecraft relative motion, both in the unperturbed and perturbed settings, to discuss the main control approaches for regulating relative satellite dynamics, using both impulsive and continuous maneuvers, and to present the main constituents required for relative navigation. We also discuss relative attitude dynamics, and the coupling between translational and rotational dynamics.

This book is intended for graduate students and academic researchers, for university professors teaching courses on distributed space systems, aerospace and mechanical engineers, mathematicians, astronomers, and astrophysicists. The book assumes prior knowledge of basic astrodynamics, attitude dynamics and control theory. Nevertheless, we provide introductory chapters dedicated to orbital mechanics, perturbation methods, control and estimation. To illustrate the main developments, we solve a number of examples in each chapter, and provide a sample simulation of a formation flying mission involving high-fidelity modeling, control and relative navigation.

Notation

Relative motion between satellites or spacecraft involves a number of coordinate systems and a few notational conventions. Usually, we will refer to one spacecraft as the *chief*, and to the other one as the *deputy*. An alternative terminology designates one of the satellites as *leader*, and the other as *follower*. The chief/deputy notation implies a more general setup, in which the motion of a satellite is modeled with respect to a potentially *virtual* or *fictitious* satellite, or, in other words, with respect to a *reference point* or a *reference orbit*. The leader/follower notation is reserved for in-line motion or for particular formation control and/or navigation problems. In the context of orbital rendezvous, one of the satellites is usually referred to as *target*, and the other is called *chaser* or *pursuer*. While usually only a single satellite is designated as the chief, there could be many deputies. We will usually denote quantities referred to the chief, leader or target by $(\cdot)_0$, and those related to the deputy, follower or chaser by $(\cdot)_1$.

Exceptions are made for denoting the epoch t_0, and the mean anomaly at epoch of the chief, $(M_0)_0$. We assume that $t_0 = 0$ frequently, and denote the initial condition of some time-dependent vector $\mathbf{w}(t)$ by $\mathbf{w}(0)$. During the discussion involving a single satellite, as in Chapter 2, $(M_0)_0$ is replaced by M_0, but in subsequent chapters dealing with multiple satellites, we designate M_0 as the mean anomaly of the chief. In some of the expressions for the relative motion variables, which are functions of the chief's orbital elements, the subscript 0 is dropped to simplify the presentation of the equations. Another exception pertains to perturbed variables. For example, we designate the zero-order and first-order Hamiltonians, respectively, by $\mathcal{H}_0$ and $\mathcal{H}_1$.

Throughout this book, we will resolve some vector $\mathbf{w}$ in a particular coordinate system, $\mathscr{A}$, by writing $[\mathbf{w}]_{\mathscr{A}}$. This notation will be followed only when a number of coordinate systems are involved in the analysis, and will be omitted when the context permits no ambiguities. Unless otherwise stated, we assume that all vectors are column vectors. The time derivative of $\mathbf{w}$ in frame $\mathscr{A}$ will be denoted by $\mathrm{d}^{\mathscr{A}}\mathbf{w}/\mathrm{d}t$.

To denote the p-norm of $\mathbf{w}$, we will use the notation $\|\mathbf{w}\|_p$. By default, $\|\mathbf{w}\| \equiv \|\mathbf{w}\|_2$, viz. the Euclidean norm.

We will use calligraphic fonts to denote quantities measured in units of energy per unit mass; for example, $\mathcal{E}$ will denote energy, and $\mathcal{H}$ will denote a Hamiltonian. Calligraphic fonts will also be used for denoting cost functions (arising in optimization problems throughout the book), which are most often related to energy measures. In a frame of reference centered at one of the orbiting satellites, we distinguish between *radial*, *along-track* and *cross-track* components of the position vector. The latter component is also referred to as *out-of-plane* or *normal*.

Depending upon context, we will use $\bar{x}$ to denote a non-dimensional (normalized) x, and $\bar{x}'$ to denote differentiation of $\bar{x}$ with respect to a non-dimensional variable; alternatively, $\overline{\infty}$ will denote the mean orbital element ∞; for example, $\bar{e}$ is the mean value of the eccentricity, e.

Finally, we elaborate below the main abbreviations used throughout this book:

$$
\begin{aligned}
\mathrm{BL} &= \text{Baseline} \\
\mathrm{CDGPS} &= \text{Carrier-Phase Differential GPS} \\
\mathrm{CLF} &= \text{Control Lyapunov Function} \\
\mathrm{CM} &= \text{Center-of-Mass} \\
\mathrm{CRD} &= \text{Cartesian Rectangular Dextral} \\
\mathrm{CW} &= \text{Clohessy–Wiltshire}
\end{aligned}
$$

DCM = Directional Cosines Matrix
DLQR = Discrete-Time Linear Quadratic Regulator
DOF = Degrees-of-Freedom
ECEF = Earth-Centered, Earth-Fixed
ECI = Earth-Centered Inertial
EKF = Extended Kalman Filter
GA = Gim–Alfriend
GCO = General Circular Orbit
GM = Geometric Method
GVE = Gauss' Variational Equations
IC = Initial Condition
KF = Kalman Filter
LCJ = Lee–Cochran–Jo
LON = Line-of-Nodes
LP = Linear Programming
LPE = Lagrange's Planetary Equations
LQR = Linear Quadratic Regulator
LTI = Linear Time-Invariant
LVLH = Local-Vertical, Local–Horizontal
MPC = Model Predicative Control
PCO = Projected Circular Orbit
RAAN = Right Ascension of the Ascending Node
RHS = Right-Hand Side
STM = State Transition Matrix
TH = Tschauner–Hempel
UKF = Unscented Kalman Filter
YA = Yamanaka–Ankersen

Book Organization

This book's organization reflects our commitment to create a complete, stand-alone text. The book is divided into 14 chapters. Chapter 1 is an introduction, wherein we raise the fundamental questions, issues, and approaches relating to spacecraft formation flying; this chapter therefore serves to provide a broad overview of the topics covered in the book. Chapter 2 is devoted to some notions in orbital mechanics. In this chapter, we concisely explain some key coordinate systems and discuss the Keplerian two-body problem. Chapter 3 contains diverse topics in mechanics, optimization, control and estimation, including Lagrangian and Hamiltonian mechanics, a discussion of static optimization, feedback control, and filtering methods, which are all pertinent to the devolvement of relative spacecraft control, measurement and navigation methods. Chapters 1–3 are all introductory chapters, constituting the first part of the book, intended to create a foundation on which we construct methods and tools for the analysis and design of spacecraft formation flying.

Chapter 4 is the first of six chapters devoted to modeling relative spacecraft dynamics. This chapter presents commonly used nonlinear models for relative motion, for both perturbed and unperturbed motion. Chapter 5 presents a variety of linear differential equations for modeling relative motion under the two-body assumptions, including satellite rendezvous. The focus in this chapter is on the physical-coordinate description of relative motion; motion modeling using orbital elements is treated extensively in Chapter 6, where we present an important modification of relative spacecraft motion modeling – the use of orbital elements as constants of motion instead of the initial conditions. Chapter 7 builds on Chapter 6 and extends its results to include orbital perturbations using a number of approaches; methods to mitigate the effect of perturbations on relative motion is the subject of Chapter 8. Chapter 9, which concludes the second part of the book, is focused on the coupling between rotation and translation; it provides equations for relative spacecraft translation as well as translational equations taking into account kinematic coupling.

Chapter 10 opens the third part of the book, which is focused on the development of spacecraft formation controllers and their practical implementation. Chapter 10 contains a brief discussion of methods for continuous and impulsive formation control: establishment, maintenance, and reconfiguration. Chapter 11 is concerned with the impact of the maneuver acceleration implementation error on the performance of spacecraft formation flying, rendezvous and docking. It also explores how accelerometers can be used to improve performance by providing accurate measurements of the applied maneuver acceleration.

The final part of this book is devoted to relative navigation and high-fidelity simulation of spacecraft formations, incorporating actual relative navigation modeling. Chapter 12 discusses an application of various estimation algorithms to the relative orbital navigation problem. Chapter 13 illustrates most of the previously introduced material by performing a series of nonlinear simulations of portions of a formation flying reference mission.

Chapter 14 discusses some future prospects. It raises an important question: given the advantages that can be gained by flying spacecraft in formations, why are there no spacecraft formations on orbit currently?

Acknowledgments

Much of the impetus for research on perturbation mitigation, control, and relative navigation for formation flying began as a means to address the biggest challenge facing the TechSat21 program proposed by AFRL more than ten years ago: how to keep small satellites in close proximity in LEO with very little fuel. TechSat21 was the first proposed mission involving satellites in close proximity and in orbits with different inclinations. This challenge has mushroomed into a body of literature consisting of more than a hundred journal articles and doctoral dissertations. Innovative solutions for formation maintenance, fuel balancing, and relative navigation developed so far can directly or indirectly be attributed to the technical challenges set by the TechSat21 mission.

The authors were supported by grants from AFOSR, AFRL, NASA GSFC, the Galileo Supervisory Authority, the European Sixth Framework Program, Israel Ministry of Science and Technology, the Asher Space Research Institute, the Gutwirth Fund for the Promotion of Research, the Rosenthal Fund for

Aerospace Research, the National Science Foundation Graduate Research Fellowship Program, NASA Grants #NAG3-2839, #NAG5-10440, #NAG5-11349, #NCC5-704, and #NCC5-729, the National Defense Science and Engineering Graduate Fellowship, administered by the American Society for Engineering Education, and Mitsubishi Electric Corporation.

The authors wish to acknowledge Alok Das (AFRL), Mark Jacobs (AFOSR), Jesse Leitner (GSFC), Russell Carpenter (GSFC), and Richard Burns (GSFC), as well as the following individuals, without whose contributions this manuscript would not have formed: John L. Junkins, Oliver Montenbruck, Sesha S. Vaddi, Prasenjit Sengupta, Hui Yan, Hanspeter Schaub, Dario Izzo, Marco Sabatini, Dong-Woo Gim, Jeremy Kasdin, David Mishne, Konstantin V. Kholshevnikov, Shay Segal and Egemen Kolemen.

In particular, we wish to acknowledge the instrumental contributions of Matthew Jeffrey to Chapter 11 and Megan Mitchell to Chapter 12. We also want to thank Kim Luu, whose contribution to the preparation of the Appendices was invaluable.

Chapter | one

Introduction

The road became a channel running flocks
Of glossy birds like ripples over rocks.

Robert Frost (1874–1963)

In this chapter we introduce the readers to some of the fundamental questions, issues, and approaches related to spacecraft formation flying. In a sense, this chapter serves to provide a broad overview of the topics covered in this book. We also list some representative formation flying missions planned by NASA and ESA.

1.1 WHAT IS SPACECRAFT FORMATION FLYING?

The definition of *spacecraft formation flying* is not very precise or universally agreed upon. Most of the space community, however, would agree to the following definition, proposed by NASA's Goddard Space Flight Center (GSFC):

> The tracking or maintenance of a desired relative separation, orientation or position between or among spacecraft.

Formation flying spacecraft are therefore a particular case of a more general category, termed *distributed space systems*, defined by NASA GSFC as follows:

> An end-to-end system including two or more space vehicles and a cooperative infrastructure for science measurement, data acquisition, processing, analysis and distribution.

NASA GSFC proposed a number of other important distinctions: A *constellation* is a collection of space vehicles that constitutes the space element of a distributed space system; *virtual aperture* is an effective aperture generated by physically independent spacecraft; and *virtual platform* is a spatially distributed network of individual vehicles collaborating as a single functional unit, and exhibiting a common system-wide capability to accomplish a shared objective. **1**

Spacecraft Formation Flying; ISBN: 9780750685337

1.2 COORDINATION APPROACHES

1.2.1 Orbit tracking

Single satellite missions are typically designed to occupy a particular orbit (as opposed to a particular location on the orbit) about which periodic stationkeeping maneuvers are conducted. This same approach can be extended to formations in which each satellite in the formation is controlled to a particular predetermined desired orbit. This approach is attractive from the perspective of allowing little or no cooperation or coordination between the spacecraft; however it is expected to require more maneuvers than methods that exploit the coupled nature of mission objectives.

1.2.2 Leader/Follower

In leader/follower coordination methods one *leader* spacecraft is controlled to a reference orbit and the other *follower* spacecraft in the formation control their relative states to that leader. This approach allows traditional periodic maneuvers to keep the leader in a desired orbit or ground-track while the remaining satellites in the formation control their relative state with respect to the leader. This approach has the advantage that it allows most satellites in the formation to follow the natural dynamics of the absolute orbit of the leader, while only performing regular automatic control on the relative states of the formation. The principal disadvantage of leader/follower is that the leader spacecraft is by definition at its correct state and will not require as much fuel use as the followers. Fuel use can be balanced among the satellites by periodically interchanging the designations of the leader and follower [5].

1.2.3 Virtual structure

Virtual structure [6] and virtual center [5] approaches fit a set of desired states to a formation in a way that minimizes the overall state error of the formation. The chief advantage of this approach over the leader/follower method is that state error will pertain to all the spacecraft in the formation. Adding fuel-weighting, as in Ref. [7], allows fuel use to be balanced methodically. The implementation of a virtual structure approach requires coordinated inter-spacecraft communication. An approach for coordination, formation maintenance and fuel balancing, developed from the fundamental astrodynamics principles of secular perturbations is presented in Ref. [156], and also discussed in Chapters 8 and 10 of this book.

1.2.4 Swarming

A number of researchers [8,9] have proposed simple heuristic control laws for arranging arbitrarily large numbers of vehicles into regular arrangements based on local information. These swarming methods have the advantage that they easily scale to large numbers of vehicles without incurring large communication or computation burdens. However, they are typically not fuel-optimal and rarely include provisions guaranteeing collision avoidance.

1.3 FUEL-USE DRIVERS

Minimizing fuel use is critical in any space mission, because fuel is expensive to launch into orbit and non-replenishable. Fuel use is expected to be a primary concern in the design of any spacecraft formation controller, because the task of keeping a formation from drifting apart and achieving science requirements is expected to require significantly more fuel than stationkeeping a single spacecraft. This section describes each of the expected drivers of fuel use in a spacecraft formation and details their expected impact on several types of control approaches.

1.3.1 Mission requirements

Several types of formation flying missions have been proposed: Interferometry, passive-aperture radar, and repeat ground-track observations. Each of these missions requires varying degrees of control to achieve observation objectives. Proposed interferometry missions require control of the spacecraft relative position with fine precision and to specific geometric templates. When the required geometry of a formation dictates that natural orbital perturbations be cancelled, significant amounts of fuel can be required to achieve mission objectives.

1.3.2 Initial conditions

The initial conditions (ICs), or the initial states, of the spacecraft in a formation dictate the future coasting behavior of the spacecraft formation. Choosing the ICs to make the future coasting behavior meet mission science requirements and minimize drift between spacecraft in the formation should minimize the need for future corrective maneuvering. In practice, it is rarely possible to choose initial conditions that completely eliminate inter-spacecraft drift, because of perturbations arising from Earth oblateness and drag. However, the ICs can be picked in such a way as to minimize the need for corrective action.

1.3.3 Navigation uncertainty

A significant source of fuel use is expected to be navigation uncertainty. In monolithic satellite systems, navigation would typically be handled by a series of ground-based observations from which a guidance maneuver would be designed. For a formation, control bandwidth is expected to be sufficiently high that on-orbit navigation will be required. Errors in navigation will lead to errors in state corrections, guaranteeing that some level of drift will always be present in the system, even if initial conditions are chosen that should completely eliminate drift. Correcting the drift in the system will require fuel.

1.3.4 Atmospheric drag

Although atmospheric drag can be considered in the dynamic models of the formation [10,11], in practice the effects of drag are partly stochastic. This is because the attitude estimation errors of each spacecraft in the formation will couple into the accelerations produced by drag for almost all spacecraft shapes. The drag equation [12]

$$\mathbf{a}_{\text{drag}} = -\frac{1}{2}\frac{c_D A}{m}\rho v^2 \hat{\mathbf{v}} = -\frac{1}{2}B^* \rho v^2 \hat{\mathbf{v}} \tag{1.1}$$

where m is the mass, c_D is the drag coefficient, ρ is the atmospheric density, $\hat{\mathbf{v}}$ is a unit vector lying along the velocity vector relative to the atmosphere and v is the magnitude of the velocity relative to the atmosphere, shows that the acceleration due to aerodynamic drag, $\mathbf{a}_{\text{drag}}$, is directly related to the cross-sectional area of the satellite A, which is determined by the attitude of the spacecraft, as well as to the *ballistic coefficient*, denoted by $B^* = c_D A/m$. An imperfect estimate of the attitude of the spacecraft will lead to errors in any trajectory predictions or corrections based on expected future drag effects.

1.3.5 Thrusting errors

Another source of uncertainty in any spacecraft mission is thrusting error. Thrusting errors arise from attitude estimation errors (i.e., thrusting in an unintended direction) and hardware design. A typical near-impulsive thruster can be modeled as a source of thrust within some percentage tolerance of a nominal force in a direction which is aimed correctly to within the knowledge of the attitude estimator. Both sources of thrusting errors couple to ensure that any attempt to cancel real drift resulting from system dynamics or perceived drift produced by navigation error will, in fact, also result in an additional source of drift, which will ultimately need to be corrected.

1.3.6 Dynamical process noise

In addition to thrusting errors, there will also be uncertainty in the dynamical model. This dynamical *process noise* will be a function of the fidelity of the dynamical model used for the controller implementation. Although the space environment is very well known, approximate linear models are often used because of their computational simplicity and ease of implementation. These models usually account for the effects of Earth's gravity modeled as a point mass, but may also take into account higher-order gravity terms (e.g., J_2). In addition, some linear models also include terms for drag [10]. However, in practice, dynamical models used for computing the control must always truncate or ignore some perturbations in order to be feasible. Typically, at least some of the higher-order gravitational terms, advanced atmospheric models, solar wind, and third-body effects of the Moon, the Sun and planets are excluded. These exclusions combine to form the dynamical model uncertainty.

1.4 CONTROL OF SPACECRAFT FORMATIONS

Formation flying spacecraft pose several control challenges beyond the problem of controlling a monolithic spacecraft or a constellation [13–16]. In a typical single-spacecraft mission, the term *control* would refer to maintaining and altering the attitude of the spacecraft, whereas *guidance* would encompass the maintenance and manipulation of the trajectory on the scale of an orbit. After launch and initial correcting maneuvers, adjusting a spacecraft's orbit would be an occasional activity planned from the ground. A constellation of spacecraft is operated in much the same way [17,18], because the constituent spacecraft operate in widely-spaced orbits (or spaced phase angles while in the same orbit), with short-term decoupled performance objectives. A formation of spacecraft is defined by the need for inter-satellite control cooperation [19]. The satellites in a formation are typically represented as sharing a common *reference orbit*, that is, being close enough in terms of their position and velocity in a central body frame that their long-term, large-scale motion can be modeled using the dynamics of a single orbit. This proximity, while typical for rendezvous missions, is uncommon for satellite missions where there is an expectation for long-term collision-free operation. Formation flying is expected to require a level of autonomous onboard guidance that in most applications would be classified as automatic control [19–22].

Many formation control approaches have been presented in the literature [6,14,19,23–30]. These papers cover a variety of approaches, including proportional-derivative (PD), linear quadratic regulation (LQR), linear matrix inequalities (LMI), nonlinear, Lyapunov, impulsive, rapidly exploring random trees (RRT), and model predictive. Typically, it is assumed that a formation is initialized to a stable orbit and deviations caused by perturbations such as differential drag and/or differential J_2 must be corrected. Some approaches, such as Lyapunov controllers and PD controllers [28,31], require that control be applied continuously, a strategy both prone to high fuel use and difficult to implement when thrusting requires attitude adjustment. Other approaches, such as the impulsive thrusting scheme introduced in Ref. [32], require spacecraft to thrust at previously-specified times and directions in the orbit, ensuring many potential maneuvers will not be fuel-optimal.

1.5 CONTROL APPROACHES

Many approaches for controlling formations exist in the formation flying literature. These approaches consider many different aspects of the formation flying spacecraft problem, including relative spacecraft control, coupled mission objectives, and global fuel minimization. The following subsections review some of the most commonly proposed methods of formation control.

1.5.1 State transition matrix inversion

One approach to spacecraft formation control is the use of a Battin matrix method [14,33]. This approach calculates the velocity change (Δv) required at the current time to achieve some desired state in the future. Similar approaches have been in the literature for decades [33] for use in correcting orbits for monolithic spacecraft.

1.5.2 Impulsive/Heuristic

Classic orbital maneuver Control for a formation of spacecraft can be treated as individual control of many individual spacecraft using traditional methods. Ref. [33] contains mathematical descriptions of the Lambert problem for finding an orbit connecting two states in a specified time, as well as methods for finding optimal one- and two-impulse burns to transfer orbits. These approaches have the advantage of low risk, because the methods have decades of established usage in the field. However, the methods will still need to be modified from their current usage to account for issues of collision avoidance and drift prevention.

Four-impulse method The algorithm in Ref. [34] can be summarized in four steps to be taken over the course of an orbit. When the argument of latitude, θ, is 0 or π radians, implement a velocity change (impulsive thrust), $\Delta v_{h_i} = [h/(r \cos \theta)]\delta i$, in the cross-track direction of an LVLH frame centered on the spacecraft to cancel the inclination error δi, with r being the orbital radius and h being the angular momentum. When the argument of latitude, θ, is $\pi/2$ radians, implement a velocity change, $\Delta v_{h_\Omega} = [h \sin i/(r \sin \theta)]\delta \Omega$ in the cross-track direction to cancel the ascending node error, $\delta \Omega$. At perigee and apogee, implement Δv_{r_p} and Δv_{r_a}, respectively, in the radial direction to cancel the argument of perigee and mean anomaly errors, denoted by $\delta \omega$ and δM, respectively:

$$\Delta v_{r_p} = -\frac{na}{4} \left(\frac{(1+e)^2}{\eta}(\delta \omega + \delta \Omega \cos i) + \delta M \right) \tag{1.2}$$

$$\Delta v_{r_a} = \frac{na}{4} \left(\frac{(1-e)^2}{\eta}(\delta \omega + \delta \Omega \cos i) + \delta M \right) \tag{1.3}$$

where a is the semimajor axis and e is the eccentricity. Also at perigee implement Δv_{θ_p} and at apogee implement Δv_{θ_a} in the along-track direction, to cancel the semimajor axis and eccentricity:

$$\Delta v_{\theta_p} = \frac{na\eta}{4} \left(\frac{\delta a}{a} + \frac{\delta e}{1+e} \right) \tag{1.4}$$

$$\Delta v_{\theta_a} = \frac{na\eta}{4} \left(\frac{\delta a}{a} - \frac{\delta e}{1-e} \right) \tag{1.5}$$

where $\eta = \sqrt{1 - e^2}$.

1.5.3 Continuous linear control

A straightforward approach to controlling the relative states of satellites in a formation when the dynamics of those states can be modeled linearly is to use a Linear Quadratic Regulator (LQR) [24]. This approach has the advantage that all the tools of linear control can be used to analyze stability and properties and should incur minimal risk. However, an LQR controller would be expected to fire continuously in response to system uncertainty, incurring a fuel penalty over maneuver-planning controllers. To prevent continuous firing, an LQR controller would likely be combined with a deadband.

1.5.4 Nonlinear control

Many formation control approaches have used nonlinear, continuous controllers [28–30,35–37] with stability guarantees based on Lyapunov proofs. Due to the use of Lyapunov-based control, these approaches give rise to Control Lyapunov Functions (CLFs), to be discussed in Chapters 3 and 10. Although these controllers eliminate concerns that nonlinearities in the dynamics could cause instability, they are still affected by many of the same issues which are expected to be problematic in continuous linear control. The form of the control input vector, $\mathbf{u}$, for an absolute state $\mathbf{x}$ and a state error ζ is [29]

$$\mathbf{u} = -K(\mathbf{x})\zeta \qquad (1.6)$$

where K is some nonlinear function of the state. A specific form shown to be asymptotically stable under nominal conditions [29] is

$$\mathbf{u} = -[B(\mathbf{x})]^T P \zeta \qquad (1.7)$$

and P is a positive definite matrix and B is the matrix representation of Gauss' variational equations [33], to be discussed in Chapter 2.

1.5.5 Model predictive control

Model Predictive Control (MPC) can be used to generate optimized plans that satisfy performance constraints [26,38–40]. MPC using linear programming (LP) has a number of other advantages for spacecraft formation flying: It easily incorporates realistic constraints on thrusting and control performance; it generates plans that closely approximate fuel-optimal "bang-off-bang" solutions rather than the continuous thrusting plans that inevitably arise from LQR, H_∞, and Lyapunov controllers; and it allows for piecewise-linear cost functions, such as the 1-norm of fuel use.

1.6 SPACE NAVIGATION AND THE GLOBAL POSITIONING SYSTEM

The NAVSTAR Global Positioning System (GPS) is a space-based radio navigation system developed, owned, and operated by the United States Department of Defense. The GPS satellites transmit signals on two carrier frequencies. The civilian L1 frequency, 1575.42 MHz, carries a pseudo-random code for timing and contains a navigation message with ephemeris data [41]. GPS positioning is based on the principle of trilateration, which is the process of ranging off at least three objects with known position to determine a local position. The clocks that are used in the GPS ranging process are of low quality, so the time is added as a fourth dimension. Because of the time uncertainty, the four required ranges are not true measures of position, and are thus called pseudoranges.

The standard method of obtaining a pseudorange involves using the navigation information on the code message. Code-based pseudorange measurements typically produce differential accuracies of several meters, which are not sufficient for formation flying missions. Carrier techniques offer much higher accuracy by calculating pseudoranges from the phase measurement of the radio-frequency (RF) carrier wave. If carrier measurements from a mobile receiver and a base station are differenced, forming a carrier-phase differential GPS (CDGPS) measurement, the motion can be observed accurately. If the base is also moving, as in the case of chief and deputy vehicles in a satellite formation, the CDGPS observable leads to relative position and velocity [42]. CDGPS measurements will be used in Chapter 12 to provide highly accurate relative navigation.

In addition to the code and carrier pseudoranges, a Doppler measurement, which can be related to velocity, is available from the GPS receiver. The Doppler measurement is created inside the receiver by differencing carrier phase measurements. Because this is not a truly independent measurement, previous research has found that adding Doppler measurements does not significantly improve the state estimate [43].

1.7 FORMATION FLYING MISSIONS

Over the past decade, numerous formation flying missions have been conceived. These missions were driven by scientific and programmatic objectives ranging from sparse-aperture imaging of extra-solar planets to lunar gravimetry. In particular, the *TechSat-21* concept[1] was a revolutionary space architecture of collaborating clusters of similar, agile, capable microsatellites that could be adapted on-demand to perform a variety of missions. In addition, it was envisioned that system performance could be increased over time through "phased" deployment, or tailored to meet evolving mission needs. The advantage of such a system is that the loss of one or more satellites in the formation has only limited impact on the cluster's performance, as remaining satellites could "absorb" the lost satellite's responsibilities.

The increased scientific return and the potential adaptability of formation flying satellites to changing mission goals have created new opportunities for the scientific endeavor. However, the current control, measurement and modeling challenges of spacecraft formation flying have rendered some of the proposed missions too costly. Thus, recent years have seen many proposed formation flying missions cancelled or transformed into technology-demonstration missions.

Nevertheless, it is safe to say that operational formation flying satellites, performing tracking or maintenance of a desired relative state (cf. our definition of formation flying in Section 1.1), will be launched in the near future. While there are currently no formation flying satellites in orbit, some operational missions certainly implement technologies and methods required to maintain satellites in formation. Two such missions are ESA's CLUSTER mission and the ESA/NASA GRACE mission.

---

[1] See http://en.wikipedia.org/wiki/TechSat-21, accessed April 30, 2009.

CLUSTER[2] comprises four identical spacecraft launched into large, highly elliptical polar orbits around the Earth, with perigee and apogee altitudes of 19,000 km and 119,000 km, respectively. These satellites fly in pre-determined relative orbits designed so as to allow scientists to measure subtle changes in the interaction between the Earth and the Sun. The four spacecraft examine how particles from the Sun interact with the Earth's magnetic field. CLUSTER observes the magnetic and electrical interactions between the Earth and the Sun by making direct measurements of the three-dimensional fields. The CLUSTER satellites were launched in August 2000 for a nine-year (extended) mission.

Another mission that implements a formation flying technology – measurement of inter-spacecraft range – is the *GRACE* mission.[3] The GRACE mission features two identical satellites in a leader/follower formation (GRACE A and GRACE B) orbiting the Earth on the same orbital plane. The purpose of this mission is to generate high-fidelity modeling of Earth's gravitational field; it is expected to yield an improvement of several orders of magnitude in gravity measurements and allow much improved resolution of the broad to fine-scale features of Earth's gravitational field over both land and sea. A secondary experiment that GRACE performs is examining how the atmosphere affects GPS signals. The initial altitude of GRACE A and GRACE B above the Earth was close to 500 km. Due to atmospheric drag, it will decrease to about 300 km towards the end of the mission. The mean inter-satellite separation varies between 170 and 270 km. Originally funded for a five-year period (2002–2007), the mission has been further extended to 2009. As the orbit decay has been slower than initially thought and the satellites' current fuel supply is expected to last another few years at the very least, the mission is likely to continue past 2012.

Two ESA formation flying missions are currently in advanced development stages: PRISMA and PROBA-3.

PRISMA[4] is a Swedish-led satellite project with the objective to develop and qualify new technology necessary for future formation flying science missions. The PRISMA project is Europe's first step to demonstrate new formation flying technology – both hardware and software. The technology of PRISMA is developed mainly in Sweden, Germany, Denmark and France. PRISMA consists of two spacecraft, with a total mass of about 200 kg. It contains several new technologies within autonomous formation flying and rendezvous, small rocket engines and Micro Electro Mechanical Systems (MEMS). PRISMA is scheduled to be launched on a Dnepr launcher in early 2010.

PROBA-3 is the ESA formation flying demonstration mission[5] preparing for future formation flying missions, such as *XEUS*[6] and *DARWIN*.[7]

---

[2] See http://sci.esa.int/science-e/www/area/index.cfm?fareaid=8, accessed March 26, 2009.

[3] See http://www.csr.utexas.edu/grace/, accessed March 26, 2009.

[4] See http://www.prismasatellites.se/?id=9033, accessed March 26, 2009.

[5] See http://www.ssc.se/?id=7613, accessed March 26, 2009.

[6] See http://sci.esa.int/science-e/www/object/index.cfm?fobjectid=39306, accessed March 26, 2009.

[7] See http://www.esa.int/esaSC/120382_index_0_m.html, accessed March 26, 2009.

The PROBA-3 mission will demonstrate algorithms, sensors, propulsion systems and other technologies needed for formation flying.

An important joint NASA/ESA mission, implementing a number of critical formation flying technologies (very precise formation design and metrology, but no formation control), is the *Laser Interferometer Space Antenna* (LISA).[8] LISA is designed to detect "ripples" in space-time, as predicted by Einstein's general theory of relativity. LISA's three spacecraft will form an equilateral triangle with an arm length of about 5 million km. Each spacecraft houses two free-floating cubes made of a gold–platinum alloy inside the spacecraft, shielded from adverse effects of being in interplanetary space. The distance between the cubes in different spacecraft is monitored using highly accurate laser-based techniques.

NASA has proposed many formation flying missions. Some of these missions are currently under development, while others are in conceptual stages. One of NASA's main formation flying missions is the *Magnetospheric Multiscale Mission* (MMS).[9] MMS includes four identical spacecraft in a variably spaced tetrahedron (1 km to several Earth radii), with a planned two-year mission lifetime. The system includes inter-spacecraft ranging, communication and instrumentation, designed to measure magnetic and electric fields using electron and ion plasma spectrometers, providing high temporal and spatial resolution. MMS is currently in the preliminary design stage. Mission confirmation is targeted for July 2009; launch is planned for 2014.

Another interesting formation flying mission is the *New Worlds Observer* (NWO).[10] NWO consists of a large telescope and an occulter spacecraft in tandem at about 50,000 km apart. The purpose of NWO is to discover and analyze terrestrial extra-solar planets. The NWO concept features two spacecraft flying at the Earth–Sun L2 (Lagrangian) point or in a drift-away solar orbit. One craft carries a 4-meter aperture-diameter diffraction-limited telescope optimized to work in the visible band. The other occulter craft would each carry a starshade. Operating 70,000 kilometers from the telescope, one starshade would be maneuvered into the telescope's line-of-sight to a nearby star, blocking starlight while passing planet light. The NWO planned launch date is circa 2018.

Other future NASA formation flying missions are the *Stellar Imager* (SI),[11] *Milli-Arc-Second Structure Imager* (MASSIM),[12] and *Black Hole Imager*,[13] all planned for the third decade of the 21st century.

The SI mission is a space-based ultraviolet (UV)/optical interferometer with over 200 times the resolution of the Hubble space telescope. The purpose of this mission is to understand, by using high angular-resolution spectral imaging,

---

[8] See http://www.esa.int/esaSC/120376_index_0_m.html, accessed April 30, 2009.

[9] See http://stp.gsfc.nasa.gov/missions/mms/mms.htm, accessed March 26, 2009.

[10] See http://newworlds.colorado.edu/implementation/index.htm, accessed March 26, 2009.

[11] See http://hires.gsfc.nasa.gov/si/, accessed March 25, 2009.

[12] See http://asd.gsfc.nasa.gov/Gerald.Skinner/massim_proposal.pdf, accessed March 25, 2009.

[13] See http://bhi.gsfc.nasa.gov/, accessed March 25, 2009.

the details and dynamics of heretofore unresolved objects and processes: Solar/stellar magnetic activity and their impact on the climates and habitability of planets and life; and magnetic and accretion processes and their roles in the evolution of structure and transport of matter throughout the Universe. The mission consists of a 0.5-km diameter space-based UV-optical interferometer located near the Sun–Earth L2 point to enable precision formation flying. The interferometer consists of 30 primary mirror elements focusing on a beam-combining hub.

The proposed MASSIM mission will image in X-rays the structure of astrophysical objects with an angular resolution three orders of magnitude better than the present state of the art. An optics spacecraft carrying an array of diffractive/refractive lenses focuses X-rays onto detectors on a spacecraft 1000 km behind.

Chapter two

Fundamental Astrodynamics

'Twas noontide of summer,
And midtime of night,
And stars, in their orbits,
Shone pale, through the light.

Edgar Allan Poe (1809–1849)

In this chapter, we review some notions in orbital mechanics, assuming that the reader has some previous knowledge of astrodynamics; a good source for acquiring such knowledge is, e.g., Ref. [44]. We will briefly explain the main coordinate systems used in the following chapters, and discuss the Keplerian *two-body problem*. We start by recalling the following terminology:

- *Orbital plane* – Newton's inverse-square law of gravitation implies that in the absence of any additional internal and/or external forces, two perfectly-spherical masses, whose shape remains unchanged under mutual gravitation, will perform planar motion. The plane of motion is the *orbital plane*. This plane contains the position and velocity vectors of the orbiting bodies. Planar motion can also be obtained in the presence of some perturbations.
- *Ecliptic plane* – A plane containing the mean orbit of the Earth around the Sun.
- *Periapsis* – When viewed from one of the gravitating masses, the *primary*, the point on the orbit of an orbiting body closest to the primary is the periapsis.
- *Orbital angular momentum* – A vector quantity related to the rotation of an orbiting body about a gravitating primary, defined as the cross product of the position and linear angular momentum vectors.
- *Vernal equinox* – The date when night and day are nearly the same length, and Sun crosses the celestial equator moving northward. The vernal equinox marks the first day of spring, and is used as the reference line for inertial measurements (with a proper fixture of some

13

Spacecraft Formation Flying; ISBN: 9780750685337

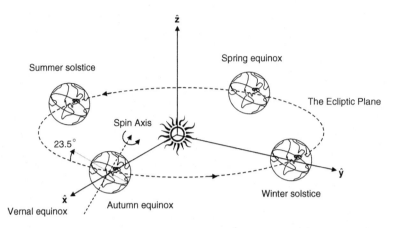

FIGURE 2.1 The heliocentric-ecliptic inertial coordinate system. Seasons refer to the northern hemisphere. Also shown is the vernal equinox vector, used as the inertial $\hat{\mathbf{x}}$-axis.

reference date); it is usually denoted by the symbol Υ, since the vernal equinox once pointed to the constellation Aries. For example, the vernal equinox of 2009 occurred on March 20, 11 hr 44 m (Universal Time).

2.1 COORDINATE SYSTEMS

Problems that involve *kinematics*, or *rates of change*, of physical quantities, require a definition of *reference frames*, giving rise to *coordinate systems*, which the rates can be referred to.

The study of problems in orbital mechanics usually requires the definition of a few coordinate systems. The most common coordinate systems are:

- $\mathscr{I}$, a Cartesian, rectangular, dextral (CRD) *inertial coordinate system*, centered at the gravitational body (primary).
    - A *heliocentric* system is centered at the Sun, the fundamental plane is the ecliptic plane, the unit vector $\hat{\mathbf{x}}$ is directed from the Sun's center along the vernal equinox, $\hat{\mathbf{z}}$ is normal to the fundamental plane, positive in the direction of the celestial north, and $\hat{\mathbf{y}}$ completes the setup (see Fig. 2.1).
    - A *geocentric* system is centered at the Earth, the fundamental plane is the equator, the unit vector $\hat{\mathbf{x}}$ is directed from the Earth's center along the vernal equinox, $\hat{\mathbf{z}}$ is normal to the fundamental plane, positive in the direction of the geographic north pole, and $\hat{\mathbf{y}}$ completes the setup (see Fig. 2.2). This reference frame is usually referred to as *Earth-centered inertial* (ECI).

    An inertial system is used to define a satellite's position and velocity vectors, denoted by $\mathbf{r}$ and $\mathbf{v}$, respectively, as well as the right ascension, α_r, and the declination, δ_d (see Fig. 2.2). The angle α_r is measured from the vernal equinox to the projection of $\mathbf{r}$ onto the

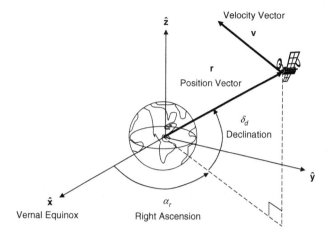

FIGURE 2.2 The Earth-centered inertial (ECI) coordinate system. Also shown are **r** and **v**, the position and velocity vectors, respectively, as well as the right ascension, α_r, and the declination, δ_d.

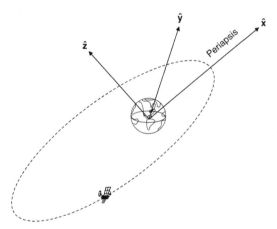

FIGURE 2.3 A perifocal coordinate system. The $\hat{\mathbf{x}}$-axis points to the (instantaneous) periapsis.

equatorial plane, whereas δ_r is measured from the same projection to **r**.

- $\mathscr{P}$, a CRD *perifocal* coordinate system, centered at the primary. The fundamental plane is the (instantaneous) orbital plane. The unit vector $\hat{\mathbf{x}}$ is directed from the primary's center to the (instantaneous) periapsis, $\hat{\mathbf{z}}$ is normal to the fundamental plane, positive in the direction of the (instantaneous) orbital angular momentum vector, and $\hat{\mathbf{y}}$ completes the setup (see Fig. 2.3). Note that in the presence of orbital perturbations, this coordinate system is non-inertial.
- $\mathscr{F}$, a CRD, *Earth-centered Earth-fixed* (ECEF) rotating coordinate system, centered at the Earth. The unit vector $\hat{\mathbf{z}}$ is parallel to the

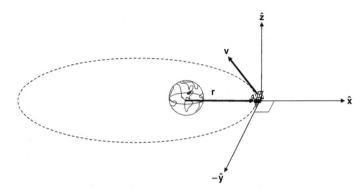

FIGURE 2.4 Local-vertical, local-horizon rotating coordinate system, centered at the spacecraft. The unit vector $\hat{\mathbf{x}}$ is directed from the spacecraft radially outward.

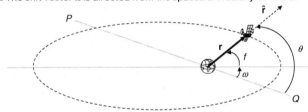

FIGURE 2.5 A polar coordinate system centered at the primary. The unit vector $\hat{\mathbf{r}}$ points radially outward. The angle θ is measured counterclockwise from some reference line, PQ, to the instantaneous radius-vector, $\mathbf{r}$. Also shown are the angles f and ω.

Earth's geographic north, $\hat{\mathbf{x}}$ intersects the Earth's sphere at the $0°$ latitude, $0°$ longitude, and $\hat{\mathbf{y}}$ completes the setup.

- $\mathscr{L}$, a CRD *local-vertical, local-horizontal* (LVLH) rotating coordinate system, centered at the spacecraft. The fundamental plane is the orbital plane. The unit vector $\hat{\mathbf{x}}$ is directed from the spacecraft radially outward, $\hat{\mathbf{z}}$ is normal to the fundamental plane, positive in the direction of the (instantaneous) angular momentum vector, and $\hat{\mathbf{y}}$ completes the setup (see Fig. 2.4).

  Occasionally, $\mathscr{L}$ is also called an RSW frame. This name is used in analysis of orbital perturbations in a frame co-rotating with the satellite. In this context, the unit vector $\hat{\mathbf{R}}$ is directed radially outwards, $\hat{\mathbf{S}}$ is perpendicular to $\hat{\mathbf{R}}$ in the direction of the instantaneous velocity, and $\hat{\mathbf{W}}$ completes the right-hand triad.

- $\mathscr{R}$, a *polar rotating coordinate system*, centered at the primary. The fundamental plane is the orbital plane. The unit vector $\hat{\mathbf{r}}$ is directed from the primary radially outward, and the angle θ is measured in the counterclockwise direction from some reference line, PQ, to $\mathbf{r}$ (see Fig. 2.5).

2.2 THE KEPLERIAN TWO-BODY PROBLEM

The equations of motion of the Keplerian two-body problem are obtained under the following assumptions:

- There are no external or internal forces except gravity.
- The gravitating bodies are spherical.
- There are no tidal forces.
- The primary's mass is much larger than the orbiting body's mass.
- The gravitational force is Newtonian.

Under these assumptions, the Keplerian two-body equations of motion can be written as follows:[1]

$$\ddot{\mathbf{r}} + \frac{\mu \mathbf{r}}{r^3} = \mathbf{0} \tag{2.1}$$

where $\mathbf{r} = [X, Y, Z]^T$ is the position vector in the ECI frame $\mathscr{I}$, μ is the gravitational constant, and $r = \|\mathbf{r}\|$. In order to solve Eq. (2.1), it is useful to transform the inertial equations of motion into polar coordinates, $\mathscr{R}$. To that end, we recall that

$$\mathbf{r} = r\hat{\mathbf{r}} \tag{2.2}$$
$$\dot{\mathbf{r}} = \dot{r}\hat{\mathbf{r}} + r\dot{\theta}\hat{\boldsymbol{\theta}} \tag{2.3}$$
$$\ddot{\mathbf{r}} = (\ddot{r} - r\dot{\theta}^2)\hat{\mathbf{r}} + (2\dot{r}\dot{\theta} + r\ddot{\theta})\hat{\boldsymbol{\theta}} \tag{2.4}$$

where θ is the *argument of latitude*. Substituting Eq. (2.1) into Eq. (2.4) yields the equations of motion in $\mathscr{R}$:

$$\ddot{r} = r\dot{\theta}^2 - \frac{\mu}{r^2} \tag{2.5}$$

$$\ddot{\theta} = -\frac{2\dot{r}\dot{\theta}}{r} \tag{2.6}$$

It is immediately seen from Eq. (2.6) that

$$\frac{d}{dt}(r^2\dot{\theta}) = r(r\ddot{\theta} + 2\dot{r}\dot{\theta}) = 0 \tag{2.7}$$

Let us show that $r^2\dot{\theta}$ is the magnitude of the orbital angular momentum vector per unit mass, $\mathbf{h}$:

$$\mathbf{h} = \mathbf{r} \times \dot{\mathbf{r}} = \mathbf{r} \times \mathbf{v} = \begin{bmatrix} r \\ 0 \\ 0 \end{bmatrix} \times \begin{bmatrix} \dot{r} \\ r\dot{\theta} \\ 0 \end{bmatrix} = r^2\dot{\theta}\hat{\mathbf{z}} = h\hat{\mathbf{z}} \tag{2.8}$$

where $\hat{\mathbf{z}}$ is a unit vector normal to the orbital plane.

---

[1] This version of the Keplerian two-body problem is sometimes referred to as the *restricted two-body problem*, implying that the particular shape factors of each body are neglected, and hence the inter-gravitation does not induce gravitational torques.

Hence, Eq. (2.7) implies *conservation of angular momentum*:

$$\mathbf{h} = \text{const.} \tag{2.9}$$

In fact, $\mathbf{h}$ is conserved as a vector quantity, meaning that not only is its magnitude, but also each of its components in the inertial space is constant. To see this, we can differentiate $\mathbf{h}$ to get

$$\dot{\mathbf{h}} = \dot{\mathbf{r}} \times \dot{\mathbf{r}} + \mathbf{r} \times \ddot{\mathbf{r}} = -\frac{\mu}{r^3}\mathbf{r} \times \mathbf{r} = 0 \tag{2.10}$$

We now examine Eq. (2.5). Note that

$$\ddot{r} = \frac{d}{dt}\left(\frac{dr}{dt}\right) = \frac{dr}{dt}\frac{d}{dr}\left(\frac{dr}{dt}\right) = \dot{r}\frac{d}{dr}(\dot{r}) = d\left(\frac{\dot{r}^2}{2}\right) \tag{2.11}$$

Substituting into Eq. (2.5) yields

$$d\left(\frac{\dot{r}^2}{2}\right) = \left(\frac{h^2}{r^3} - \frac{\mu}{r^2}\right) \tag{2.12}$$

Integrating both sides of Eq. (2.12), we obtain

$$\mathcal{E} = \frac{\dot{r}^2}{2} + \frac{h^2}{2r^2} - \frac{\mu}{r} = \underbrace{\frac{\dot{r}^2}{2} + \frac{(r\dot{\theta}^2)}{2}}_{\text{kinetic energy}} \underbrace{-\frac{\mu}{r}}_{\text{potential energy}} = \text{const.} \tag{2.13}$$

where $\mathcal{E}$, the constant of integration, is the *total energy* per unit mass, viz. the *kinetic energy* plus the *potential energy*. Equation (2.13) thus implies *conservation of energy*. It is sometime written in the *vis-viva*[2] form

$$\mathcal{E} = \frac{v^2}{2} - \frac{\mu}{r} \tag{2.14}$$

where v is the magnitude of the velocity vector.

We are now able to solve the equations of motion in polar coordinates. Utilizing both constants of motion, we write

$$\dot{r} = \sqrt{2\left(\mathcal{E} + \frac{\mu}{r}\right) - \frac{h^2}{r^2}} \tag{2.15}$$

$$\dot{\theta} = \frac{h}{r^2} \tag{2.16}$$

Although Eqs. (2.15) and (2.16) appear with the time as the independent variable, it is more convenient to obtain a relationship of the form $r = r(\theta)$.

---

[2]Latin for "living force".

To that end, we divide Eq. (2.15) by Eq. (2.16) to obtain

$$\frac{dr}{d\theta} = \frac{r^2\sqrt{2\left(\mathcal{E} + \mu/r\right) - h^2/r^2}}{h} \tag{2.17}$$

Equation (2.17) is a separable differential equation which can be solved by direct integration utilizing the initial condition $\theta_0 = \omega$, where ω, shown in Fig. 2.5, is the *argument of periapsis*:

$$\theta = \int \frac{h\,dr}{r^2\sqrt{2(\mathcal{E} + \mu/r) - h^2/r^2}} + \omega = \cos^{-1}\frac{1/r - \mu/h^2}{\sqrt{2\mathcal{E}/h + \mu^2/h^4}} + \omega \tag{2.18}$$

Solving for r yields

$$r = \frac{h^2/\mu}{1 + \sqrt{1 + 2\mathcal{E}h^2/\mu^2}\cos(\theta - \omega)} \tag{2.19}$$

This is the well-known equation of *conic sections* in polar coordinates, and is sometimes referred to as the *conic equation*, as elaborated below. The orbits determined by Eq. (2.19) are called *Keplerian orbits*. An alternative formulation of Eq. (2.19) is

$$r = \frac{p}{1 + e\cos f} \tag{2.20}$$

where

$$p = h^2/\mu \tag{2.21}$$

is called the *semilatus rectum* or *the parameter*,

$$e = \sqrt{1 + 2\mathcal{E}h^2/\mu^2} \tag{2.22}$$

is the *eccentricity*, and

$$f = \theta - \omega \tag{2.23}$$

is the *true anomaly*, shown in Fig. 2.5.

Equation (2.20) yields two families of non-degenerate orbits, shown in Fig. 2.6. By defining

$$a = \frac{p}{1 - e^2} \tag{2.24}$$

these families can be categorized as follows:

- An *ellipse*, which is a *closed orbit* constituting the locus of all points whose sum of distances from two fixed points, called the *foci*, is

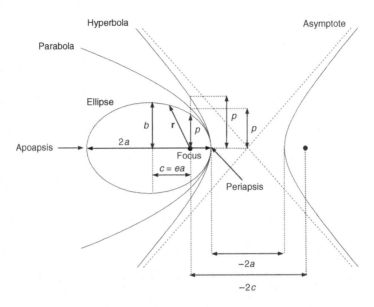

FIGURE 2.6 The conic sections: Ellipse, parabola and hyperbola.

constant. In this case, $a > 0$ is the *semimajor axis*, $0 < e < 1$, and $b = a\sqrt{1 - e^2}$ is the *semiminor axis*. We distinguish between the following special cases of an ellipse:
- The case $e = 0$ is a closed orbit called a *circle*, and
- The case $e = 1, a = \infty$ is an *open orbit* called a *parabola*
• A *hyperbola*, which is an open orbit constituting the locus of all points whose difference of distances from two fixed points, the foci, is constant. In this case, the semimajor axis is negative, $a < 0$, and equals half the minimum distance between the two hyperbola branches. The eccentricity satisfies $e > 1$.

In orbital mechanics, the gravitating body is located at one of the foci and the additional focus is vacant. The closest point to the gravitating body is called periapsis (see above) and the farthest point is called *apoapsis*. The line connecting the periapsis and apoapsis is called the *line of apsides*. Open orbits do not posses an apoapsis.

The Keplerian orbits can be similarly categorized in terms of the *total orbital energy*, $\mathcal{E}$. Substitution of Eqs. (2.21), (2.22) and (2.24) into Eq. (2.13) yields

$$\mathcal{E} = -\frac{\mu}{2a} \tag{2.25}$$

Thus, for an ellipse $(0 < a < \infty)$ $\mathcal{E} < 0$, for a parabola $(a = \infty)$ $\mathcal{E} = 0$ and for a hyperbola $(-\infty < a < 0)$ $\mathcal{E} > 0$.

Expression (2.20) provides an elegant closed-form expression for the radius vector as a function of the true anomaly. However, it is sometimes required to express the position as a function of time. To that end, a relationship between the

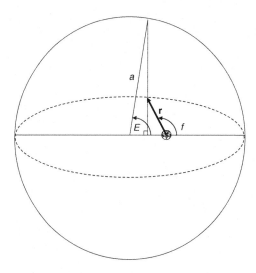

FIGURE 2.7 The eccentric anomaly. A bounding circle of radius a is plotted around the elliptic orbit. At a given point on the orbit, a line perpendicular to the ellipse major axis is plotted. Connecting the circle–line intersection with the center of the circle will yield a line whose angle with respect to the major axis is the eccentric anomaly, E.

true anomaly and time must be found. However, a closed-form relationship of the form $f = f(t)$ does not exist. Time is therefore introduced into the problem by using an auxiliary variable called the *eccentric anomaly*, $E(f)$, defined as the angle between the perifocal unit vector $\hat{\mathbf{x}}$ and the radius of a bounding circle at a point normal to the line of apsides at a given f, as depicted in Fig. 2.7. *Kepler's equation* states that

$$M = M_0 + n(t - t_0) = E - e \sin E \tag{2.26}$$

where $n = \sqrt{\mu/a^3}$ is the *mean motion*, M is the *mean anomaly*, t_0 is the *epoch* and M_0 is the *mean anomaly at epoch*. We note that due to Kepler's equation, the true anomaly is a function of a, e, t and M_0.

We say that the mean anomaly at epoch, M_0, is a *constant of motion* for the Keplerian two-body problem. It joins the other constants of motion: The total energy, $\mathcal{E}$, and the three components of the angular momentum in inertial space, $\mathbf{h}$; so we are one constant-of-motion short of solving the three-degrees-of-freedom equations (2.1). In fact, we can find three such constants of motion, by recalling that motion in a conservative field yields a constant vector called the *Laplace–Runge–Lenz vector*. In the Keplerian two-body problem, the Laplace–Runge–Lenz vector is usually referred to as the *eccentricity vector*. For $\mathbf{v} = \dot{\mathbf{r}}, v = \|\mathbf{v}\|$, this vector is defined as follows:

$$\mathbf{e} = \frac{v^2 \mathbf{r}}{\mu} - \frac{(\mathbf{r} \cdot \mathbf{v})\mathbf{v}}{\mu} - \frac{\mathbf{r}}{r} \tag{2.27}$$

or, alternatively,

$$\mathbf{e} = \frac{\mathbf{v} \times \mathbf{h}}{\mu} - \frac{\mathbf{r}}{r} \tag{2.28}$$

By taking the time derivative of (2.28), keeping in mind that $\mathbf{h} = $ const. and recalling that $\mathbf{r} \times (\mathbf{r} \times \mathbf{v}) = \mathbf{r}(\mathbf{r} \cdot \mathbf{v}) - \mathbf{v}(\mathbf{r} \cdot \mathbf{r})$, it can be shown that $\dot{\mathbf{e}} = \mathbf{0}$. Furthermore, $\mathbf{r} \cdot \mathbf{e} = r e \cos f$, and $e = \|\mathbf{e}\|$. Thus, the unit vector $\hat{\mathbf{e}} = \mathbf{e}/e$ points to the periapsis, and is therefore identical to the $\hat{\mathbf{x}}$-axis in our perifocal frame, $\mathscr{P}$.

To conclude this section, we will mention that the angular velocity along a Keplerian orbit is obtained by differentiating the true anomaly with respect to time. This operation yields [33]:

$$\dot{f} = \sqrt{\frac{\mu}{a^3(1 - e^2)^3}} (1 + e \cos f)^2 \tag{2.29}$$

For Keplerian orbits $\omega = $ const. and hence

$$\dot{f} = \dot{\theta} \tag{2.30}$$

Based on Eq. (2.8), Eq. (2.30) implies that $h = r^2 \dot{f}$, or, equivalently, $dt = r^2/h\,df$. Letting the radius-vector of an elliptic orbit sweep an element area $dA = r^2 df/2$, we are convinced that $dt = 2dA/h$. Hence, if T stands for the orbital period and πab is the ellipse's area,

$$\int_0^T dt = \frac{2}{h} \int_0^{\pi ab} dA \tag{2.31}$$

which, upon substitution of $h = \sqrt{\mu a(1 - e^2)}$ and $b = a\sqrt{1 - e^2}$, provides us with an expression for the orbital period on elliptic (and circular) orbits:

$$T = 2\pi \sqrt{\frac{a^3}{\mu}} \tag{2.32}$$

Substituting from Eq. (2.25) yields an expression for the orbital period in terms of the total energy, $\mathcal{E}$:

$$T = \frac{\pi \mu}{\sqrt{2(-\mathcal{E})^3}} \tag{2.33}$$

2.3 SOLUTION OF THE INERTIAL EQUATIONS OF MOTION

We have obtained a solution to the differential equations of motion written in polar coordinates. Recall, however, that the original equations of motion (2.1) were formulated in inertial coordinates. In order to solve the inertial differential equations, it is customary to express the position vector in the perifocal coordinate system, $\mathscr{P}$:

$$[\mathbf{r}]_{\mathscr{P}} = \begin{bmatrix} r\cos f \\ r\sin f \\ 0 \end{bmatrix} \qquad (2.34)$$

where r is given in Eq. (2.20) with $p = a(1 - e^2)$. An alternative expression for the perifocal position vector can be written in terms of the eccentric anomaly, E [33]:

$$[\mathbf{r}]_{\mathscr{P}} = \begin{bmatrix} a(\cos E - e) \\ b\sin E \\ 0 \end{bmatrix} \qquad (2.35)$$

This expression will be used in Chapter 6.

The velocity vector in the perifocal system, $\mathscr{P}$, can be calculated by writing

$$[\dot{\mathbf{r}}]_{\mathscr{P}} = \dot{f}\left[\frac{d\mathbf{r}}{df}\right]_{\mathscr{P}} = \sqrt{\frac{\mu}{a^3(1-e^2)^3}}(1 + e\cos f)^2\left[\frac{d\mathbf{r}}{df}\right]_{\mathscr{P}} \qquad (2.36)$$

which yields

$$[\dot{\mathbf{r}}]_{\mathscr{P}} = \sqrt{\frac{\mu}{a(1-e^2)}}\begin{bmatrix} -\sin f \\ e + \cos f \\ 0 \end{bmatrix} \qquad (2.37)$$

In order to obtain $\mathbf{r}$, the inertial position vector, and $\dot{\mathbf{r}}$, the inertial velocity vector, we need to find a rotation matrix from perifocal to inertial coordinates, $T_{\mathscr{P}}^{\mathscr{I}}$. Recall that frame $\mathscr{P}$ is defined by the triad $\hat{\mathbf{e}}$, pointing to the periapsis; $\hat{\mathbf{h}}$, coinciding with the angular momentum vector; and a unit vector $\hat{\mathbf{p}}$, which completes the right-hand system (this vector actually lies along the orbit's semilatus rectum). Both $\mathscr{P}$ and $\mathscr{I}$ are shown in Fig. 2.8.

Now, assume that the spacecraft is rotating counterclockwise in its orbit when viewed from the primary's north pole. If the orbital plane of the spacecraft and the fundamental plane of frame $\mathscr{I}$ (which coincides with the equatorial plane if the inertial system is geocentric) intersect, then we define two points of interest on the line of intersection, as shown in Fig. 2.8: The first is the *ascending node*, denoted by Ω. This point marks the spacecraft's location on the line of intersection when moving to the north; the second is the *descending node*, denoted by $\mho$. This point marks the spacecraft's location on the line of intersection when moving to the south. The line connecting $\mho$ to Ω is called the *line of nodes* (LON); we will use the notation $\hat{\mathbf{l}}$ to denote a unit vector that lines along the LON.

We can now transform from $\mathscr{P}$ to $\mathscr{I}$ using three consecutive clockwise rotations by Euler angles conforming to the 3–1–3 sequence. The rotation

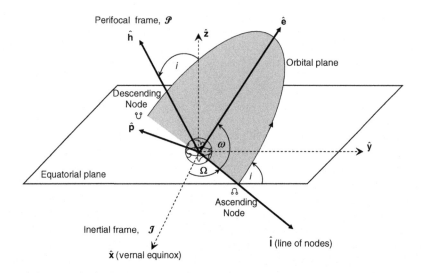

FIGURE 2.8 Definition of the classical orbital elements: The right ascension of the ascending node, Ω, the argument of periapsis, ω, and the inclination, i. These three Euler angles are used when transforming from a perifocal frame, $\mathscr{P}$, to an inertial frame, $\mathscr{I}$.

sequence is as follows:

- $T_1(\omega, \hat{\mathbf{h}})$, a rotation about $\hat{\mathbf{h}}$ by $0 \leq \omega \leq 2\pi$, mapping $\hat{\mathbf{e}}$ onto $\hat{\mathbf{I}}$.
- $T_2(i, \hat{\mathbf{I}})$, a rotation about $\hat{\mathbf{I}}$ by $0 \leq i \leq \pi$, mapping $\hat{\mathbf{h}}$ onto $\hat{\mathbf{z}}$.
- $T_3(\Omega, \hat{\mathbf{z}})$, a rotation about $\hat{\mathbf{z}}$ by $0 \leq \Omega \leq 2\pi$, mapping $\hat{\mathbf{I}}$ onto $\hat{\mathbf{x}}$.

The composite rotation, $T = T_3 \circ T_2 \circ T_1 \in SO(3)$, transforming any vector in $\mathscr{P}$ into the inertial frame, is given by[3]

$$T_{\mathscr{P}}^{\mathscr{I}}(\omega, i, \Omega) = T_3(\Omega, \hat{\mathbf{z}})T_2(i, \hat{\mathbf{I}})T_1(\omega, \hat{\mathbf{h}}) \tag{2.38}$$

Evaluating Eq. (2.38) gives the directional cosines matrix (DCM)

$$T_{\mathscr{P}}^{\mathscr{I}}(\omega, i, \Omega) = \begin{bmatrix} c_\Omega c_\omega - s_\Omega s_\omega c_i & -c_\Omega s_\omega - s_\Omega c_\omega c_i & s_\Omega s_i \\ s_\Omega c_\omega + c_\Omega s_\omega c_i & -s_\Omega s_\omega + c_\Omega c_\omega c_i & -c_\Omega s_i \\ s_\omega s_i & c_\omega s_i & c_i \end{bmatrix} \tag{2.39}$$

where we used the compact notation $c_x = \cos x$, $s_x = \sin x$. The Euler angles used for the transformation are

- Ω, the *right ascension of the ascending node*, an angle measured from the vernal equinox to the LON;
- ω, the argument of periapsis, which we already used in Eq. (2.18), an angle measured from the LON to the eccentricity vector; and

---

[3] The subgroup of 3×3 orthogonal matrices with determinant $+1$ is called the *special orthogonal group*, denoted SO(3).

- i, the *inclination*, an angle measured from $\hat{\mathbf{z}}$ to $\hat{\mathbf{h}}$.

These angles are shown in Fig. 2.8.

Transforming into inertial coordinates using Eqs. (2.34) and (2.39), we obtain the general solution to Eq. (2.1):[4]

$$
\mathbf{r} = T_{\mathscr{P}}^{\mathscr{I}}(\omega, i, \Omega)\mathbf{r}_{\mathscr{P}}(a, e, M_0, t) = \mathbf{r}(a, e, i, \Omega, \omega, M_0, t)
$$
$$
= \frac{a(1 - e^2)}{1 + e \cos f}
\begin{bmatrix}
c_{f+\omega}c_{\Omega} - c_i s_{f+\omega}s_{\Omega} \\
c_i c_{\Omega}s_{f+\omega} + c_{f+\omega}s_{\Omega} \\
s_i s_{f+\omega}
\end{bmatrix}
\tag{2.40}
$$

In a similar fashion, the expression for the inertial velocity is given by

$$
\mathbf{v} = \dot{\mathbf{r}} = T_{\mathscr{P}}^{\mathscr{I}}(\omega, i, \Omega)\dot{\mathbf{r}}_{\mathscr{P}}(a, e, M_0, t) = \mathbf{v}(a, e, i, \Omega, \omega, M_0, t)
$$
$$
= \sqrt{\frac{\mu}{a(1 - e^2)}}
\begin{bmatrix}
-c_{\Omega}s_{f+\omega} - s_{\Omega}c_i c_{f+\omega} - e(c_{\Omega}s_{\omega} + s_{\Omega}c_{\omega}c_i) \\
c_{\Omega}c_i c_{f+\omega} - s_{\Omega}s_{f+\omega} - e(s_{\Omega}s_{\omega} - c_{\Omega}c_{\omega}c_i) \\
s_i(c_{f+\omega} + ec_{\omega})
\end{bmatrix}
\tag{2.41}
$$

Thus, the inertial position and velocity depend upon time and the *classical orbital elements*, given by

$$
œ = \{a, e, i, \Omega, \omega, M_0\}
\tag{2.42}
$$

While the Euler angles Ω, i, ω may become degenerate in some cases (for instance, ω is undefined for circular orbits; both ω and Ω are undefined for equatorial orbits), the position and velocity vectors are always well-defined. However, occasionally alternative orbital elements are used to alleviate these deficiencies. These alternative elements are collectively referred to as *nonsingular orbital elements*. A good survey of these elements can be found in Ref. [45]. The next section provides a brief overview of the nonsingular elements used in this book (e.g. in Chapter 10).

2.4 NONSINGULAR ORBITAL ELEMENTS

To alleviate the singularity of the classical elements for circular orbits, Deprit and Rom [46] suggested replacing e, M and ω by the following elements:

$$
q_1 = e \cos \omega, \quad q_2 = e \sin \omega, \quad \lambda = \omega + M
\tag{2.43}
$$

Here, λ is the *mean argument of latitude*. However, the set $\{a, q_1, q_2, i, \Omega, \lambda\}$ will still be singular for equatorial orbits. Removal of all singularities associated with the classical elements is possible by defining the *equinoctial elements* [47–49]:

$$
\left\{ a, e \sin(\omega + \Omega), e \cos(\omega + \Omega), \tan\frac{i}{2}\sin\Omega, \tan\frac{i}{2}\cos\Omega, \omega + \Omega + M \right\}
\tag{2.44}
$$

---

[4]When resolving some vector $\mathbf{w}$ in inertial coordinates, we will usually write $\mathbf{w}$ instead of $[\mathbf{w}]_{\mathscr{I}}$.

In Eqs. (2.43) and (2.44), M may be replaced by M_0.

Another set of nonsingular elements, to be used in Section 7.2, was proposed by Gurfil [50]. The main idea is to replace the three Euler angles Ω, i, ω, transforming from the perifocal to the ECI frame, by *Euler parameters*, denoted by $\beta_1, \beta_2, \beta_3, \beta_0$, subject to the constraint

$$\sum_{i=0}^{3} \beta_i^2 = 1 \tag{2.45}$$

Another variation of such nonsingular representations can be found in Sengupta and Vadali [31]. Note that the Euler parameters, presented herein as a means for regularizing orbital motion, are more common in the context of attitude dynamics; indeed, they are used in this context in Chapter 9. An analogy between orbital motion and rigid body attitude dynamics is drawn by using the Euler parameters as in Ref. [51].

The new set of elements is given by

$$\mathbf{œ} = \left\{ a, \sqrt{1 - e^2}, \beta_1, \beta_2, \beta_3, M_0 \right\} \tag{2.46}$$

where

$$\beta_1 = \sin \frac{i}{2} \cos \frac{\Omega - \omega}{2} \tag{2.47}$$

$$\beta_2 = \sin \frac{i}{2} \sin \frac{\Omega - \omega}{2} \tag{2.48}$$

$$\beta_3 = \cos \frac{i}{2} \sin \frac{\Omega + \omega}{2} \tag{2.49}$$

and

$$\beta_0 = \cos \frac{i}{2} \cos \frac{\Omega + \omega}{2} \tag{2.50}$$

The Euler parameters form a *quaternion*,[5] given by $\beta_0 + i\beta_1 + j\beta_2 + k\beta_3$, where i, j, k satisfy the relations

$$ij = -ji = k, \, jk = -kj = i, \, ki = -ik = j, \, i^2 = j^2 = k^2 = -1$$

The transformation from the perifocal frame to the ECI frame can be written as [52]

$$ix + jy + kz = (\beta_0 + i\beta_1 + j\beta_2 + k\beta_3)(iX + jY + kZ)$$
$$\times (\beta_0 - i\beta_1 - j\beta_2 - k\beta_3) \tag{2.51}$$

---

[5]Quaternions form a number system that extends the complex numbers. The quaternions were first described by Hamilton in 1843.

This transformation represents a single rotation of magnitude $\wp$ about the *Euler vector*. The relationships between the Euler parameters and the rotation angle $\wp$ are given by [52]

$$\beta_1 = \sin \frac{\wp}{2} \cos \phi_1 \tag{2.52}$$

$$\beta_2 = \sin \frac{\wp}{2} \cos \phi_2 \tag{2.53}$$

$$\beta_3 = \sin \frac{\wp}{2} \cos \phi_3 \tag{2.54}$$

where ϕ_i, $i = 1, 2, 3$ is the angle between the Euler vector and the inertial unit vectors $\hat{\mathbf{x}}$, $\hat{\mathbf{y}}$, $\hat{\mathbf{z}}$, respectively, and

$$\beta_0 = \cos \frac{\wp}{2} \tag{2.55}$$

The DCM rotating from $\mathscr{P}$ to $\mathscr{I}$ can now be written in terms of Euler parameters instead of the Euler angles. This DCM, replacing the Euler angles-based DCM of Eq. (2.39), is [52]

$$
\begin{aligned}
&T_{\mathscr{P}}^{\mathscr{I}}(\beta_1, \beta_2, \beta_3) \\
&= \begin{bmatrix} \beta_1^2 - \beta_2^2 - \beta_3^2 + \beta_0^2 & 2(\beta_1\beta_2 - \beta_3\beta_0) & 2(\beta_1\beta_3 + \beta_2\beta_0) \\ 2(\beta_1\beta_2 + \beta_3\beta_0) & \beta_2^2 + \beta_0^2 - \beta_1^2 - \beta_3^2 & 2(\beta_2\beta_3 - \beta_1\beta_0) \\ 2(\beta_1\beta_3 - \beta_2\beta_0) & 2(\beta_2\beta_3 + \beta_1\beta_0) & \beta_3^2 + \beta_0^2 - \beta_1^2 - \beta_2^2 \end{bmatrix}
\end{aligned} \tag{2.56}
$$

where β_0 is given by Eq. (2.45).

Using the new DCM of Eq. (2.56), the expressions for the inertial position and velocity read

$$
\begin{aligned}
\mathbf{r} &= T_{\mathscr{P}}^{\mathscr{I}}(\beta_1, \beta_2, \beta_3)\mathbf{r}_{\mathscr{P}}(a, e, M_0, t) = \mathbf{r}(a, e, \beta_1, \beta_2, \beta_3, M_0, t) \\
&= \frac{a(1 - e^2)}{1 + e \cos f} \begin{bmatrix} (\beta_1^2 - \beta_2^2 - \beta_3^2 + \beta_0^2)c_f + 2(\beta_1\beta_2 - \beta_3\beta_0)s_f \\ 2(\beta_1\beta_2 + \beta_3\beta_0)c_f + (\beta_2^2 + \beta_0^2 - \beta_1^2 - \beta_3^2)s_f \\ 2(\beta_1\beta_3 - \beta_2\beta_0)c_f + 2(\beta_2\beta_3 + \beta_1\beta_0)s_f \end{bmatrix}
\end{aligned} \tag{2.57}
$$

and

$$
\begin{aligned}
\mathbf{v} &= \dot{\mathbf{r}} = T_{\mathscr{P}}^{\mathscr{I}}(\omega, i, \Omega)\mathbf{v}_{\mathscr{P}}(a, e, M_0, t) = \mathbf{v}(a, e, i, \Omega, \omega, M_0, t) \\
&= \sqrt{\frac{\mu}{a(1 - e^2)}} \\
&\times \begin{bmatrix} 2(\beta_1\beta_2 - \beta_3\beta_0)c_f + (2\beta_2^2 + 2\beta_3^2 - 1)s_f + 2e(\beta_1\beta_2 - \beta_3\beta_0) \\ (1 - 2\beta_1^2 - 2\beta_3^2)c_f - 2(\beta_1\beta_2 + \beta_3\beta_0)s_f - e(2\beta_1^2 + 2\beta_3^2 - 1) \\ 2(\beta_2\beta_3 + \beta_1\beta_0)c_f + 2(\beta_2\beta_0 - \beta_1\beta_3)s_f + 2e(\beta_2\beta_3 + \beta_1\beta_0) \end{bmatrix}
\end{aligned}
$$
$$\tag{2.58}$$

A set of nonsingular elements that also form a quaternion, or a *spinor*, are the *Kustaanheimo–Stiefel orbital elements* [53,54], which are designed to regularize collision orbits in addition to the common singularities using a Hamiltonian framework [55].

2.5 NON-KEPLERIAN MOTION AND ORBITAL PERTURBATIONS

The solutions in Section 2.3 were obtained for the nominal, undisturbed Keplerian motion. When perturbations act upon the body, the motion is no longer Keplerian. In order to solve for the resulting non-Keplerian motion, Euler [56] and Lagrange [57] have developed the *variation-of-parameters* (VOP) procedure, a general and powerful method for the solution of nonlinear differential equations. Before applying this method to non-Keplerian motion, we will illustrate it in the next subsection.

2.5.1 Variation of parameters

In essence, the VOP method suggests to turn the constants of the unperturbed motion, resulting from the homogenous solution of a given differential equation, into functions of time. In order to illustrate the VOP method, consider the following example:

Example 2.1 (*Newman's example [58]*). *Solve the differential equation*

$$\ddot{x} + x = f(t), \quad x(0) = x_0, \quad \dot{x}(0) = \dot{x}_0 \tag{2.59}$$

using the VOP method.

The homogenous solution of Eq. (2.59) is given by

$$x = s \sin t + c \cos t \tag{2.60}$$

where s and c are constants. According to the VOP method, we need to seek a solution of the form

$$x = s(t) \sin t + c(t) \cos t \tag{2.61}$$

Differentiation of Eq. (2.61) results in

$$\dot{x} = \dot{s}(t) \sin t + \dot{c}(t) \cos t + s(t) \cos t - c(t) \sin(t) \tag{2.62}$$

Note that an additional differentiation, required to substitute for $\ddot{x}$, will yield a fourth-order system whereas the original differential equation is only of second order. Obviously, there is an excess of freedom in the system stemming from the transformation to the new state variables, $(x, \dot{x}) \mapsto (s, \dot{s}, c, \dot{c})$.

In order to solve for the excess freedom, it is customary to impose the constraint

$$\dot{s}(t) \sin t + \dot{c}(t) \cos t = 0 \tag{2.63}$$

This constraint will simplify the resulting differential equations, but is otherwise completely arbitrary. A more general form is

$$\dot{s}(t)\sin t + \dot{c}(t)\cos t = \Xi(t) \tag{2.64}$$

where $\Xi(t)$ is arbitrary. Substituting Eq. (2.64) into Eq. (2.62) and differentiating the resulting expression yields

$$\ddot{x} = \dot{\Xi} + \dot{s}(t)\cos(t) - \dot{c}(t)\sin t - s(t)\sin(t) - c(t)\cos(t) \tag{2.65}$$

Hence

$$\ddot{x} + x = \dot{\Xi} + \dot{s}(t)\cos(t) - \dot{c}(t)\sin(t) = f(t) \tag{2.66}$$

The resulting system of differential equations are

$$\dot{\Xi} + \dot{s}(t)\cos(t) - \dot{c}(t)\sin t = f(t) \tag{2.67}$$
$$\dot{s}(t)\sin(t) + \dot{c}(t)\cos(t) = \Xi(t) \tag{2.68}$$

This system can be re-written in the form

$$\dot{s}(t) = f(t)\cos(t) - \frac{d}{dt}(\Xi\cos t) \tag{2.69}$$

$$\dot{c}(t) = -f(t)\sin t + \frac{d}{dt}(\Xi\sin t) \tag{2.70}$$

Integration of Eqs. (2.69)–(2.70) yields

$$s(t) = \int_0^t f(\tau)\cos\tau d\tau - \Xi\cos t + s(0) \tag{2.71}$$

$$c(t) = \int_0^t f(\tau)\sin\tau d\tau + \Xi\sin t + c(0) \tag{2.72}$$

Substituting Eqs. (2.71)–(2.72) into Eq. (2.61) entails

$$x = -\cos t \int_0^t f(\tau)\sin\tau d\tau + \sin t \int_0^t f(\tau)\cos\tau d\tau$$
$$+ s(0)\sin t + c(0)\cos t \tag{2.73}$$

Not unexpectedly, the Ξ-dependent terms cancel out. Thus, while the state-space representation using the new state variables s and c depends upon the constraint function Ξ, the solution in terms of x is invariant to a particular selection of Ξ. This phenomenon is called *symmetry*. The analogy to orbital elements is straightforward, and will be discussed next.

2.5.2 Lagrange's planetary equations

When a perturbing specific force, $\mathbf{d}$, is introduced into Eq. (2.1), we have[6]

$$\ddot{\mathbf{r}} + \frac{\mu \mathbf{r}}{r^3} = \mathbf{d} \tag{2.74}$$

Applying the VOP formalism requires re-defining the classical orbital elements as functions of time, yielding a modified solution of the form

$$\mathbf{r} = \mathbf{f}[\mathbf{œ}(t), t] \tag{2.75}$$

Taking the time derivative of Eq. (2.75) yields the relationship

$$\dot{\mathbf{r}} = \frac{\partial \mathbf{f}}{\partial t} + \frac{\partial \mathbf{f}}{\partial \mathbf{œ}} \dot{\mathbf{œ}} \tag{2.76}$$

The differential equations describing the temporal change of the classical orbital elements for a conservative, position-only dependent perturbing potential are known as *Lagrange's planetary equations* (LPE). The *Gauss variational equations* (GVE) model the time evolution of the orbital elements due to an arbitrary perturbing acceleration resolved in the spacecraft's own LVLH frame.

To derive the LPE, Eq. (2.76) is differentiated and substituted into Eq. (2.74). This operation results in a 12-dimensional system of differential equations for $\mathbf{œ}$ and $\dot{\mathbf{œ}}$. However, there are only three degrees of freedom. Hence, the resulting system will be underdetermined, meaning that three extra conditions can be imposed. Lagrange chose to impose the constraint

$$\frac{\partial \mathbf{f}}{\partial \mathbf{œ}} \dot{\mathbf{œ}} = \mathbf{0} \tag{2.77}$$

which is also known as the *Lagrange constraint* or *osculation constraint*. Mathematically, this restriction confines the dynamics of the orbital state space to a 9-dimensional submanifold of the 12-dimensional manifold spanned by the orbital elements and their time derivatives. More importantly, this freedom reflects an internal symmetry in the mapping $(\mathbf{r}, \dot{\mathbf{r}}) \mapsto (\mathbf{œ}, \dot{\mathbf{œ}})$, which is inherent to Lagrange's planetary equations.

Physically, the Lagrange constraint postulates that the trajectory in the inertial configuration space is always tangential to an "instantaneous" ellipse (or hyperbola) defined by the "instantaneous" values of the time-varying orbital elements $\mathbf{œ}(t)$, meaning that the perturbed physical trajectory would coincide with the Keplerian orbit that the body would follow if the perturbing force was to cease instantaneously. This instantaneous orbit is called the *osculating orbit*. Accordingly, the orbital elements which satisfy the Lagrange constraint

---

[6]The external specific force can also be a control (thrust) input. In this case we will denote it by $\mathbf{u}$.

are called *osculating orbital elements*. The Lagrange constraint, however, is completely arbitrary. The generalized form of the Lagrange constraint is

$$\frac{\partial \mathbf{f}}{\partial \mathbf{œ}} \dot{\mathbf{œ}} = \mathbf{v}_g(\mathbf{œ}, \dot{\mathbf{œ}}, t) \tag{2.78}$$

where the velocity $\mathbf{v}_g$ is an arbitrary, user-defined function of the classical orbital elements, their time derivatives and possibly time. Equation (2.76) then becomes

$$\dot{\mathbf{r}} = \frac{\partial \mathbf{f}}{\partial t} + \mathbf{v}_g \tag{2.79}$$

where

$$\frac{\partial \mathbf{f}}{\partial t} = \mathbf{g}[\mathbf{œ}(t), t] = \mathbf{v} \tag{2.80}$$

This important observation has been made by a few researchers [59–61]. Recently, Efroimsky et al. have published key works on planetary equations with a generalized Lagrange constraint [58,62–64]. They termed the constraint function $\mathbf{v}_g$ *gage function* or *gage velocity*, which are terms taken from the field of electrodynamics.[7] The zero gage $\mathbf{v}_g = 0$ was termed the *Lagrange gage*.

The use of a generalized Lagrange constraint gives rise to *non-osculating orbital elements*. Thus, although the description of the physical orbit in the inertial Cartesian configuration space remains invariant to a particular selection of a gage velocity, its description in the orbital elements space depends on whether osculating or non-osculating orbital elements are used. However, to avoid complications, we will adhere to the Lagrange gage, and assume from now on that the orbital elements are osculating, i.e., $\mathbf{v}_g \equiv 0$. In this case

$$\dot{\mathbf{r}} = \mathbf{v} \tag{2.81}$$

We can now derive the differential equations for the orbital elements utilizing the VOP method as illustrated in Subsection 2.5.1. The time derivative of Eq. (2.79), using Eq. (2.81), is

$$\ddot{\mathbf{r}} = \frac{d\dot{\mathbf{r}}}{dt} = \frac{\partial}{\partial t} \frac{\partial \mathbf{r}}{\partial t} + \frac{\partial \mathbf{v}}{\partial \mathbf{œ}} \dot{\mathbf{œ}} = \frac{\partial^2 \mathbf{r}}{\partial^2 t} + \frac{\partial \mathbf{v}}{\partial \mathbf{œ}} \dot{\mathbf{œ}} \tag{2.82}$$

Substituting for $\ddot{\mathbf{r}}$ from Eq. (2.74) yields

$$\frac{\partial^2 \mathbf{r}}{\partial^2 t} + \frac{\mu}{r^3} \mathbf{r} + \frac{\partial \mathbf{v}}{\partial \mathbf{œ}} \dot{\mathbf{œ}} = \mathbf{d} \tag{2.83}$$

---

[7]Gage theory is an approach for developing a unified theory of the fundamental forces based on the concept of symmetry. Gage theory gets its name from the fact that measurements can be "re-gaged", yielding the same results. Gage theory is of great importance in electromagnetic interactions; mathematical models of gage symmetry have facilitated the discovery of *bosons*, which are counterparts of the photon. Gage symmetry is also used in cosmology to establish the theory of inflation.

Since the Keplerian solution satisfies

$$\frac{\partial^2 \mathbf{r}}{\partial t^2} + \frac{\mu}{r^3}\mathbf{r} = \mathbf{0} \tag{2.84}$$

Thus, Eq. (2.83) reduces to the perturbed Keplerian motion equation:

$$\frac{\partial \mathbf{v}}{\partial \text{œ}}\dot{\text{œ}} = \mathbf{d} \tag{2.85}$$

Together with Eq. (2.77), we have six differential equations for the classical osculating orbital elements. These equations can be written in a compact form utilizing the following formalism, originally conceived by Lagrange: Multiply Eq. (2.77) by $[\partial \mathbf{v}/\partial \text{œ}]^T$, multiply Eq. (2.85) by $[\partial \mathbf{r}/\partial \text{œ}]^T$, and subtract the two expressions. This procedure results in

$$\mathfrak{L}\dot{\text{œ}} = \left[\frac{\partial \mathbf{r}}{\partial \text{œ}}\right]^T \mathbf{d} \tag{2.86}$$

where $\mathfrak{L}$ is called the *Lagrange matrix*,

$$\mathfrak{L} \triangleq \left[\frac{\partial \mathbf{r}}{\partial \text{œ}}\right]^T \frac{\partial \mathbf{v}}{\partial \text{œ}} - \left[\frac{\partial \mathbf{v}}{\partial \text{œ}}\right]^T \frac{\partial \mathbf{r}}{\partial \text{œ}} \tag{2.87}$$

The entries of this matrix are called *Lagrange brackets*,

$$[\text{œ}_i, \text{œ}_j] \triangleq \left[\frac{\partial \mathbf{r}}{\partial \text{œ}_i}\right]^T \frac{\partial \mathbf{v}}{\partial \text{œ}_j} - \left[\frac{\partial \mathbf{r}}{\partial \text{œ}_j}\right]^T \frac{\partial \mathbf{v}}{\partial \text{œ}_i} \tag{2.88}$$

It can be shown that [33] $[\text{œ}_i, \text{œ}_j] = -[\text{œ}_j, \text{œ}_i]$ and $[\text{œ}_i, \text{œ}_i] = 0$, so the Lagrange matrix is skew symmetric,

$$\mathfrak{L}^T = -\mathfrak{L} \tag{2.89}$$

Also, $\partial[\text{œ}_i, \text{œ}_j]/\partial t = 0$, meaning that the Lagrange matrix is time-invariant,

$$\frac{\partial \mathfrak{L}}{\partial t} = 0 \tag{2.90}$$

The Lagrange matrix is nonsingular $\forall \text{œ} \setminus \{e = 0, i = 0\}$, and therefore can be inverted to yield the explicit differential equation

$$\dot{\text{œ}} = \mathfrak{L}^{-1}\left[\frac{\partial \mathbf{r}}{\partial \text{œ}}\right]^T \mathbf{d} \tag{2.91}$$

The inverse of the Lagrange matrix is denoted by

$$\mathfrak{P}^T = \mathfrak{L}^{-1} \tag{2.92}$$

where $\mathfrak{P}$ is a skew symmetric matrix called the *Poisson matrix*. Its entries are called the *Poisson brackets*,

$$\{\text{œ}_i, \text{œ}_j\} \triangleq \frac{\partial \text{œ}_i}{\partial \mathbf{r}} \left[\frac{\partial \text{œ}_j}{\partial \mathbf{v}}\right]^T - \frac{\partial \text{œ}_i}{\partial \mathbf{v}} \left[\frac{\partial \text{œ}_j}{\partial \mathbf{r}}\right]^T \tag{2.93}$$

The Poisson matrix is given by

$$\mathfrak{P}^T = \begin{bmatrix} 0 & 0 & 0 & 0 & 0 & \frac{2}{na} \\ 0 & 0 & 0 & 0 & -\frac{\sqrt{1-e^2}}{na^2 e} & \frac{1-e^2}{na^2 e} \\ 0 & 0 & 0 & -\frac{1}{na^2\sqrt{1-e^2}\sin i} & \frac{\cot i}{na^2\sqrt{1-e^2}} & 0 \\ 0 & 0 & \frac{1}{na^2\sqrt{1-e^2}\sin i} & 0 & 0 & 0 \\ 0 & \frac{\sqrt{1-e^2}}{na^2 e} & -\frac{\cot i}{na^2\sqrt{1-e^2}} & 0 & 0 & 0 \\ -\frac{2}{na} & -\frac{1-e^2}{na^2 e} & 0 & 0 & 0 & 0 \end{bmatrix} \tag{2.94}$$

An important special case arises when the orbital perturbations are conservative and depend only upon position, viz.

$$\mathbf{d} = \nabla_{\mathbf{r}} \mathcal{R} = \frac{\partial \mathcal{R}}{\partial \mathbf{r}} \tag{2.95}$$

where $\mathcal{R}$ is a perturbing potential. Substituting

$$\left[\frac{\partial \mathbf{r}}{\partial \text{œ}}\right]^T \mathbf{d} = \left[\frac{\partial \mathbf{r}}{\partial \text{œ}}\right]^T \frac{\partial \mathcal{R}}{\partial \mathbf{r}} = \frac{\partial \mathcal{R}}{\partial \text{œ}} \tag{2.96}$$

into Eq. (2.91) yields *Lagrange's planetary equations* (LPE),

$$\dot{\text{œ}} = \mathfrak{L}^{-1} \frac{\partial \mathcal{R}}{\partial \text{œ}} \tag{2.97}$$

2.5.3 Zonal harmonics

The perturbing potential $\mathcal{R}$ appearing in Eq. (2.97) is due to any conservative perturbation; for example, *zonal gravitational harmonics*, accounting for an axially-symmetric non-spherical primary with equatorial radius R_e, yield the perturbing potential[8]

$$\mathcal{R} = -\frac{\mu}{r} \sum_{k=2}^{\infty} J_k \left(\frac{R_e}{r} \right)^k P_k(\cos \phi) \tag{2.98}$$

where ϕ is the *colatitude angle*, satisfying

$$\cos \phi = \sin i \sin(f + \omega) \tag{2.99}$$

and P_k is a *Legendre polynomial* of the first kind of order k. The Legendre polynomials, denoted for some argument x by $P_k(x)$, are useful when expanding functions such as

$$\frac{1}{\|\mathbf{r} - \mathbf{r}'\|} = \frac{1}{\sqrt{r^2 + r'^2 - 2rr' \cos \gamma}} = \sum_{k=0}^{\infty} \frac{(r')^k}{r^{k+1}} P_k(\cos \gamma) \tag{2.100}$$

where $\|\mathbf{r}\|$ and $\|\mathbf{r}'\|$ are the Euclidean norms of the vectors $\mathbf{r}$ and $\mathbf{r}'$, respectively, so that $\mathbf{r} > \mathbf{r}'$, and γ is the angle between those two vectors. Each Legendre polynomial, $P_k(x)$, is an k th-degree polynomial. It may be expressed using the *Rodrigues formula*:

$$P_k(x) = \frac{1}{2^k k!} \frac{d^k}{dx^k} \left[(x^2 - 1)^k \right] \tag{2.101}$$

For example, $P_0(x) = 1$, $P_1(x) = x$, $P_2(x) = 0.5(3x^2 - 1)$ and $P_3(x) = 0.5(5x^3 - 3x)$. It is worth noting that the Legendre polynomials are orthogonal, and that the most dominant term in Eq. (2.98) is the J_2 term. The values of the second and third zonal harmonics for the Earth are $J_2 = 1082.63 \times 10^{-6}$ and $J_3 = -2.52 \times 10^{-6}$.

2.5.4 Gauss' variational equations

Gauss' variational equations (GVE), describing the time evolution of the orbital elements in the presence of perturbations or control inputs, can be obtained by the use of the chain rule:

$$\dot{\boldsymbol{\alpha}} = \frac{\partial \boldsymbol{\alpha}}{\partial t} + \frac{\partial \boldsymbol{\alpha}}{\partial \mathbf{r}} \left(\frac{\partial \mathbf{r}}{\partial t} + \frac{\partial \mathbf{r}}{\partial \boldsymbol{\alpha}} \dot{\boldsymbol{\alpha}} \right) + \frac{\partial \boldsymbol{\alpha}}{\partial \mathbf{v}} \left(\frac{\partial \mathbf{v}}{\partial t} + \frac{\partial \mathbf{v}}{\partial \boldsymbol{\alpha}} \dot{\boldsymbol{\alpha}} \right) \tag{2.102}$$

Substituting Eq. (2.77) into Eq. (2.102) using the fact that for the unperturbed problem

$$\frac{\partial \boldsymbol{\alpha}}{\partial \mathbf{r}} \frac{\partial \mathbf{r}}{\partial t} + \frac{\partial \boldsymbol{\alpha}}{\partial \mathbf{v}} \frac{\partial \mathbf{v}}{\partial t} = 0 \tag{2.103}$$

---

[8]Equation (2.98) is written using the *Vinti representation* [65]. Brouwer [66] and Kozai [67, 68] use different representations by introducing functions of both J_k and R_e as the zonal series constants.

and that α_i lacks direct time-dependence will yield

$$\dot{\boldsymbol{\alpha}} = \frac{\partial \boldsymbol{\alpha}}{\partial \mathbf{v}} \mathbf{d} \tag{2.104}$$

The second step is to express $\mathbf{r}$, $\mathbf{v}$, $\partial \boldsymbol{\alpha}/\partial \mathbf{v}$ and $\mathbf{d}$ in an LVLH coordinate system fixed to the spacecraft. In this coordinate system, denoted by $\mathscr{L}$ in Section 2.1, the unit vector $\hat{\mathbf{R}}$ is directed radially outwards, $\hat{\mathbf{S}}$ is perpendicular to $\hat{\mathbf{R}}$ in the direction of the instantaneous velocity, and $\hat{\mathbf{W}}$ completes the right-hand triad.

An explicit form of the GVE is derived by choosing the classical orbital elements as the state variables, i.e., $\boldsymbol{\alpha} = [a, e, i, \Omega, \omega, M_0]^T$, where, as before, a is the semimajor axis, e is the eccentricity, i is the inclination, Ω is the right ascension of the ascending node, ω is the argument of periapsis, and M_0 is defined as [69]

$$M_0 = M - \int_{t_0}^{t} n \, dt \tag{2.105}$$

where M is the mean anomaly, t_0 is a reference time (which may differ from the periapsis passage time) and $n = \sqrt{\mu/a^3}$ is the mean motion.

In addition, we write the position and velocity vectors in frame $\mathscr{L}$ using a polar representation, so that

$$\mathbf{r} = r\hat{\mathbf{R}}, \quad \mathbf{v} = \frac{\partial r}{\partial t}\hat{\mathbf{R}} + r\frac{\partial f}{\partial t}\hat{\mathbf{S}} \tag{2.106}$$

where f is the true anomaly. By collecting all expressions and substituting into Eq. (2.104), we can obtain GVE. Writing $\mathbf{d} = d_r\hat{\mathbf{R}} + d_\theta\hat{\mathbf{S}} + d_h\hat{\mathbf{W}} = [d_r, d_\theta, d_h]^T$, provides the following equations:

$$\frac{da}{dt} = 2\frac{d_r a^2 e \sin f}{h} + 2\frac{d_\theta a^2 p}{hr} \tag{2.107a}$$

$$\frac{de}{dt} = \frac{d_r p \sin f}{h} + \frac{d_\theta \left[(p + r)\cos f + re\right]}{h} \tag{2.107b}$$

$$\frac{di}{dt} = \frac{d_h r \cos(f + \omega)}{h} \tag{2.107c}$$

$$\frac{d\Omega}{dt} = \frac{d_h r \sin(f + \omega)}{h \sin i} \tag{2.107d}$$

$$\frac{d\omega}{dt} = -\frac{d_r p \cos f}{he} + \frac{d_\theta (p + r)\sin f}{he}$$
$$- \frac{d_h r \sin(f + \omega)\cos i}{h \sin i} \tag{2.107e}$$

$$\frac{dM_0}{dt} = d_r \left[\frac{\left(-2e + \cos f + e \cos^2 f\right)\left(1 - e^2\right)}{e(1 + e \cos f)na} \right]$$
$$+ d_\theta \left[\frac{\left(e^2 - 1\right)(e \cos f + 2)\sin f}{e(1 + e \cos f)na} \right] \tag{2.107f}$$

where as before $p = a(1 - e^2)$ is the semilatus rectum and $h = \sqrt{\mu p}$ is the magnitude of the angular momentum vector. Another useful relationship is the variational equation for the true anomaly, obtained by using Eqs. (2.16), (2.23) and (2.107e):

$$\frac{df}{dt} = \frac{h}{r^2} + \frac{1}{eh} [d_r p \cos f - d_\theta (p + r) \sin f] \qquad (2.108)$$

2.6 AVERAGING THEORY

A very powerful and useful tool for dealing with the orbital elements variational equations is *averaging*. Classical averaging theory was originally developed in order to simplify nonlinear nonautonomous systems. The standard form of the equations of motion for averaging is

$$\dot{\mathbf{x}} = \epsilon \mathbf{F}(\mathbf{x}, t), \quad \mathbf{x}(0) = \mathbf{x}_0 \qquad (2.109)$$

where $\mathbf{x} \in \mathbb{M} \subset \mathbb{R}^n$, $\mathbf{F}$ is T-periodic, $\mathbf{F}(\mathbf{x}, t) = \mathbf{F}(\mathbf{x}, t + T)$. The *averaging operator* on $\mathbf{F}$, denoted by $\langle \mathbf{F} \rangle$ yields a *mean value*, $\bar{\mathbf{F}}$, via the quadrature

$$\bar{\mathbf{F}} = \langle \mathbf{F} \rangle \triangleq \frac{1}{T} \int_t^{t+T} \mathbf{F}(\mathbf{x}, \tau) d\tau \qquad (2.110)$$

The averaged differential equations are

$$\dot{\bar{\mathbf{x}}} = \epsilon \mathbf{F}(\bar{\mathbf{x}}) \qquad (2.111)$$

The primary goal of averaging is removing the time-dependence of the original differential equations, thus permitting a considerable simplification of the resulting dynamics. However, one should determine under what conditions the averaged and the original systems coincide, in order for the averaging to be meaningful. To that end, we consider the following theorem.

Theorem 2.1 (*First-order averaging [70]*). *Consider the dynamical systems* (2.110) *and* (2.111). *If* $\mathbf{F}(\mathbf{x}, t)$ *is continuous in* $\mathbf{x}$ *and* t *and in addition* $\bar{\mathbf{x}} \in \mathbb{M}, \forall t_0 \leq t \leq t_0 + T/\epsilon$, *then* $\mathbf{x} = \bar{\mathbf{x}} + \mathcal{O}(\epsilon), \forall t_0 \leq t \leq t_0 + T/\epsilon$ *as* $\epsilon \to 0$.

Theorem 2.1 determines the conditions under which the averaged and original dynamical systems are identical to first-order, meaning that taking $\mathbf{x} = \bar{\mathbf{x}}$ and $\dot{\mathbf{x}} = \dot{\bar{\mathbf{x}}}$ is correct to first-order in ϵ. In this case, the averaging error will be $\mathcal{O}(\epsilon^2)$.

Application of averaging to the orbital variational equations is straightforward. For some perturbing potential $\mathcal{R}$, we average using [33]:

$$\bar{\mathcal{R}} = \langle \mathcal{R} \rangle = \frac{1}{2\pi} \int_0^{2\pi} \mathcal{R} dM \qquad (2.112)$$

or, alternatively,

$$\bar{\mathcal{R}} = \frac{1}{2\pi} \int_0^{2\pi} \frac{n}{h} \mathcal{R} r^2 df \tag{2.113}$$

Battin [33] provides an expression for the first-order averaged potential due to J_2 (cf. Subsection 2.5.3):

$$\bar{\mathcal{R}} = \frac{\bar{n}^2 J_2 R_e^2}{4(1 - \bar{e}^2)^{\frac{3}{2}}} (3 \cos^2 \bar{i} - 1) \tag{2.114}$$

Upon substitution into the LPE (2.97), one obtains the linear differential equations for the *mean classical orbital elements*:

$$\frac{d\bar{a}}{dt} = 0 \tag{2.115a}$$

$$\frac{d\bar{e}}{dt} = 0 \tag{2.115b}$$

$$\frac{d\bar{i}}{dt} = 0 \tag{2.115c}$$

$$\frac{d\bar{\Omega}}{dt} = -\frac{3}{2} J_2 \left(\frac{R_e}{\bar{p}}\right)^2 \bar{n} \cos \bar{i} \tag{2.115d}$$

$$\frac{d\bar{\omega}}{dt} = \frac{3}{4} J_2 \left(\frac{R_e}{\bar{p}}\right)^2 \bar{n}(5 \cos^2 \bar{i} - 1) \tag{2.115e}$$

$$\frac{d\bar{M}_0}{dt} = \frac{3}{4} J_2 \left(\frac{R_e}{\bar{p}}\right)^2 \bar{n}\bar{\eta}(3 \cos^2 \bar{i} - 1) \tag{2.115f}$$

where $\bar{n} = \sqrt{\frac{\mu}{\bar{a}^3}}$ and $\bar{\eta} = \sqrt{1 - \bar{e}^2}$. These equations predict secular growths of Ω, ω and M_0. To first order in J_2, there are no long-periodic and no secular variations in the remaining elements.

Gurfil [50] presented the first-order averaged LPE for the Euler-parameters-based elements defined by Eqs. (2.46) due to the J_2-perturbation as an alternative for Eqs. (2.115). The alternative equations are given below:

$$\dot{\bar{a}} = 0 \tag{2.116a}$$

$$\dot{\bar{\eta}} = 0 \tag{2.116b}$$

$$\dot{\bar{\beta}}_1 = \frac{3}{4} J_2 \left(\frac{R_e}{\bar{p}}\right)^2 \bar{n}\bar{\beta}_2[3 + 10(\bar{\beta}_1^2 + \bar{\beta}_2^2)^2 - 12(\bar{\beta}_1^2 + \bar{\beta}_2^2)] \tag{2.116c}$$

$$\dot{\bar{\beta}}_2 = -\frac{3}{4} J_2 \left(\frac{R_e}{\bar{p}}\right)^2 \bar{n}\bar{\beta}_1[3 + 10(\bar{\beta}_1^2 + \bar{\beta}_2^2)^2 - 12(\bar{\beta}_1^2 + \bar{\beta}_2^2)] \tag{2.116d}$$

$$\dot{\bar{\beta}}_3 = \frac{3}{4} J_2 \left(\frac{R_e}{\bar{p}}\right)^2 \bar{n}\bar{\beta}_4[1 + 10(\bar{\beta}_1^2 + \bar{\beta}_2^2)^2 - 8(\bar{\beta}_1^2 + \bar{\beta}_2^2)] \tag{2.116e}$$

$$\dot{\bar{M}}_0 = \frac{3}{2} J_2 \left(\frac{R_e}{\bar{p}} \right)^2 \bar{n} \bar{\eta} [1 - 6\bar{\beta}_1^2 - 6\bar{\beta}_2^2 + 6(\bar{\beta}_1^2 + \bar{\beta}_2^2)^2] \qquad (2.116f)$$

Example 2.2. *Determine the impulse required to suppress the nodal precession accumulated over one orbit period for a mean circular orbit with $\bar{a} = 7100$ km and $\bar{i} = 70°$.*

Assuming that $R_e = 6378.1363$ km and $\mu = 3.98604415 \times 10^5$ km^3/s^2, the nodal precession accumulated over one orbit period can be calculated from Eq. (2.115d) to be $\delta\Omega = -0.00282$ rad. The cross-track impulse required to cancel this accumulation is obtained from Eq. (2.107d) using the impulsive thrust assumption as

$$\Delta v_h = \left| \frac{\delta\Omega \sqrt{\mu/\bar{a}} \sin \bar{i}}{\sin \theta} \right| = 19.829 \text{ m/s} \qquad (2.117)$$

We made the tacit assumption in performing the above calculation that the impulse is applied at $\theta = \pi/2$. This example also illustrates the expense involved in "fighting with nature".

Example 2.3. *Determine the impulse required to suppress the differential nodal precession accumulated over one orbit period for a mean circular orbit with $\bar{a} = 7100$ km and $\bar{i} = 70°$ due to a differential inclination $\delta i = 1/7100$ rad.*

The differential nodal precession for the example of a circular orbit over one period, using Eq. (2.115d), is $\delta\Omega = 3\pi J_2 (R_e/\bar{a})^2 (\sin \bar{i})\delta i$. Substituting for the given data into Eqs. (2.107d) and (2.115d), we obtain

$$\Delta v_h = 7.6732 \times 10^{-3} \text{ m/s} \qquad (2.118)$$

Over a period of one year the impulse required to perform this operation is 41 m/s.

SUMMARY

This chapter presented a brief treatment of the vast field of astrodynamics. We touched upon the important topics that serve as the building blocks for the material to follow in the subsequent chapters. The orbital mechanics of Keplerian motion as well as perturbed motion were introduced. Of prime significance are the Gauss variational equations, the concept of averaging, first-order secular rates in the orbital elements due to the J_2 perturbation, and the estimates of impulse requirements to mitigate the absolute and differential effects of the perturbations.

Chapter three

The Basics of Analytical Mechanics, Optimization, Control and Estimation

making tapes together, discs together,
sweating for applause
they read basically to and for
each other.

Charles Bukowski (1920–1994)

In this chapter, we review the basics of *analytical mechanics* as applied to orbit theory, optimization, control and estimation. We start by presenting some notions in Lagrangian and Hamiltonian mechanics, including *canonical transformations*. We introduce the *Delaunay variables* and show their applications in *Brouwer's satellite theory*. This material is followed by a discussion of static optimization, control, and filtering methods, to be applied in subsequent chapters to the development of relative spacecraft control and navigation algorithms.

3.1 LAGRANGIAN AND HAMILTONIAN MECHANICS

Lagrangian and Hamiltonian mechanics are two main approaches to mechanics that constitute a generalization of the classical Newtonian mechanics. The rationale behind the Lagrangian and Hamiltonian formalisms, often termed analytical mechanics, is related to *variational principles*, which we will briefly discuss in this section. Hamiltonian mechanics is based on the energy concept, and is directly related to symmetry, one of the most powerful tools in modern physics. The subject of Lagrangian and Hamiltonian mechanics is vast. The reader is referred to, e.g., Marsden and Ratiu [71] for a rigorous treatment of these subjects.

39

Spacecraft Formation Flying; ISBN: 9780750685337

Lagrangian and Hamiltonian mechanics have played a fundamental role in astrodynamics during the past two and a half centuries. Hamiltonian mechanics was applied to some of the most fundamental astrodynamical theories, such as Brouwer's artificial satellite theory [59,66], to be discussed in Chapters 7 and 8. Our interest in Lagrangian and Hamiltonian mechanics stems from the important applications of these methods to the development of spacecraft formation flying models. We will subsequently use Lagrangian and Hamiltonian methods in Chapter 4, where we show how to develop nonlinear models of relative motion using analytical mechanics; in Chapter 5, where we show how to derive linear relative motion models based on energy methods; and in Chapter 8, where we develop perturbation theories.

Lagrangian mechanics deals with a *configuration space*, $\mathbb{Q}$, parameterized by a set of *generalized coordinates*, $\mathbf{q} \in \mathbb{Q}$, and *generalized velocities*, $\dot{\mathbf{q}}$. The structure $T_s(\mathbb{Q}) = \mathbb{Q} \times \mathbb{Q}$ is called the *tangent space*. Coordinates on $T_s(\mathbb{Q})$ are then $(\mathbf{q}, \dot{\mathbf{q}})$. The *Lagrangian*, $\mathcal{L} : T_s(\mathbb{Q}) \times \mathbb{R} \rightarrow \mathbb{R}$, is a function of time and the generalized coordinates and velocities, and is defined as the kinetic minus the potential energy of the system:

$$\mathcal{L}(\mathbf{q}, \dot{\mathbf{q}}, t) = \mathcal{K} - \mathcal{U} \tag{3.1}$$

Hamilton's variational principle states that

$$\delta \int_a^b \mathcal{L}(\mathbf{q}, \dot{\mathbf{q}}, t)dt = 0 \tag{3.2}$$

Using the calculus of variations, Eq. (3.2) leads to the *Euler–Lagrange equations*

$$\frac{d}{dt}\frac{\partial \mathcal{L}}{\partial \dot{\mathbf{q}}} - \frac{\partial \mathcal{L}}{\partial \mathbf{q}} = \mathbf{0} \tag{3.3}$$

To use the Hamiltonian formalism, one introduces the *conjugate momenta*, defined as

$$\mathbf{p} = \frac{\partial \mathcal{L}}{\partial \dot{\mathbf{q}}} \tag{3.4}$$

and then adopts the change of variables $(\mathbf{q}, \dot{\mathbf{q}}) \rightarrow (\mathbf{q}, \mathbf{p})$. The *Legendre transformation* provides a connection between the *Hamiltonian*, $\mathcal{H}$, and the Lagrangian:

$$\mathcal{H}(\mathbf{q}, \mathbf{p}, t) = \mathbf{p}^T \dot{\mathbf{q}} - \mathcal{L}(\mathbf{q}, \dot{\mathbf{q}}, t) \tag{3.5}$$

The structure $T_c(\mathbb{Q})$, having coordinates $(\mathbf{q}, \mathbf{p})$ is referred to as the *cotangent space*, so that $\mathcal{H} : T_c(\mathbb{Q}) \times \mathbb{R} \rightarrow \mathbb{R}$. Using the change of variables and the Legendre transformation leads to *Hamilton's equations*:

$$\dot{\mathbf{q}} = \frac{\partial \mathcal{H}}{\partial \mathbf{p}}$$

$$\dot{\mathbf{p}} = -\frac{\partial \mathcal{H}}{\partial \mathbf{q}} \tag{3.6}$$

3.2 THE DELAUNAY ELEMENTS

The Hamiltonian formalism can be applied to astrodynamical problems using the Delaunay canonical elements. These elements are obtained when one defines the transformation

$$l = M \tag{3.7a}$$
$$g = \omega \tag{3.7b}$$
$$\hbar = \Omega \tag{3.7c}$$
$$L = \sqrt{\mu a} \tag{3.7d}$$
$$G = \sqrt{\mu a (1 - e^2)} \tag{3.7e}$$
$$H = G \cos i \tag{3.7f}$$

In this case $(l, g, \hbar)$ become generalized coordinates, and (L, G, H) are the conjugate momenta.[1] Given a nominal Hamiltonian, $\mathcal{H}_0$, representing the two-body Keplerian motion, a perturbing Hamiltonian, $\mathcal{H}_1$, and the total Hamiltonian, $\mathcal{H} = \mathcal{H}_0 + \varepsilon \mathcal{H}_1$, the Delaunay variables satisfy the following Hamilton equations (cf. Eqs. (3.6)):

$$\dot{l} = \frac{\partial \mathcal{H}}{\partial L}, \quad \dot{g} = \frac{\partial \mathcal{H}}{\partial G}, \quad \dot{\hbar} = \frac{\partial \mathcal{H}}{\partial H} \tag{3.8}$$

$$\dot{L} = -\frac{\partial \mathcal{H}}{\partial l}, \quad \dot{G} = -\frac{\partial \mathcal{H}}{\partial g}, \quad \dot{H} = -\frac{\partial \mathcal{H}}{\partial \hbar} \tag{3.9}$$

Note that G is the orbital angular momentum and H is the corresponding polar component. The term $\mathcal{H}_0$ is given by

$$\mathcal{H}_0 = -\frac{\mu^2}{2L^2} \tag{3.10}$$

The consequences of $\mathcal{H}_0$ being only a function of L, for the two-body problem are that all the momenta and the coordinates g and $\hbar$ are constants and l varies linearly with time.

These results are no longer valid in the presence of the perturbing Hamiltonian. The Delaunay variables lend themselves to canonical transformations, which render the Hamiltonian system easily integrable; this is the basis for Brouwer's satellite theory [66]. We discuss such transformations in Section 3.3, before introducing Brouwer's satellite theory. We will also use the Delaunay formalism in Chapter 8 to derive perturbation mitigation methods.

---

[1] We will use the notation $\hbar$ instead of the customary h so as to avoid confusion with the orbital angular momentum which is denoted by H.

3.3 CANONICAL TRANSFORMATIONS

In general, the Hamiltonian is a function of both $\mathbf{q}$ and $\mathbf{p}$. It is advantageous in many applications to determine transformations of the generalized coordinates and momenta such that the new variables $(\widetilde{\mathbf{q}}, \widetilde{\mathbf{p}})$ also satisfy Eq. (3.6), but with a different Hamiltonian, $\widetilde{\mathcal{H}}(\widetilde{\mathbf{q}}, \widetilde{\mathbf{p}}, t)$, having some special properties. Such a transformation is called a *canonical transformation*, and it is a vast subject matter in itself. There are many advantages of finding such transformations; for example, $\widetilde{\mathcal{H}} = \widetilde{\mathcal{H}}(\widetilde{\mathbf{p}}, t)$ leads to the constants of motion $\widetilde{\mathbf{p}}$ (cf. Eq. (3.10)). We will restrict our discussion to the special case of *autonomous dynamical systems*, for which it can be shown that a canonical transformation must satisfy the Hamiltonian invariance property Ref. [59]

$$\widetilde{\mathbf{q}}^T \, d\widetilde{\mathbf{p}} - \mathbf{q}^T \, d\mathbf{p} = d\left[\widetilde{\mathbf{q}}^T \widetilde{\mathbf{p}} - W \right] \tag{3.11}$$

where $W(\widetilde{\mathbf{q}}, \mathbf{p})$ is called a *generating function* and d stands for the differential. The generating function can be of various types, but the one being discussed here depends on the new generalized coordinates and the old momentum variables such that

$$\widetilde{\mathbf{p}} = \frac{\partial W(\widetilde{\mathbf{q}}, \mathbf{p})}{\partial \widetilde{\mathbf{q}}} \tag{3.12a}$$

$$\mathbf{q} = \frac{\partial W(\widetilde{\mathbf{q}}, \mathbf{p})}{\partial \mathbf{p}} \tag{3.12b}$$

Given $(\widetilde{\mathbf{q}}, \widetilde{\mathbf{p}})$, Eqs. (3.12) can be solved for $(\mathbf{q}, \mathbf{p})$ if W is known. A variety of transformations can be generated by selecting different generating functions.

It is usually a difficult task to obtain an explicit generating function to meet the desired objectives for the transformation, since equations of the type (3.12) are implicit. However, the task is simplified for systems whose Hamiltonians can be represented as

$$\mathcal{H}(\mathbf{q}, \mathbf{p}) = \mathcal{H}_0(\mathbf{q}, \mathbf{p}) + \epsilon \mathcal{H}_1(\mathbf{q}, \mathbf{p}) + 0.5 \, \epsilon^2 \mathcal{H}_2(\mathbf{q}, \mathbf{p}) + \mathcal{O}(\epsilon^3) \tag{3.13}$$

where ϵ is a small non-dimensional parameter and the unperturbed Hamiltonian $\mathcal{H}_0$ is that of an integrable system, such as the two-body problem. The perturbation parameter ϵ, for example, can be related to J_2. There are two common methods for determining the canonical transformations of such systems: von Zeipel's method, which expands the perturbed Hamiltonian into a Taylor series [66], and Hori's method [72], which uses the Lie series expansion [73]. The von Zeipel method involves a generating function of the mixed type, as used in Eqs. (3.12). On the other hand, Hori's method determines a generating function of the new coordinates and momenta. The generating function is expanded as a series in ϵ:

$$W(\widetilde{\mathbf{q}}, \widetilde{\mathbf{p}}) = W_1(\widetilde{\mathbf{q}}, \widetilde{\mathbf{p}}) + \epsilon W_2(\widetilde{\mathbf{q}}, \widetilde{\mathbf{p}}) + \mathcal{O}(\epsilon^2) \tag{3.14}$$

and the transformed Hamiltonian, also expanded as

$$\widetilde{\mathcal{H}}(\widetilde{\mathbf{q}}, \widetilde{\mathbf{p}}) = \widetilde{\mathcal{H}}_0(\widetilde{\mathbf{q}}, \widetilde{\mathbf{p}}) + \epsilon \widetilde{\mathcal{H}}_1(\widetilde{\mathbf{q}}, \widetilde{\mathbf{p}}) + \mathcal{O}(\epsilon^2) \tag{3.15}$$

The Lie theorem provides a relationship between the Hamiltonian and the generating function as follows:

$$\mathcal{H}(\mathbf{q}, \mathbf{p}) = \mathcal{H}_0(\widetilde{\mathbf{q}}, \widetilde{\mathbf{p}}) + \epsilon \{\mathcal{H}, W\} + 0.5 \, \epsilon^2 \{\{\mathcal{H}, W\}, W\} + \mathcal{O}(\epsilon^3) \tag{3.16}$$

where $\{\cdot, \cdot\}$ is the Poisson bracket, defined by Eq. (2.93).

The coefficients of the like powers of ϵ can be equated between Eqs. (3.15) and (3.16) to obtain a series of equations, the first two of which are

$$\widetilde{\mathcal{H}}_0(\widetilde{\mathbf{q}}, \widetilde{\mathbf{p}}) = \mathcal{H}_0(\widetilde{\mathbf{q}}, \widetilde{\mathbf{p}}) \tag{3.17a}$$

$$\widetilde{\mathcal{H}}_1(\widetilde{\mathbf{q}}, \widetilde{\mathbf{p}}) = \mathcal{H}_1(\widetilde{\mathbf{q}}, \widetilde{\mathbf{p}}) + \{\mathcal{H}_0(\widetilde{\mathbf{q}}, \widetilde{\mathbf{p}}), W_1\} \tag{3.17b}$$

Thus, $\widetilde{\mathcal{H}}_0$ is obtained by substituting the new variables $(\widetilde{\mathbf{q}}, \widetilde{\mathbf{p}})$ for the respective old variables $(\mathbf{q}, \mathbf{p})$ in $\mathcal{H}_0$. Since Eq. (3.17b) is a single equation in two unknowns, $\widetilde{\mathcal{H}}_1$ and W_1, it can be solved in a variety of ways. Generally, the secular part of $\mathcal{H}_1(\widetilde{\mathbf{q}}, \widetilde{\mathbf{p}})$ is absorbed into $\widetilde{\mathcal{H}}_1(\widetilde{\mathbf{q}}, \widetilde{\mathbf{p}})$ and the periodic part is absorbed into the solution for W_1 via quadratures. This process is called averaging. The detailed treatments of the Brouwer and Hori methods are given in Vinti [65].

A modified version of Hori's method has also been implemented for a higher-order satellite theory by Coffey and Deprit [74] with the help of a special-purpose symbolic algebra program.

3.4 BROUWER THEORY

Brouwer [66] developed his satellite theory based on the conversion of the mean Delaunay variables into the corresponding osculating variables, a process known as the *Brouwer transformation*. Brouwer writes Eq. (3.13) for the perturbed Hamiltonian with $\epsilon = -J_2$. Our development of the Brouwer theory uses only the J_2 terms. The higher-order geopotential terms are ignored. The canonical variables used are the Delaunay variables, introduced in Section 3.2. The first term of the Hamiltonian is as defined by Eq. (3.10), and the second term is

$$\mathcal{H}_1 = \frac{\mu^4 R_e^2}{4L^6} \left(\frac{a}{r}\right)^3 \left[\left(3\frac{H^2}{G^2} - 1\right) + 3\left(1 - \frac{H^2}{G^2}\right) \cos\theta \right] \tag{3.18}$$

where r is the orbit radius and θ is the argument of latitude. Note that $\mathcal{H}$ is a function of all the variables except $\hbar$; consequently, none of them will be constant except H, the polar component of the angular momentum.[2] Using Lie

---

[2]With J_2 as the only perturbation, the right ascension, $\hbar$, does not appear; however, it would appear at the second-order if the time-dependent terms of the geopotential were included.

series [73] or the von Zeipel method as done by Brouwer [66], the Hamiltonian is averaged first with respect to the mean anomaly l to remove the short-periodic (of the order of the orbit period) terms, and then with respect to the argument of perigee g to remove the long-periodic (approximately one order higher than the orbit period) terms. In the remainder of this development, the notation $(\bar{\cdot})$ will be dropped and the elements (l, g, h, L, G, H) will be assumed to be mean elements. The algebra is minimized if dimensionless variables are used, so distances will be normalized by the Earth's radius, and time will be normalized by the mean motion of a satellite at one Earth radius. This gives $\mu = 1$. In the dimensionless variables, the averaged Hamiltonians become

$$\bar{\mathcal{H}}_0 = -\frac{1}{2L^2} \tag{3.19}$$

$$\bar{\mathcal{H}}_1 = -\frac{1}{4L^6} \left(\frac{L}{G}\right)^3 \left(1 - 3\frac{H^2}{G^2}\right) \tag{3.20}$$

The first-order long-periodic and short-periodic generating functions of Brouwer are

$$W^{(lp)} = -\left(\frac{1}{32G^3}\right)\left(1 - \frac{G^2}{L^2}\right)\left(1 - 5\frac{H^2}{G^2}\right)^{-1}$$
$$\times \left(1 - 16\frac{H^2}{G^2} + 15\frac{H^4}{G^4}\right) \sin 2g \tag{3.21}$$

and

$$W^{(sp)} = -\frac{1}{4G^3}\left(1 - 3\frac{H^2}{G^2}\right)(f - l + e\sin f) + \frac{3}{8G^3}\left(1 - \frac{H^2}{G^2}\right)$$
$$\times \left[\sin(2f + 2g) + e\sin(f + 2g) + \frac{e}{3}\sin(3f + 2g)\right] \tag{3.22}$$

For the sake of completeness and for reference in Chapter 7, the second-order averaged Hamiltonian is also given below:

$$\bar{\mathcal{H}}_2 = -\frac{15}{64L^{10}}\left(\frac{L}{G}\right)^5\left[\left(1 - \frac{18}{5}\frac{H^2}{G^2} + \frac{H^4}{G^4}\right)\right.$$
$$+ \frac{4}{5}\left(\frac{L}{G}\right)\left(1 - 6\frac{H^2}{G^2} + 9\frac{H^4}{G^4}\right)$$
$$\left. - \left(\frac{L}{G}\right)^2\left(1 - 2\frac{H^2}{G^2} - 7\frac{H^4}{G^4}\right)\right] \tag{3.23}$$

Note that after averaging, the Hamiltonian is a function of only the momenta, i.e., the angles are ignorable. Consequently, the angle rates and momenta are constant. As will be shown in Chapter 8, this property makes the design of formations in mean element space easier than that with osculating elements. In the osculating space, the angles are not linear in time and the momenta are not constant. Hence, finding the initial values of the elements for a specific formation is much more difficult than when using mean elements.

Brouwer's transformation in its original form accepts mean elements as inputs and produces the osculating elements as the outputs. The exact inverse transformation is nontrivial to write, since it is nonlinear. However, as we are dealing with a perturbed system with a small perturbation parameter, the first-order approximate inverse transformation can be obtained by replacing J_2 by $-J_2$ in the Brouwer transformation and treating the osculating elements as the inputs and the mean elements as the outputs. This is very convenient, but it may not be acceptable for accurate simulations. If such is the case, then an iterative solution to the inverse transformation may be pursued.

The first-order transformation between the mean and osculating nonsingular orbital elements [75] presented in Chapter 7 forms an important part of the material presented in this book and it is based on generating functions obtained by Brouwer. The detailed expressions of this transformation are provided in Appendix E. Lyddane's modification [76] of Brouwer theory to eliminate singularities associated with small eccentricities and inclinations can be found in Ref. [29]. We present an example of the forward and inverse Brouwer transformation next.

Example 3.1 (*Mean-to-osculating transformation*). *Let the given mean elements be:*

$$\bar{a} = 7100 \text{ km}, \quad \bar{\theta} = 0 \text{ rad}, \quad \bar{i} = 70^\circ$$
$$\bar{q}_1 = 0.05, \quad \bar{q}_2 = 0.05, \quad \bar{\Omega} = 45^\circ \tag{3.24}$$

Obtain the osculating elements.

The first-order Brouwer transformation results in the following osculating elements:

$$a = 7109.31795 \text{ km}, \quad \theta = 0.00005 \text{ rad}, \quad i = 1.22196 \text{ rad}$$
$$q_1 = 0.05063, \quad q_2 = 0.05003, \quad \Omega = 0.78547 \text{ rad} \tag{3.25}$$

Note that there is approximately a 10 km difference between the mean and osculating semimajor axis values.

Example 3.2 (*Osculating-to-mean transformation*). *Now consider the osculating elements of the previous example as given and obtain the mean elements via the first-order inverse transformation.*

The first-order inverse transformation provides the following mean elements:

$$\bar{a} = 7099.996055 \text{ km}, \quad \bar{\theta} = 0.000008 \text{ rad}, \quad \bar{i} = 1.221731 \text{ rad}$$
$$\bar{q}_1 = 0.0500006, \quad \bar{q}_2 = 0.04999994, \quad \bar{\Omega} = 0.7853984 \text{ rad} \tag{3.26}$$

Note that there are minor differences between the mean elements obtained for this example and their respective counterparts from the previous example.

3.4.1 Osculating-to-mean Iterative Solution

If the mean-to-osculating transformation is represented as a nonlinear mapping

$$œ = œ(\overline{œ}) \tag{3.27}$$

then an iterative procedure can be devised to solve Eq. (3.27) for $\overline{œ}$, given $œ$. This procedure will require an initial guess for $\overline{œ}$ and the information on the Jacobian $D = \partial œ / \partial \overline{œ}$. The matrix D is provided in Appendix F.

3.5 CONSTRAINED STATIC OPTIMIZATION

In cases where the control of relative motion is applied using impulsive thrusters, the relative dynamics and the associated kinematic constraints may be described by algebraic equations. In this case, derivation of relative spacecraft control methods can be based on simple *static optimization* procedures; by static optimization we mean that differential equations are absent. We will elaborate on this topic in Section 4.3. The purpose of the following discussion is to introduce a method for solving constrained static optimization problems using the formalism of *Lagrange multipliers*.

To that end, consider the following static optimization problem:

$$\text{minimize} \quad \mathcal{J}(\mathbf{x})$$

s. t.

$$g_k(\mathbf{x}) = 0, \quad k = 1, \ldots, m \tag{3.28}$$

In Eq. (3.27), $\mathcal{J}$ is the *objective function*, $\mathbf{x}$ is an n-dimensional vector of optimization variables, $[x_1, x_2, \ldots, x_n]^T$, and g_k are m equality constraints.[3] Now, we define the Lagrangian, $\mathcal{L}$, as follows:

$$\mathcal{L}(\mathbf{x}, \mathbf{\Lambda}) = \mathcal{J}(\mathbf{x}) + \sum_{k=1}^{m} \lambda_k g_k(\mathbf{x}) \tag{3.29}$$

where $\mathbf{\Lambda} = [\lambda_1, \lambda_2, \ldots, \lambda_m]^T$, $m < n$, is a vector of Lagrange multipliers. If $\mathcal{J}$ and the g_k have continuous first partial derivatives and the gradients of the g_k do not vanish on the domain of $\mathcal{J}$ (the domain of $\mathcal{J}$ is assumed an open set containing all points satisfying the constraints), then the stationary points of the Lagrangian are determined by

$$\nabla_{\mathbf{x}} \mathcal{L} = \mathbf{0} \tag{3.30}$$

$$\nabla_{\mathbf{\Lambda}} \mathcal{L} = \mathbf{0} \tag{3.31}$$

---

[3] The method of Lagrange multipliers is generalized by the Karush–Kuhn–Tucker conditions [77], which can also take into account inequality constraints.

Equivalently, the stationary points of the Lagrangian must satisfy

$$\nabla \mathcal{L} = \mathbf{0} \tag{3.32}$$

Equation (3.32) gives $m + n$ unique equations for the unknowns $\mathbf{x}^\star$ and $\mathbf{\Lambda}^\star$, constituting the solutions for the optimization parameters and Lagrange multipliers at the critical point.

To determine whether some stationary point $\{\mathbf{x}^\star, \mathbf{\Lambda}^\star\}$ is a minimum, one must examine the *Hessian* [4] of the Lagrangian, also referred to as the *bordered Hessian*. If there exist vectors $\mathbf{x}^\star \in \mathbb{R}^n$ and $\mathbf{\Lambda}^\star \in \mathbb{R}^m$ such that

$$\nabla \mathcal{L}(\mathbf{x}^\star, \mathbf{\Lambda}^\star) = \mathbf{0} \tag{3.33}$$

and if, for twice continuously-differentiable $\mathcal{J}$ and g_k,

$$(-1)^m \det \begin{bmatrix} \dfrac{\partial^2 \mathcal{L}(\mathbf{x}^\star, \mathbf{\Lambda}^\star)}{\partial x_1 \partial x_1} & \cdots & \dfrac{\partial^2 \mathcal{L}(\mathbf{x}^\star, \mathbf{\Lambda}^\star)}{\partial x_1 \partial x_p} & \dfrac{\partial g_1(\mathbf{x}^\star)}{\partial x_1} & \cdots & \dfrac{\partial g_m(\mathbf{x}^\star)}{\partial x_1} \\ \vdots & \vdots & \vdots & \vdots & \vdots & \vdots \\ \dfrac{\partial^2 \mathcal{L}(\mathbf{x}^\star, \mathbf{\Lambda}^\star)}{\partial x_p \partial x_1} & \cdots & \dfrac{\partial^2 \mathcal{L}(\mathbf{x}^\star, \mathbf{\Lambda}^\star)}{\partial x_p \partial x_p} & \dfrac{\partial g_1(\mathbf{x}^\star)}{\partial x_p} & \cdots & \dfrac{\partial g_m(\mathbf{x}^\star)}{\partial x_p} \\ \dfrac{\partial g_1(\mathbf{x}^\star)}{\partial x_1} & \cdots & \dfrac{\partial g_1(\mathbf{x}^\star)}{\partial x_p} & 0 & \cdots & 0 \\ \vdots & \vdots & \vdots & \vdots & \vdots & \vdots \\ \dfrac{\partial g_m(\mathbf{x}^\star)}{\partial x_1} & \cdots & \dfrac{\partial g_m(\mathbf{x}^\star)}{\partial x_p} & 0 & \cdots & 0 \end{bmatrix} > 0 \tag{3.34}$$

for $p = m + 1, \ldots, n$, then $\mathcal{J}$ has a local minimum at $\mathbf{x}^\star$ such that

$$g_k(\mathbf{x}^\star) = 0, \quad k = 1, \ldots, m \tag{3.35}$$

3.6 CONTROL LYAPUNOV FUNCTIONS

Our interest in *Control Lyapunov Functions* (CLFs) stems from their application to relative spacecraft control, to be discussed in Chapter 10. We consider autonomous nonlinear systems of the form

$$\dot{\mathbf{x}} = \mathbf{f}(\mathbf{x}) \tag{3.36}$$

to provide an elementary treatment of the CLF approach. It is assumed that the system of Eq. (3.36) satisfies the property

$$\mathbf{f}(\mathbf{0}) = \mathbf{0}, \tag{3.37}$$

---

[4]Named after the 19th century German mathematician Ludwig Otto Hesse.

i.e., the origin is an equilibrium point. The stability of the system of Eq. (3.36) at the origin can be verified by using *Lyapunov's theorem* [78], which can be stated as follows.

Theorem 3.1 (*Lyapunov's theorem [78]*). *If V(x) is a continuously differentiable, positive definite function, defined in a domain $\mathbb{D}$ containing the origin (i.e., V(0) = 0 and V(x) > 0 in $\mathbb{D}$, except at x = 0) and furthermore,*

$$\dot{V}(0) = 0 \quad and \quad \dot{V}(x) < 0; \quad x \neq 0 \tag{3.38}$$

then the equilibrium point x = 0 of Eq. (3.36) is asymptotically stable and V(x) is a Lyapunov function.

For our discussion, Eq. (3.36) can be considered to be the representation of a dynamical system under the action of a feedback control law, $u = u(x)$. The CLF approach begins with the selection of a positive definite Lyapunov test function and the feedback control law is determined to satisfy Eq. (3.38) along the trajectories of Eq. (3.36), thus ensuring closed-loop stability. The choice of a CLF is not unique, but in many cases it may resemble the energy or the Hamiltonian of a closely-related open-loop (unforced) system.

3.7 LINEAR QUADRATIC REGULATION

As mentioned in the Chapter 1, the *Linear Quadratic Regulator* (LQR) is an optimal control approach based on a linear approximation of a dynamical system of the form

$$\dot{x} = Ax + Bu + N(x) \tag{3.39}$$

where A and B are matrices of appropriate dimensions and **N** denotes the effects due to nonlinearities and unmodeled dynamics. The control law for **u** is obtained in this approach by ignoring the term **N** in Eq. (3.39) and minimizing the following *performance index*:

$$\mathcal{J} = \frac{1}{2} \int_0^{t_f} (x^T Q x + u^T R u) dt \tag{3.40}$$

where t_f is the final time and $Q \geq 0$ and $R > 0$ are, respectively, the state and control *weight matrices*. For an autonomous system, constant weight matrices, and $t_f \to \infty$, the minimization of Eq. (3.40) is achieved by the following control law:

$$u = -Kx \tag{3.41}$$

where $K = R^{-1}B^T Sx$ and S satisfies the *Algebraic Riccati Equation* (ARE)

$$SA + A^T S - SBR^{-1}B^T S + Q = 0 \tag{3.42}$$

The solution to the ARE is positive definite if the pair (A, B) is controllable and the pair $(A, Q^{\frac{1}{2}})$ is observable [79]. Positive definiteness of S guarantees closed-loop stability, i.e., asymptotic stability of the system

$$\dot{\mathbf{x}} = (A - BK)\mathbf{x} \tag{3.43}$$

Numerical solvers of the ARE are standard in software packages such as MATLAB®.

The *Discrete-time Linear Quadratic Regulator* (DLQR) formulation is a discretized version of the continuous LQR, and is defined as follows: Minimize

$$\mathcal{J} = \sum_{i=0}^{N-1} \mathbf{x}^T(i) Q \mathbf{x}(i) + \mathbf{u}^T(i) R \mathbf{u}(i) \tag{3.44}$$

where N indicates the total number of time steps, and $\mathbf{x}(i)$ and $\mathbf{u}(i)$ are, respectively, the state and control vectors at the ith time instant, subject to the state equation

$$\mathbf{x}(i + 1) = A_d \mathbf{x}(i) + B_d \mathbf{u}(i) \tag{3.45}$$

As for the continuous-time, autonomous problem, a discrete-time algebraic Riccati equation results when N is large. The discrete-time Riccati equation is

$$A_d^T S A_d - S - A_d^T S B_d R^{-1} B_d^T S A_d + Q = 0 \tag{3.46}$$

and the feedback control law is obtained in the form

$$\mathbf{u}(i) = -K\mathbf{x}(i) = -R^{-1} B_d^T S \mathbf{x}(i) \tag{3.47}$$

3.8 KALMAN FILTERING

In 1960, Robert Kalman introduced a new approach for minimum mean-square error filtering that used state-space methods [80]. The *Kalman Filter* (KF) is a recursive scheme that propagates a current estimate of a state and the error covariance matrix of that state forward in time. The filter optimally blends the new information introduced by the measurements with old information embodied in the prior state with a Kalman gain matrix. The gain matrix balances uncertainty in the measurements with the uncertainty in the dynamics model. The KF is guaranteed to be the optimal filter (in the sense of minimizing the 2-norm-squared of the estimation error) for a linear system with linear measurements [81]. However, few systems can be accurately modeled with linear dynamics. Shortly after its inception, improvements on the Kalman filter to handle nonlinear systems were proposed. One of the most popular choices, the *Extended Kalman Filter* (EKF), was applied to relative navigation filters in LEO [43]. We will demonstrate how to use the EKF for relative spacecraft state estimation in Chapter 12.

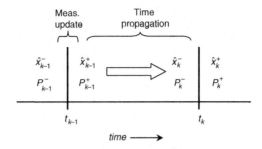

FIGURE 3.1 A Kalman filter process.

The discrete EKF is as a state estimator for systems whose state dynamics model, measurement model, or both may be nonlinear, as in Eqs. (3.48) and (3.53) [81]. The dynamics model provides the equations to propagate $\hat{\mathbf{x}}_k$, the estimate of the state $\mathbf{x}$ at time k, to time step $k + 1$, producing $\hat{\mathbf{x}}_{k+1}$. The measurement model then incorporates the new sensor information to update this estimate, updating the a priori estimate $\hat{\mathbf{x}}_{k+1}^-$ to the a posteriori estimate, $\hat{\mathbf{x}}_{k+1}^+$. This process is illustrated in Fig. 3.1.

The continuous state $\mathbf{x}$ is governed by the dynamics

$$\dot{\mathbf{x}}(t_k) = \mathbf{f}(\mathbf{x}, \mathbf{u}, t_k) + \mathbf{w}(t_k) \tag{3.48}$$

where $\mathbf{u}$ is a known control input, and $\mathbf{w}(t)$ is an additive white noise that models the error accumulated by uncertainty in the dynamics during the time step. The power spectral density of this zero mean, white noise process is

$$Q = E[\mathbf{w}(t)\,\mathbf{w}(t)^T] \tag{3.49}$$

To proceed, linear expressions for the dynamics and measurement equations must be formed. In general, this requires knowledge of the probability density function [81], but the EKF approximates the nonlinear function by expanding it in a Taylor series, at each time step, about the current estimate,

$$F_k = \left.\frac{\partial \mathbf{f}}{\partial \mathbf{x}}\right|_{\mathbf{x}=\hat{\mathbf{x}}_k} \tag{3.50}$$

The dynamics are discretized with time step Δt by forming the state transition matrix,

$$\Phi_k = e^{F_k \Delta t} \tag{3.51}$$

The cumulative effect of the white noise process $\mathbf{w}(t)$ over the time step is captured in the discrete process noise covariance matrix

$$Q_k = \int_0^{\Delta t} e^{F_k \tau} Q (e^{F_k \tau})^T \, d\tau \tag{3.52}$$

The vector of measurements, **y**,

$$\mathbf{y} = \mathbf{h}(\mathbf{x}, t) + \mathbf{v}_k \tag{3.53}$$

is modeled as a nonlinear function of the state and time, with an additive white noise process $v(t)$ that accounts for uncertainty in the sensors and their models. The measurement noise covariance matrix is defined by

$$R_k = E[\mathbf{v}_k \; \mathbf{v}_k^T] \tag{3.54}$$

The nonlinear measurement equation is also linearized about the current estimate,

$$H_k = \left. \frac{\partial \mathbf{h}}{\partial \mathbf{x}} \right|_{\mathbf{x} = \hat{\mathbf{x}}_k^-} \tag{3.55}$$

Because approximations must be made in the linearization, the EKF is a suboptimal filter, in the sense that its stability and performance are not guaranteed. Fortunately, the dynamics of orbital motion are fairly simple, and the EKF can have very good performance in space navigation applications. The discrete, linear representation of the system dynamics are

$$\mathbf{x}_k = \Phi_{k-1}\mathbf{x}_{k-1} + \mathbf{w}_{k-1} + \mathbf{u}_{k-1} \tag{3.56}$$

The confidence in the current estimate is captured in the state error covariance matrix, P,

$$P_k = \mathrm{E}[\tilde{\mathbf{x}}_k \tilde{\mathbf{x}}_k^T] = \mathrm{E}[(\hat{\mathbf{x}}_k - \mathbf{x}_k)(\hat{\mathbf{x}}_k - \mathbf{x}_k)^T] \tag{3.57}$$

where $\tilde{\mathbf{x}}_k = \hat{\mathbf{x}}_k - \mathbf{x}_k$ is the estimation error. The first step in the EKF involves propagating the state and error covariance forward in time. Equation (3.56), with zero process noise, is used to propagate the state estimate. The error covariance is propagated forward using

$$P_k^- = \Phi_{k-1}P_{k-1}^+\Phi_{k-1}^T + Q_{k-1} \tag{3.58}$$

An alternate approach to the time propagation step involves using the nonlinear dynamics equations to propagate the state. A 4th-order Runge–Kutta integration scheme uses the nonlinear state dynamics equation

$$\dot{\hat{\mathbf{x}}}(t) = \mathbf{f}(\hat{\mathbf{x}}(t), \mathbf{u}(t)) \quad \text{for } t = t_{k-1} \to t_k \tag{3.59}$$

to find $\hat{\mathbf{x}}_k$. The state covariance is still propagated with Eq. (3.58), so the state transition matrix Φ_{k-1} must be calculated regardless of whether the linear or nonlinear state propagation is chosen.

The second step of the filter uses the measurement equation to update the a priori state $\hat{\mathbf{x}}_k^-$ to the a posteriori state $\hat{\mathbf{x}}_k^+$. When a measurement becomes available, the new information provided by the measurement and the previous information captured in the state estimate are combined to form an updated state estimate. The Kalman gain K is the blending gain matrix that is used to weight the importance of the old and new information. The optimum gain matrix is formulated by minimizing the trace of the a posteriori state error covariance matrix P_k^+, which essentially minimizes the estimation error vector at each time step [81]. The terms in the gain matrix equation include the previous state estimate, the linearized measurement matrix, and the expected noise of the new measurements,

$$K_k = P_k^- H_k^T (H_k P_k^- H_k^T + R_k)^{-1} \tag{3.60}$$

The nonlinear measurement equation is used to update the state estimate

$$\hat{\mathbf{x}}_k^+ = \hat{\mathbf{x}}_k^- - K_k(\mathbf{y}_k - \mathbf{h}_k(\hat{\mathbf{x}}_k^-)) \tag{3.61}$$

Note that the computation of the gain matrix K_k requires the linear measurement matrix H_k. The covariance is updated after the measurement with

$$P_k^+ = (I - K_k H_k) P_k^- (I - K_k H_k)^T + K_k R_k K_k^T \tag{3.62}$$

which is the Joseph form of the covariance update whose inherent symmetry makes it numerically stable [82].

3.9 THE UNSCENTED KALMAN FILTER

The development of the EKF propagation and update equations requires ignoring terms which may not be small for many nonlinear systems [81,83]. The inherent linearization in the process typically introduces significant biases in the estimation results. Although the EKF has been widely used for many years, experience has shown that it is only reliable for systems that are almost linear on the time scale of the update intervals [84]. In contrast, the *Unscented Kalman Filter* (UKF) does not require the linearization of any nonlinear functions.[5] Instead, the UKF uses a set of points, called *sigma points*, that are distributed around the current estimate. These sigma points are chosen so that their mean matches the current estimate and the covariance of the distribution of sigma points matches the current covariance of the estimate [87]. The often-stated premise of this approach is "it is easier to approximate a Gaussian distribution than it is to approximate an arbitrary nonlinear function or transformation" [87].

During the propagation step of the filter, each sigma point is propagated forward through the actual nonlinear dynamics equation. After the propagation, the set of sigma points is condensed back to a single state estimate using a weighted sum of the propagated sigma points. The propagated state covariance is also set to be the calculated covariance of the propagated sigma points.

---

[5]Ref. [85] presents an opposing view, countered by Ref. [86].

Similarly, the measurement update step requires updating the set of sigma points using the nonlinear measurement equation. The updated sigma points are then condensed back into a single state and their covariances used for the updated state estimate covariance.

While the EKF handles the nonlinearities by approximating them (i.e., by linearization) in the measurement update and time propagation steps, the UKF approximates the *distribution* of $\hat{\mathbf{x}}$, with the sigma points. The mean and covariance of the original state estimate are represented precisely in the distribution of the sigma points. The mean and covariance of the propagated states (i.e., either time update or measurement update) are correct to second order as well [87], which means that the UKF calculates the mean with more accuracy than the EKF, and the covariance with accuracy of at least the EKF [87]. In fact, Theorem 2 of Ref. [84] strengthens this statement to *the prediction algorithm introduces errors in estimating the mean and covariance at the fourth and higher orders in the Taylor series.*

3.9.1 The standard form of the UKF

The original form of the UKF requires augmenting the state with process and measurement noise variables. However, as shown in the equations included in the next subsection, the number of sigma points required is determined by the length of the state (or augmented state). Since the nonlinear state propagation and measurement update is performed for each sigma point, a larger state can require many more calculations. Fortunately, if the measurement and process noises are purely additive, the standard form can be reduced to what is called the additive form of the UKF (UKF-A), which does not require state augmentation [88]. The UKF-A has a smaller state and a reduced computational burden.

In the relative orbital navigation problem, detailed in Chapter 12, the process and measurement noises filter are additive, so the reduced additive form can be used for this application. The additive form of the UKF is presented in the following subsection.

3.9.2 The additive form of the UKF

The additive form of the UKF (UKF-A) is used for systems whose process and measurement noises are purely additive, as in the case discussed in Chapter 12. This form is preferred over the standard form of the UKF because it has a smaller state vector, resulting in fewer sigma points and less computation. Similar to the standard form, the UKF-A uses the nonlinear dynamics and measurement equations and employs a set of sigma points in each time propagation and measurement update step.

At each step, $2n + 1$ sigma points are required, where n indicates the length of the state vector. The sigma points that are used for the time propagation and measurement update steps are chosen to have a cumulative mean and standard deviation identical to the prior estimate. The sigma points are formed by adding and subtracting scaled columns of the matrix square root of the covariance matrix to the original state estimate. Adding these $2n$ columns to the original state vector produces the required $2n + 1$ sigma points [83,88]. This procedure is shown below in the presentation of the UKF-A algorithm.

Several weights and constants are used in the UKF algorithm. The set of scalar weights used to recombine the sigma points into the a posteriori state mean and covariance estimates, $\{W_i\}$, are defined by

$$W_0^{(m)} = \frac{\lambda_{\text{UKF}}}{n + \lambda_{\text{UKF}}} \tag{3.63}$$

$$W_0^{(c)} = \frac{\lambda_{\text{UKF}}}{n + \lambda_{\text{UKF}}} + 1 - \alpha_{\text{UKF}}^2 + \beta_{\text{UKF}} \tag{3.64}$$

$$W_i^{(m)} = \frac{1}{2(n + \lambda_{\text{UKF}})}, \quad i = 1, \dots, 2n \tag{3.65}$$

$$W_i^{(c)} = \frac{1}{2(n + \lambda_{\text{UKF}})}, \quad i = 1, \dots, 2n \tag{3.66}$$

The scaling parameter λ_{UKF} is $\lambda_{\text{UKF}} = \alpha_{\text{UKF}}^2(n+\kappa) - n$. The related parameter, $\gamma = \sqrt{n + \lambda_{\text{UKF}}}$, is used in the sigma point calculation. The parameter α_{UKF}, set between 10^{-4} and 1, determines the spread of points around the original estimate. The parameter β_{UKF} influences how much the prior estimate of the covariance is weighted and is typically set to 2 for state estimation problems.

The algorithm for the UKF-A is presented in Table 3.1. In several places, the nonlinear dynamics equation **f** or measurement equations **h** is applied to a matrix of vectors, producing matrix output. First the sigma points are formed around the $\hat{\mathbf{x}}_{k-1}$ state estimate (Eq. (3.69)). The sigma points are propagated (Eq. (3.70)) and weighted sums of the sigma points are used to calculate propagated state estimate and covariance, $\hat{\mathbf{x}}_k^-$ and P_k^- (Eqs. (3.71) and (3.72)). A new set of sigma points is calculated for the measurement update step (Eq. (3.73)). The measurement equation is used to create an expected measurement for each of the sigma points (Eq. (3.74)), which is then condensed with a weighted sum to create an overall, composite expected measurement, $\hat{\mathbf{y}}_k^-$ (Eq. (3.75)). Covariance matrices are calculated with the expected measurements (Eqs. (3.76) and (3.77)), and a Kalman gain matrix is formed (Eq. (3.78)). Finally, the updated state and covariance, $\hat{\mathbf{x}}_k$ and P_k, are calculated (Eqs. (3.79) and (3.80)) [88].

As a practical note, it was observed that numerical errors would sometimes introduce very small imaginary components when the matrix square root of the covariance was computed. This tends to disrupt the algorithm, so preventative measures should be included.

Another form of the UKF, called the square root form (UKF-S), has been proposed as an alternative to the additive form. The square root form makes use of *Cholesky factorization*, and in some applications may increase numerical stability and performance. A comparison of the UKF-A and the UKF-S for the relative spacecraft navigation problem is discussed in Chapter 12.

Table 3.1 UKF-A algorithm

1. The UKF-A is initialized with

$$\hat{\mathbf{x}}_0 = E[\mathbf{x}_0] \tag{3.67}$$
$$P_0 = E[(\mathbf{x}_0 - \hat{\mathbf{x}}_0)(\mathbf{x}_0 - \hat{\mathbf{x}}_0)^T] \tag{3.68}$$

For $k \in \{1, \ldots, \infty\}$,
2. Calculate sigma points:

$$\mathcal{X}_{k-1} = \left[\hat{\mathbf{x}}_{k-1}, \ \hat{\mathbf{x}}_{k-1} + \gamma\sqrt{P_{k-1}}, \ \hat{\mathbf{x}}_{k-1} - \gamma\sqrt{P_{k-1}}\right] \tag{3.69}$$

3. Propagate the sigma points with the nonlinear dynamics equation $\hat{\mathbf{x}}_{k+1} = \mathbf{f}(\hat{\mathbf{x}}_k)$ and find the covariance:

$$\mathcal{X}^*_{k|k-1} = \mathbf{f}(\mathcal{X}_{k-1}, \mathbf{u}_{k-1}) \tag{3.70}$$

$$\hat{\mathbf{x}}^-_k = \sum_{i=0}^{2n} W_i^{(m)} \mathcal{X}^*_{i,k|k-1} \tag{3.71}$$

$$P^-_k = \sum_{i=0}^{2n} W_i^{(c)} (\mathcal{X}^*_{i,k|k-1} - \hat{\mathbf{x}}^-_k)(\mathcal{X}^*_{i,k|k-1} - \hat{\mathbf{x}}^-_k)^T + Q \tag{3.72}$$

4. Create a new set of sigma points and, with the nonlinear function measurement equation $\hat{\mathbf{y}}_k = \mathbf{h}(\hat{\mathbf{x}}_k)$, create expected measurements:

$$\mathcal{X}_{k|k-1} = \left[\hat{\mathbf{x}}^-_k, \ \hat{\mathbf{x}}^-_k + \gamma\sqrt{P^-_k}, \ \hat{\mathbf{x}}^-_k - \gamma\sqrt{P^-_k}\right] \tag{3.73}$$
$$\mathcal{Y}_{k|k-1} = \mathbf{h}(\mathcal{X}_{k|k-1}) \tag{3.74}$$
$$\hat{\mathbf{y}}^-_k = \sum_{i=0}^{2n} W_i^{(m)} \mathcal{Y}_{i,k|k-1} \tag{3.75}$$

5. Create the Kalman gain and perform a measurement update:

$$P_{\tilde{\mathbf{y}}_k \tilde{\mathbf{y}}_k} = \sum_{i=0}^{2n} W_i^{(c)} (\mathcal{Y}_{i,k|k-1} - \hat{\mathbf{y}}^-_k)(\mathcal{Y}_{i,k|k-1} - \hat{\mathbf{y}}^-_k)^T + R \tag{3.76}$$

$$P_{x_k y_k} = \sum_{i=0}^{2n} W_i^{(c)} (\mathcal{X}_{i,k|k-1} - \hat{\mathbf{x}}^-_k)(\mathcal{Y}_{i,k|k-1} - \hat{\mathbf{y}}^-_k)^T \tag{3.77}$$

$$K_k = P_{x_k y_k} P_{\tilde{\mathbf{y}}_k \tilde{\mathbf{y}}_k}^{-1} \tag{3.78}$$
$$\hat{\mathbf{x}}_k = \hat{\mathbf{x}}^-_k + K_k(\mathbf{y}_k - \hat{\mathbf{y}}^-_k) \tag{3.79}$$
$$P_k = P^-_k - K_k P_{\tilde{\mathbf{y}}_k \tilde{\mathbf{y}}_k} K_k^T \tag{3.80}$$

3.9.3 The square root form of the UKF

Van der Merwe and Wan introduced the Square Root form of the UKF (UKF-S) [89] to handle some of the numerical issues referred to in the previous subsection. The computational burden of the UKF-S is expected to be similar to that of the UKF-A. However, the UKF-S may provide improved numerical stability and it also guarantees that the covariance matrix will be positive semidefinite, which is required for filter stability.

The main innovation introduced in the UKF-S is the use of the Cholesky factor, S, of the covariance matrix. The Cholesky factor, S, is initialized taking

Table 3.2 UKF-S algorithm

1. The UKF-S is initialized with

$$\hat{\mathbf{x}}_0 = E[\mathbf{x}_0] \tag{3.81}$$

$$S_0 = \texttt{chol}\left(E[(\mathbf{x}_0 - \hat{\mathbf{x}}_0)(\mathbf{x}_0 - \hat{\mathbf{x}}_0)^T]\right) \tag{3.82}$$

For $k \in \{1, \dots, \infty\}$,
2. Calculate sigma points

$$\mathcal{X}_{k-1} = [\hat{\mathbf{x}}_{k-1}, \ \hat{\mathbf{x}}_{k-1} + \gamma S_k, \ \hat{\mathbf{x}}_{k-1} - \gamma S_k] \tag{3.83}$$

3. Propagate sigma points using nonlinear dynamics equation and find covariance

$$\mathcal{X}_{k|k-1}^* = \mathbf{f}(\mathcal{X}_{k-1}, \mathbf{u}_{k-1}) \tag{3.84}$$

$$\hat{\mathbf{x}}_k^- = \sum_{i=0}^{2n} W_i^{(m)} \mathcal{X}_{i,k|k-1}^* \tag{3.85}$$

$$S_k^- = \texttt{qr}\left(\left[\sqrt{W_1^{(c)}}(\mathcal{X}_{1:2n,k|k-1}^* - \hat{\mathbf{x}}_k^-), \ \sqrt{Q}\right]\right) \tag{3.86}$$

$$S_k^- = \texttt{cholupdate}\left(S_k^-, \ \mathcal{X}_{0,k}^* - \hat{\mathbf{x}}_k^-, \ W_0^{(c)}\right) \tag{3.87}$$

4. Create new set of sigma points and, using the nonlinear measurement equation, and perform the Cholesky update

$$\mathcal{X}_{k|k-1} = [\hat{\mathbf{x}}_k^-, \ \hat{\mathbf{x}}_k^- + \gamma S_k^-, \ \hat{\mathbf{x}}_k^- - \gamma S_k^-] \tag{3.88}$$

$$\mathcal{Y}_{k|k-1} = \mathbf{h}(\mathcal{X}_{k|k-1}) \tag{3.89}$$

$$\hat{\mathbf{y}}_k^- = \sum_{i=0}^{2n} W_i^{(m)} \mathcal{Y}_{i,k|k-1} \tag{3.90}$$

$$S_{\tilde{y}_k} = \texttt{qr}\left(\left[\sqrt{W_1^{(c)}}(\mathcal{Y}_{1:2n,k} - \hat{\mathbf{y}}_k), \ \sqrt{R}\right]\right) \tag{3.91}$$

$$S_{\tilde{y}_k} = \texttt{cholupdate}\left(S_{\tilde{y}_k}, \ \mathcal{Y}_{0,k} - \hat{\mathbf{y}}_k, \ W_0^{(c)}\right) \tag{3.92}$$

5. Create the Kalman gain and perform the measurement update

$$P_{x_k y_k} = \sum_{i=0}^{2n} W_i^{(c)} (\mathcal{X}_{i,k|k-1} - \hat{\mathbf{x}}_k^-)(\mathcal{Y}_{i,k|k-1} - \hat{\mathbf{y}}_k^-)^T \tag{3.93}$$

$$K_k = (P_{x_k y_k}/S_{\tilde{y}_k}^T)/S_{\tilde{y}_k} \tag{3.94}$$

$$\hat{\mathbf{x}}_k = \hat{\mathbf{x}}_k^- + K_k(\mathbf{y}_k - \hat{\mathbf{y}}_k^-) \tag{3.95}$$

$$U = K_k S_{\tilde{y}_k} \tag{3.96}$$

$$S_k = \texttt{cholupdate}(S_k^-, U, -1) \tag{3.97}$$

the matrix square root of the initial P with a Cholesky factorization. Three linear algebra techniques are recommended by van der Merwe for use in the UKF-S [89]:

- **QR decomposition** The QR decomposition factors a matrix A into the product of an orthogonal matrix, Q, and an upper triangular matrix, R, so $A^T = QR$. The upper triangular part of R is the transpose of the Cholesky factor of a matrix P, defined such that $P = AA^T$. The MATLAB® function qr returns the *lower* Cholesky factor. The transpose of the output of qr should be used to get the *upper* Cholesky factor required in this algorithm.

- **Cholesky factor updating** Consider a matrix $P = AA^T$ that has a Cholesky factor of S. If the matrix P is updated so that $P^+ = P \pm \sqrt{\nu}\mathbf{u}\mathbf{u}^T$, the Cholesky factor of the updated P^+ is found using the cholupdate MATLAB® function, $\text{chol}(P^+) = \text{cholupdate}(S, \mathbf{u}, \pm \nu)$. When $\mathbf{u}$ is a matrix, the function performs a Cholesky update for each of the columns of $\mathbf{u}$.
- **Efficient least squares** The MATLAB® function "/" finds the solution to $(AA^T)\mathbf{v} = A^T\mathbf{b}$ using an efficient QR decomposition method.

These three techniques were incorporated into the Square Root Unscented Kalman Filter used in Chapter 12. The UKF-S algorithm is summarized in Table 3.2.

SUMMARY

We presented some key tools of analytical mechanics, optimization, control and estimation, which supplement the introductory material on astrodynamics presented in Chapter 2. The tools provided in the chapter will be extensively used in subsequent discussions. Hamiltonian and Lagrangian mechanics will be used for perturbation modeling, relative motion modeling, mean-to-osculating element conversion using the Brouwer transformation, and averaging; optimization theory, including the concept of bordered Hessians, will be used for developing optimal formation-keeping maneuvers; Linear Quadratic Regulators and Control Lyapunov Functions will be used to control spacecraft formations; and Kalman filtering will be extensively applied to the problem of relative navigation.

Chapter | four

Nonlinear Models of Relative Dynamics

wheels repeating the same gesture
remain relatively stationary
rails forever parallel
return on themselves infinitely.
The dance is sure.

William Carlos Williams (1883–1963)

We present commonly used nonlinear models for relative motion in this chapter, for both perturbed and unperturbed motion. For unperturbed Keplerian motion, this chapter suggests a rather simple method for generating bounded relative motion between any two spacecraft [90]. The underlying methodology relies on the concept of orbital-period *commensurability*: Two nonzero real numbers c and d are said to be commensurable if and only if c/d is a rational number; in other words, there exists some real number g, and integers m and n, such that $c = mg$ and $d = ng$. Thus, two elliptic orbits with orbital periods T_1 and T_2 are said to be $m : n$ commensurable if

$$\frac{T_1}{T_2} = \frac{m}{n}, \tag{4.1}$$

Based on our discussion of Keplerian orbits in Chapter 2, it is obvious that the corresponding relationship between the orbital energies, $\mathcal{E}_1$ and $\mathcal{E}_2$, and the semimajor axes, a_1 and a_2, derived based on Eq. (2.33), satisfies

$$\frac{T_1}{T_2} = \left(\frac{\mathcal{E}_2}{\mathcal{E}_1}\right)^{3/2} = \left(\frac{a_1}{a_2}\right)^{3/2}, \tag{4.2}$$

which implies that

$$\frac{\mathcal{E}_2}{\mathcal{E}_1} = \frac{a_1}{a_2} = \left(\frac{m}{n}\right)^{2/3} \tag{4.3}$$

59

Spacecraft Formation Flying; ISBN: 9780750685337

Thus, although m/n is rational, $\mathcal{E}_2/\mathcal{E}_1$ and a_1/a_2 may not be so. Nevertheless, orbital commensurability necessarily constrains the ratio between orbital energies and between the semimajor axes of the orbits. This constraint is therefore referred to as the *energy matching condition*. When $m = n = 1$, which is of a particular interest in our forthcoming discussion, we have $a_1 = a_2$ and $\mathcal{E}_1 = \mathcal{E}_2$.

The concept of commensurability will help us model the general problem of Keplerian relative motion. We will do so by formulating the energy matching condition in the rotating reference-orbit-fixed frame. The formulation using orbital commensurability will give us a single, simple, algebraic constraint on initial conditions guaranteeing bounded relative motion between spacecraft flying on elliptic, perturbation-free orbits.

The approach for generating bounded relative motion presented in this chapter is a convenient stepping stone for developing an optimal *formation-keeping* scheme. In real-world scenarios, spacecraft formation initialization will almost always entail an initialization error [91]. A formation-keeping maneuver is required for arresting the relative spacecraft drift. As relative spacecraft position control is sometimes performed utilizing relative measurements, it is important to design formation-keeping maneuvers utilizing the relative state variables.

We will discuss in detail formation control concepts in Chapter 10. In this chapter, we develop a *single-impulse formation-keeping* method by using the inherent freedom of the energy matching condition. We also provide an insight into the resulting formation-keeping maneuver by utilizing the classical orbital elements discussed in Chapter 2.

The bottom line is that we design a simple framework for both initialization and initialization-error correction for spacecraft formation flying. This framework yields bounded relative motion between any two spacecraft flying on arbitrary elliptic orbits – at this stage, without utilizing any simplifying assumptions regarding the relative dynamics – except that it is perturbation-free. We will relieve this assumption, however, in Section 4.7, where we present a model for relative motion under the influence of J_2.

4.1 EQUATIONS OF RELATIVE MOTION IN THE UNPERTURBED CASE

Consider two spacecraft orbiting a common primary. One of the spacecraft will be termed *chief* and the other will be referred to as *deputy* (see our discussion of the chief/deputy notation vs. the leader/follower notation on p. 2). We wish to develop the equations of relative motion under the setup of the Keplerian two-body problem. We will start with an inertial description of the relative motion in $\mathscr{I}$ and then transform the equations into a chief-fixed, LVLH rotating frame, $\mathscr{L}$, called the *Euler–Hill frame*, as shown in Fig. 4.1 (both $\mathscr{I}$ and $\mathscr{L}$ were discussed in Chapter 2). This operation is required due to the fact that control of relative position and velocity, to be discussed in Chapters 10 and 11, most often utilizes measurements taken in the chief rotating frame.

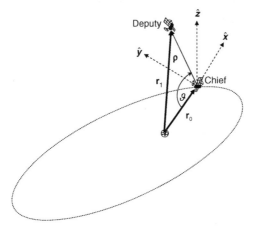

FIGURE 4.1 Rotating Euler–Hill frame, centered at the chief spacecraft. This figure also shows the deputy spacecraft, whose position vectors in the rotating and inertial reference frames are denoted by ρ and $\mathbf{r}_1$, respectively.

Recalling Eq. (2.1), the inertial equations of motion of the chief are given by

$$\ddot{\mathbf{r}}_0 = -\frac{\mu}{r_0^3}\mathbf{r}_0 \tag{4.4}$$

where

$$r_0 = \|\mathbf{r}_0\| = \frac{a_0(1 - e_0^2)}{(1 + e_0 \cos f_0)}, \tag{4.5}$$

and a_0, e_0, f_0 are the chief's orbit semimajor axis, eccentricity and true anomaly, respectively. In a similar fashion, the inertial equations of motion of the deputy are

$$\ddot{\mathbf{r}}_1 = -\frac{\mu}{r_1^3}\mathbf{r}_1 \tag{4.6}$$

where

$$r_1 = \|\mathbf{r}_1\| = \frac{a_1(1 - e_1^2)}{(1 + e_1 \cos f_1)}, \tag{4.7}$$

and a_1, e_1, f_1 are the deputy's orbit semimajor axis, eccentricity and true anomaly, respectively. Let

$$\rho = \mathbf{r}_1 - \mathbf{r}_0 \tag{4.8}$$

denote the position of the deputy relative to the chief. Subtracting Eq. (4.4) from Eq. (4.6) yields

$$\ddot{\boldsymbol{\rho}} = -\frac{\mu(\mathbf{r}_0 + \boldsymbol{\rho})}{\|\mathbf{r}_0 + \boldsymbol{\rho}\|^3} + \frac{\mu}{r_0^3}\mathbf{r}_0 \tag{4.9}$$

In order to express the relative acceleration in frame $\mathscr{L}$, we recall that

$$\ddot{\boldsymbol{\rho}} = \frac{d^{2\mathscr{L}}\boldsymbol{\rho}}{dt^2} + 2^{\mathscr{I}}\boldsymbol{\omega}^{\mathscr{L}} \times \frac{d^{\mathscr{L}}\boldsymbol{\rho}}{dt}$$
$$+ \frac{d^{\mathscr{I}}\boldsymbol{\omega}^{\mathscr{L}}}{dt} \times \boldsymbol{\rho} + {}^{\mathscr{I}}\boldsymbol{\omega}^{\mathscr{L}} \times ({}^{\mathscr{I}}\boldsymbol{\omega}^{\mathscr{L}} \times \boldsymbol{\rho}) \tag{4.10}$$

where ${}^{\mathscr{I}}\boldsymbol{\omega}^{\mathscr{L}}$ denotes the angular velocity vector of frame $\mathscr{L}$ relative to frame $\mathscr{I}$.

As ${}^{\mathscr{I}}\boldsymbol{\omega}^{\mathscr{L}}$ is normal to the orbital plane, we may write

$${}^{\mathscr{I}}\boldsymbol{\omega}^{\mathscr{L}} = [0, 0, \dot{\theta}_0]^T \tag{4.11}$$

The position vector of the chief can be written as

$$\mathbf{r}_0 = [r_0, 0, 0]^T \tag{4.12}$$

Also, let

$$[\boldsymbol{\rho}]_{\mathscr{L}} = [x, y, z]^T \tag{4.13}$$

Substituting Eqs. (4.9), (4.11), and (4.13) into Eq. (4.10) yields the following component-wise equations for relative motion:

$$\ddot{x} - 2\dot{\theta}_0\dot{y} - \ddot{\theta}_0 y - \dot{\theta}_0^2 x = -\frac{\mu(r_0 + x)}{\left[(r_0 + x)^2 + y^2 + z^2\right]^{\frac{3}{2}}} + \frac{\mu}{r_0^2} \tag{4.14}$$

$$\ddot{y} + 2\dot{\theta}_0\dot{x} + \ddot{\theta}_0 x - \dot{\theta}_0^2 y = -\frac{\mu y}{\left[(r_0 + x)^2 + y^2 + z^2\right]^{\frac{3}{2}}} \tag{4.15}$$

$$\ddot{z} = -\frac{\mu z}{\left[(r_0 + x)^2 + y^2 + z^2\right]^{\frac{3}{2}}} \tag{4.16}$$

Equations (4.14)–(4.16) together with Eqs. (2.5) and (2.6),

$$\ddot{r}_0 = r_0\dot{\theta}_0^2 - \frac{\mu}{r_0^2}, \quad \ddot{\theta}_0 = -\frac{2\dot{r}_0\dot{\theta}_0}{r_0}$$

constitute a 10-dimensional system of nonlinear differential equations. For $\ddot{\theta}_0 \neq 0$, these equations admit a single relative equilibrium at $x = y = z = 0$, meaning that the deputy spacecraft will appear stationary in the chief frame if and only if their positions coincide on a given elliptic orbit. We will later see that the single relative equilibrium is transformed into infinitely many relative equilibria if the chief is assumed to follow a circular reference orbit.

If there are external (differential) perturbations, denoted by $\mathbf{d} = [d_x, d_y, d_z]^T$, and (differential) control forces, $\mathbf{u} = [u_x, u_y, u_z]^T$, they are introduced into Eqs. (4.14)–(4.16) in the following manner:

$$\ddot{x} - 2\dot{\theta}_0\dot{y} - \ddot{\theta}_0 y - \dot{\theta}_0^2 x = -\frac{\mu(r_0 + x)}{\left[(r_0 + x)^2 + y^2 + z^2\right]^{\frac{3}{2}}} + \frac{\mu}{r_0^2} + d_x + u_x \quad (4.17)$$

$$\ddot{y} + 2\dot{\theta}_0\dot{x} + \ddot{\theta}_0 x - \dot{\theta}_0^2 y = -\frac{\mu y}{\left[(r_0 + x)^2 + y^2 + z^2\right]^{\frac{3}{2}}} + d_y + u_y \quad (4.18)$$

$$\ddot{z} = -\frac{\mu z}{\left[(r_0 + x)^2 + y^2 + z^2\right]^{\frac{3}{2}}} + d_z + u_z \quad (4.19)$$

Let us mention some mathematical formalities related to Eqs. (4.14)–(4.16). We say that the configuration space for the relative spacecraft dynamics of Eqs. (4.14)–(4.16) is $\mathbb{R}^3$, and that $T_S(\mathbb{R}^3) = \mathbb{R}^3 \times \mathbb{R}^3$ is the tangent space of $\mathbb{R}^3$. We use $(\rho, \dot{\rho})$ as coordinates for $T_S(\mathbb{R}^3)$.

To get a more compact form of Eqs. (4.14)–(4.16), we can use the formalism developed by Szebehely and Giacaglia [92] for simplifying the equations of motion obtained in the elliptic restricted three-body problem. To that end, we define normalized position components,

$$\bar{x} = x/r_0, \bar{y} = y/r_0, \bar{z} = z/r_0 \quad (4.20)$$

and use the true anomaly, f, as our independent variable. By the chain rule, derivatives with respect to f, to be denoted by $(\cdot)'$, satisfy

$$\frac{d(\cdot)}{dt} = (\cdot)'\dot{f} \quad (4.21)$$

In the Keplerian setup, we can use the auxiliary relations

$$\dot{\theta}_0 = \dot{f}_0 = \frac{h_0}{r_0^2}, h_0 = \sqrt{\mu p_0}, r_0 = \frac{p_0}{1 + e_0 \cos f_0}, r_0' = \frac{p_0 e_0 \sin f_0}{(1 + e_0 \cos f_0)^2} \quad (4.22)$$

Thus,

$$\dot{x} = \dot{f}_0(r_0'\bar{x} + r_0\bar{x}') = \frac{h_0}{p_0}[(e_0 \sin f_0)\bar{x} + (1 + e_0 \cos f_0)\bar{x}'] \quad (4.23)$$

Equivalent relations hold for $\dot{y}$ and $\dot{z}$. Application of the chain rule again provides us with the expression

$$\ddot{x} = \frac{h_0^2}{p_0^2}[(1 + e_0 \cos f_0)\bar{x}'' + e_0(\cos f_0)\bar{x}](1 + e_0 \cos f_0)^3 \quad (4.24)$$

Again, similar relations hold for $\ddot{\bar{y}}$ and $\ddot{\bar{z}}$. Now, we define the non-dimensional potential function

$$\mathcal{U} = -\frac{1}{\left[(1+\bar{x})^2 + \bar{y}^2 + \bar{z}^2\right]^{\frac{1}{2}}} + 1 - \bar{x} \tag{4.25}$$

and the *pseudo-potential* function

$$\mathcal{W} = \frac{1}{1 + e_0 \cos f_0}\left[\frac{1}{2}(\bar{x}^2 + \bar{y}^2 - e_0\bar{z}^2 \cos f_0) - \mathcal{U}\right] \tag{4.26}$$

This enables us to write

$$\bar{x}'' - 2\bar{y}' = \frac{\partial \mathcal{W}}{\partial \bar{x}} \tag{4.27}$$

$$\bar{y}'' + 2\bar{x}' = \frac{\partial \mathcal{W}}{\partial \bar{y}} \tag{4.28}$$

$$\bar{z}'' = \frac{\partial \mathcal{W}}{\partial \bar{z}} \tag{4.29}$$

4.2 THE ENERGY MATCHING CONDITION

It is natural to inquire whether Eqs. (4.14)–(4.16) provide bounded solutions. This question has a straightforward answer which is derived from the physics of the orbits: If both vehicles follow Keplerian elliptic orbits, then their separation cannot grow unboundedly. However, it is intuitively clear that if the periods of the spacecraft orbits are not commensurate, periodicity of the relative motion will not be exhibited, and hence, on shorter time scales, the relative motion may appear to be "locally" unbounded. The relative motion between two non-commensurable elliptic orbits is said to be *quasi-periodic*.

Since the periods of elliptic orbits uniquely determine the energy of the orbit, we can use the energy matching conditions to find periodic relative orbits. In spacecraft formation flying, the only interesting case is 1:1 commensurability, guaranteeing that the semimajor axes and energies are equal. Other commensurability ratios are of interest in problems such as interplanetary travel, orbital transfer and Lambert's problem [33].

To implement the energy matching condition for finding periodic relative orbits in the formation flying scenario, we will first write the velocity of the deputy in the rotating frame:

$$\mathbf{v}_1 = \frac{d^{\mathcal{L}}}{dt}\boldsymbol{\rho} + \frac{d^{\mathcal{L}}}{dt}\mathbf{r}_0 + {}^{\mathcal{I}}\boldsymbol{\omega}^{\mathcal{L}} \times \mathbf{r}_0 + {}^{\mathcal{I}}\boldsymbol{\omega}^{\mathcal{L}} \times \boldsymbol{\rho} \tag{4.30}$$

Substitution of Eqs. (4.11)–(4.13) into Eq. (4.30) yields

$$\mathbf{v}_1 = \begin{bmatrix} \dot{x} - \dot{\theta}_0 y + \dot{r}_0 \\ \dot{y} + \dot{\theta}_0(x + r_0) \\ \dot{z} \end{bmatrix} = \begin{bmatrix} v_x \\ v_y \\ v_z \end{bmatrix} \tag{4.31}$$

where

$$
\dot{r}_0 = \frac{d}{dt}\left[\frac{a_0(1 - e_0^2)}{1 + e_0 \cos f_0}\right] = \dot{f}_0 \frac{d}{df_0}\left[\frac{a_0(1 - e_0^2)}{1 + e_0 \cos f_0}\right]
$$

$$
= \dot{\theta}_0 \frac{a_0 e_0 (1 - e_0^2) \sin f_0}{(1 + e_0 \cos f_0)^2}
\tag{4.32}
$$

Substituting for $\dot{\theta}_0$ from Eq. (2.29) we obtain

$$
\dot{r}_0 = e_0 \sin f_0 \sqrt{\frac{\mu}{a_0(1 - e_0^2)}}
\tag{4.33}
$$

The total specific energy of the deputy spacecraft comprises the kinetic and potential energies,

$$
\mathcal{E}_1 = \frac{1}{2} v_1^2 - \frac{\mu}{r_1} = \frac{1}{2} v_x^2 + v_y^2 + v_z^2 - \frac{\mu}{r_1}
$$

$$
= \frac{1}{2}\{(\dot{x} - \dot{\theta}_0 y + \dot{r}_0)^2 + [\dot{y} + \dot{\theta}_0(x + r_0)]^2 + \dot{z}^2\}
$$

$$
- \frac{\mu}{\sqrt{(r_0 + x)^2 + y^2 + z^2}}
\tag{4.34}
$$

The total energy of the chief is given by (cf. Eq. (2.25))

$$
\mathcal{E}_0 = -\frac{\mu}{2a_0}
\tag{4.35}
$$

The energy matching condition, guaranteeing a 1:1 commensurable relative motion, is therefore

$$
\frac{1}{2}\{(\dot{x} - \dot{\theta}_0 y + \dot{r}_0)^2 + [\dot{y} + \dot{\theta}_0(x + r_0)]^2 + \dot{z}^2\}
$$

$$
- \frac{\mu}{\sqrt{(r_0 + x)^2 + y^2 + z^2}} = -\frac{\mu}{2a_0}
\tag{4.36}
$$

In order to design a 1:1 bounded formation, we require the following constraint on the initial conditions of Eqs. (4.14)–(4.16):

$$
\frac{1}{2}\{[\dot{x}(0) - \dot{\theta}_0(0)y(0) + \dot{r}_0(0)]^2 + \{\dot{y}(0) + \dot{\theta}_0(0)[x(0) + r_0(0)]\}^2 + \dot{z}(0)^2\}
$$

$$
- \frac{\mu}{\sqrt{[r_0(0) + x(0)]^2 + y^2(0) + z^2(0)}} = -\frac{\mu}{2a_0}
\tag{4.37}
$$

Most often, Eq. (4.37) is normalized, with the distances being measured in units of a_0 and angular velocities in units of $\sqrt{\mu/a_0^3}$. We will denote normalized quantities by $(\bar{\cdot})$ and differentiation with respect to normalized time by $(\cdot)'$.

Example 4.1. *Consider a chief spacecraft on an elliptic orbit. Normalize positions by a_0 and angular velocities by $\sqrt{\mu/a_0^3}$ so that $a_0 = \mu = 1$. Using these normalized units, let*

$$\bar{y}(0) = 0, \bar{z}(0) = 0.1, \bar{x}'(0) = 0.02$$
$$\bar{y}'(0) = 0.02, \bar{z}'(0) = 0, f_0(0) = 0, e_0 = 0.1 \qquad (4.38)$$

Find $\bar{x}(0)$ that guarantees a 1:1 bounded relative motion.

We first need to calculate $\bar{r}_0(0), \bar{r}_0'(0)$ and $\theta_0'(0)$:

$$\bar{r}_0(0) = \frac{1 - e_0^2}{1 + e_0 \cos f_0(0)} = \frac{1 - 0.1^2}{1 + 0.1} = 0.9$$

$$\bar{r}_0'(0) = e_0 \sin f_0(0) \sqrt{\frac{1}{(1 - e_0^2)}} = 0$$

$$\theta_0'(0) = \sqrt{\frac{1}{(1 - e_0^2)^3}}(1 + e_0 \cos f_0(0))^2 = 1.22838$$

Upon substitution into (4.37), we obtain a sixth-order equation for $\bar{x}_0$:

$$2.2768\bar{x}^6(0) + 12.4431\bar{x}^5(0) + 31.3762\bar{x}^4(0) + 45.46062\bar{x}^3(0)$$
$$+ 39.5905\bar{x}^2(0) + 19.5344\bar{x}(0) + 0.2151 = 0$$

There are two real solutions:

$$\bar{x}_1(0) = -0.01127, \bar{x}_2(0) = -1.8059$$

These initial conditions on the radial separation of the spacecraft will guarantee 1:1 bounded motion. To illustrate the resulting orbits, we simulated Eqs. (4.14)–(4.16) using the initial conditions (4.38) and $\bar{x}(0) = \bar{x}_1(0)$. The results are depicted in Figs. 4.2 and 4.3. Figure 4.2 shows that the relative position components are periodic, with period equal to the chief's orbital period. This is not surprising, as we have obtained the 1:1 commensurability solution. Fig. 4.3 shows the relative orbits in the configuration space. We note that the constraint (4.37) can be satisfied in many other ways, e.g., by selecting $\dot{y}(0)$, with the other initial conditions specified.

4.3 IMPULSIVE FORMATION-KEEPING

Equation (4.37) constitutes a necessary and sufficient condition for 1:1 commensurable relative motion. However, in practice, due to initialization errors, this constraint cannot be satisfied exactly. The ideal initial conditions required for 1:1 boundedness are violated, and a "drift" in relative position results. To compensate for such errors, the deputy spacecraft must maneuver. In most cases, the maneuver will be carried out using the on-board propulsion system,

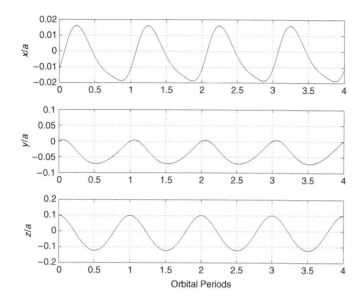

FIGURE 4.2 Time histories of the normalized relative position components show periodic behavior with period equal to the chief's orbital period.

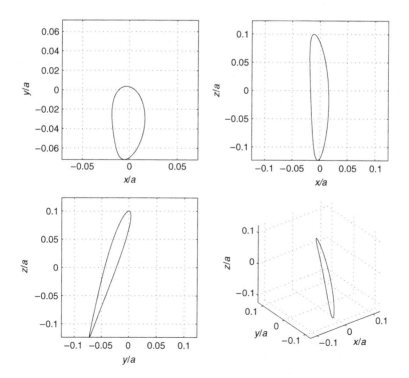

FIGURE 4.3 The relative motion in the configuration space exhibits bounded orbits.

providing short thrust pulses. Such maneuvers are called *impulsive* and the relative position regulation is called *formation-keeping*. Ideal impulsive maneuvers provide instantaneous velocity changes only; in other words, while the velocity of an impulsively maneuvering spacecraft is immediately changed, the position remains intact at impulse application.

We will develop an impulsive formation-keeping maneuver strategy aimed at correcting the relative position drift while consuming minimum fuel. To this end, let the position initialization errors be denoted by

$$\delta\rho(0) = [\delta x(0), \delta y(0), \delta z(0)]^T$$

and the velocity initialization errors be

$$\delta\mathbf{v}(0) = [\delta\dot{x}(0), \delta\dot{y}(0), \delta\dot{z}(0)]^T$$

The actual initial conditions are thus

$$\rho^\delta(0) = \rho(0) + \delta\rho(0)$$
$$\mathbf{v}^\delta(0) = \mathbf{v}(0) + \delta\mathbf{v}(0) \qquad (4.39)$$

The impulsive maneuver will be represented by a relative velocity correction performed by the deputy in the rotating Euler–Hill frame, $[\Delta\mathbf{v}]_\mathscr{L} = [\Delta v_x, \Delta v_y, \Delta v_z]^T$. We will subsequently omit the subscript $\mathscr{L}$ to facilitate the notation. If the impulsive maneuver is initiated at time t_i, then the required velocity correction components can be determined by the energy matching condition applied on the deputy's total energy after the correction, $\mathcal{E}_1^+$ (recall that under the ideal impulsive maneuver assumption, the position remains unchanged due to an impulsive correction):

$$\mathcal{E}_1^+ = \frac{1}{2}\left[(v_x^- + \Delta v_x)^2 + (v_y^- + \Delta v_y)^2 + (v_z^- + \Delta v_z)^2\right] - \frac{\mu}{r_1}$$
$$= -\frac{\mu}{2a_0} \qquad (4.40)$$

where

$$v_x^- = \dot{x}^-(t_i) - \dot{\theta}_0^-(t_i)y(t_i) + \dot{r}_0^-(t_i) \qquad (4.41)$$
$$v_y^- = \dot{y}^-(t_i) + \dot{\theta}_0^-(t_i)[x(t_i) + r_0(t_i)] \qquad (4.42)$$
$$v_z^- = \dot{z}^-(t_i) \qquad (4.43)$$
$$r_1 = \sqrt{[r_0(t_i) + x(t_i)]^2 + y^2(t_i) + z^2(t_i)} \qquad (4.44)$$

and $(\cdot)^-(t_i)$ indicates values prior to the impulsive maneuver, applied at $t = t_i$. Apparently, we have three degrees of freedom for choosing the velocity correction components but only one constraint (4.40). This means that we are free to choose two degrees of freedom. As we are interested in saving fuel consumption, the extra freedom will be used for minimizing the required fuel consumption by solving an optimization problem. We note that if t_i is given, the underlying optimization problem is static, as it involves parameters defined at a single time instant, t_i. Otherwise, a dynamic optimization problem must be solved or approximated by successive static optimization procedures for each t_i. We will assume hereafter that the t_i is pre-determined based upon operational consideration, and subsequently solve a static optimization problem. Recall that we described a method for solving such problems in Section 3.5.

Static parameter optimization problems with equality constraints can be straightforwardly solved utilizing the well-known concept of Lagrange multipliers, described in Section 3.5. In our case, the optimization problem may be stated as follows: Find an optimal impulsive maneuver, $\Delta \mathbf{v}^*$, satisfying

$$\Delta \mathbf{v}^* = \arg \min_{\Delta \mathbf{v}} \| \Delta \mathbf{v} \|^2$$

s. t.

$$\mathcal{E}_1^+ = -\frac{\mu}{2a_0} \qquad (4.45)$$

Augmenting the objective function with the equality constraint using the Lagrange multiplier λ yields the Lagrangian

$$\mathcal{L} = \| \Delta \mathbf{v} \|^2 + \lambda \left(\mathcal{E}_1^+ + \frac{\mu}{2a_0} \right) \qquad (4.46)$$

The necessary condition for the existence of a stationary point is

$$\frac{\partial \mathcal{L}}{\partial (\Delta \mathbf{v})} = \mathbf{0} \qquad (4.47)$$

This stationary point is a minimum if the bordered Hessian of $\mathcal{L}$ satisfies the test of Eq. (3.34). Equation (4.47) together with the constraint (4.40) establishes a system of four quadratic algebraic equations for the four variables Δv_x, Δv_y, Δv_z and λ. These equations are

$$2\Delta v_x + \lambda(v_x^- + \Delta v_x) = 0, \qquad (4.48)$$

$$2\Delta v_y + \lambda(v_y^- + \Delta v_y) = 0, \qquad (4.49)$$

$$2\Delta v_z + \lambda(v_z^- + \Delta v_z) = 0, \qquad (4.50)$$

$$\frac{1}{2} \left[(v_x^- + \Delta v_x)^2 + (v_y^- + \Delta v_y)^2 + (v_z^- + \Delta v_z)^2 \right] - \frac{\mu}{r_1} - \frac{\mu}{2a_0} = 0 \quad (4.51)$$

A solution thereof yields

$$\frac{\Delta v_x}{v_x^-} = \frac{\Delta v_y}{v_y^-} = \frac{\Delta v_z}{v_z^-} = -1 \pm \frac{1}{v_1^-} \sqrt{\frac{\mu(2a_0 - r_1)}{a_0 r_1}} \qquad (4.52)$$

$$\lambda = -2 \pm 2v_1^- \sqrt{\frac{a_0 r_1}{\mu(2a_0 - r_1)}} \qquad (4.53)$$

Real and finite solutions are obtained provided that $r_1 \leq 2a_0$. The bordered Hessian test of Eq. (3.34) shows that a sufficient condition for a minimum is $\lambda > -2$. Since $v_1^- > 0, r_1 > 0, \mu > 0, a > 0$, we conclude that only the first of solutions (4.53) corresponds to a minimum,

$$\lambda = -2 + 2v_1^- \sqrt{\frac{a_0 r_1}{\mu(2a_0 - r_1)}} \tag{4.54}$$

The corresponding optimal velocity corrections are therefore

$$\frac{\Delta v_x^*}{v_x^-} = \frac{\Delta v_y^*}{v_y^-} = \frac{\Delta v_z^*}{v_z^-} = -1 + \frac{1}{v_1^-}\sqrt{\frac{\mu(2a_0 - r_1)}{a_0 r_1}} \tag{4.55}$$

and the minimum velocity correction is

$$\Delta v^* = \sqrt{(\Delta v_x^*)^2 + (\Delta v_y^*)^2 + (\Delta v_z^*)^2} = v_1^- - \sqrt{\frac{\mu(2a_0 - r_1)}{a_0 r_1}} \tag{4.56}$$

Example 4.2. *Consider a chief spacecraft on an elliptic orbit. Normalize positions by a_0 and angular velocities by $\sqrt{\mu/a_0^3}$ so that $a_0 = \mu = 1$. Let the nominal normalized initial conditions be as in Eq. (4.38) with $\bar{x}(0) = -0.01127$. Assume that the initialization errors are*

$$\delta\bar{x}(0) = 0.001, \delta\bar{y}(0) = 0.001, \delta\bar{z}(0) = 0.01$$
$$\delta\bar{x}'(0) = 0, \delta\bar{y}'(0) = 0, \delta\bar{z}'(0) = 0 \tag{4.57}$$

Compute the minimum-fuel maneuver required to obtain a 1 : 1 bounded relative motion assuming that the maneuver is to be applied after one orbital period of the chief.

After a single orbital period, $\bar{t}_i = 1$ (in normalized units). Utilizing the initial conditions (4.38) and Eqs. (4.14)–(4.16) we have, at $\bar{t}_i = 1$,

$$\bar{x} = -0.015374, \bar{y} = -0.084596, \bar{z} = 0.109547$$
$$(\bar{x}')^- = 0.00994, (\bar{y}')^- = 0.021792, (\bar{z}')^- = 0.011765$$
$$(\theta_0')^- = 1.22838, \bar{r}_0 = 0.9, (\bar{r}_0')^- = 0 \tag{4.58}$$

Substituting into Eqs. (4.41)–(4.44) yields the normalized values

$$(\bar{v}_x)^- = 0.11386, (\bar{v}_y)^- = 1.10845,$$
$$(\bar{v}_z)^- = 0.01177, \bar{r}_1 = 0.89538 \tag{4.59}$$

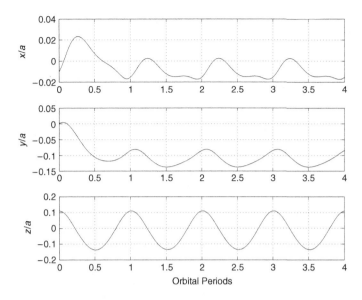

FIGURE 4.4 Time histories of the normalized relative position components show periodic behavior following the formation-keeping maneuver initiated after one orbital period.

Substituting again into Eq. (4.55) yields the optimal formation-keeping maneuver components (in normalize units)

$$\Delta \bar{v}_x^* = -0.00037144, \ \Delta \bar{v}_y^* = -0.00361606,$$
$$\Delta \bar{v}_z^* = -0.00003838 \tag{4.60}$$

which results in the total normalized Δv

$$\Delta \bar{v}^* = \sqrt{(\Delta \bar{v}_x^*)^2 + (\Delta \bar{v}_y^*)^2 + (\Delta \bar{v}_z^*)^2} = 0.0036353 \tag{4.61}$$

Figures 4.4–4.6 depict the results of a simulation performed utilizing the above values. Fig. 4.4 shows the time histories of the normalized relative position components. The time is normalized by the chief's orbital period. The impulsive formation-keeping maneuver is carried out at $\bar{t} = 1$. Following this maneuver, the position components converge to a periodic motion reflecting the 1:1 relative motion commensurability. Note the transient response leading to periodicity.

Figure 4.5 shows the three-dimensional relative orbit in the Euler–Hill frame utilizing normalized position components and the projections of the orbit on the xy, xz and yz planes. The initial relative drift is arrested at $\bar{t} = 1$, and a bounded relative motion results.

Figure 4.6 exhibits a magnification of the time history of the normalized relative velocity components $\bar{x}'$, $\bar{y}'$, $\bar{z}'$ at the vicinity of the impulsive maneuver, and the total energy of the deputy. The discontinuity in the relative velocity component is a result of the impulsive velocity change. Note that the initial

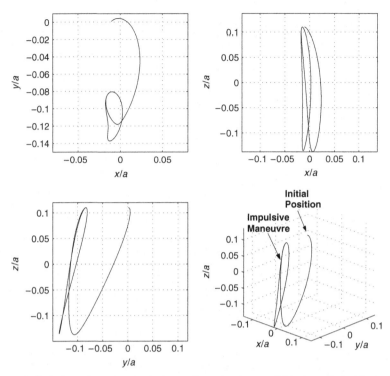

FIGURE 4.5 The relative motion in the configuration space exhibits bounded orbits following the formation-keeping maneuver.

energy of the deputy in normalized units is $\bar{\mathcal{E}}_1 = -0.496$, which differs from the total energy of the chief, $\bar{\mathcal{E}}_0 = -0.5$. The impulsive maneuver then decreases the total energy of the deputy by 0.004, matching it to the chief's energy and establishing a 1:1 bounded motion.

4.4 ANOTHER OUTLOOK ON OPTIMAL FORMATION-KEEPING

In the previous section, we developed an optimal formation-keeping maneuver for the deputy spacecraft using an impulsive velocity correction, $[\Delta\mathbf{v}]_{\mathscr{L}}$, formulated in a chief-fixed Euler–Hill frame. This maneuver has straightforward meaning if formulated in a *deputy-fixed frame*. To see this, we utilize the GVE (2.108). Denoting $[u_r, u_\theta, u_h]^T$ as the deputy's thrust acceleration components in the radial, along-track and cross-track directions, respectively, in its own $\mathscr{L}$ frame, the variational equation for the semimajor axis of the deputy is

$$\dot{a}_1 = \frac{2a_1^2}{h_1}\left(u_r e_1 \sin f_1 + \frac{p_1}{r_1}u_\theta\right) \tag{4.62}$$

where $h_1 = \sqrt{\mu a_1(1 - e_1^2)}$ is the deputy's orbital angular momentum. An

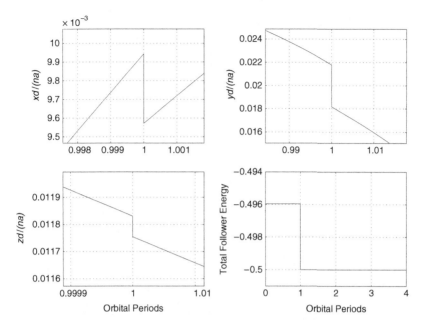

FIGURE 4.6 Magnification of the relative position components at the vicinity of the formation-keeping maneuver shows the applied velocity impulse, decreasing the total energy of the deputy to match the total energy of the chief (bottom right graph).

impulsive maneuver aimed at matching the semimajor axis of the deputy to the chief's semimajor axis can therefore be expressed as

$$\Delta a_1 = \frac{2a_1^2}{h_1}\left(\Delta v_r e_1 \sin f_1 + \frac{p_1}{r_1}\Delta v_\theta\right) \tag{4.63}$$

where $\Delta \mathbf{v_1} = [\Delta v_r, \Delta v_\theta, \Delta v_h]^T$ is the impulsive velocity correction vector in a deputy-fixed $\mathscr{L}$ frame.

Let us find a minimum-energy impulsive maneuver using an optimization problem formulated similarly to Eq. (4.45),

$$\Delta \mathbf{v_1^*} = \arg\min_{\Delta v_r, \Delta v_\theta} \Delta v_r^2 + \Delta v_\theta^2 \tag{4.64}$$

s. t.

$$\Delta a_1 = \frac{2a_1^2}{h_1}\left(\Delta v_r e_1 \sin f_1 + \frac{p_1}{r_1}\Delta v_\theta\right) \tag{4.65}$$

Using a Lagrange multiplier λ, the Lagrangian becomes

$$\mathcal{L}_1 = \Delta v_r^2 + \Delta v_\theta^2 + \lambda_1\left[\frac{2a_1^2}{h_1}\left(\Delta v_r e \sin f_1 + \frac{p_1}{r_1}\Delta v_\theta\right) - \Delta a_1\right] \tag{4.66}$$

Note that $\mathcal{L}_1$ is time-dependent through the deputy's true anomaly, f_1. The necessary conditions for an optimum are therefore

$$\frac{\partial \mathcal{L}_1}{\partial (\Delta \mathbf{v}_1)} = 0,$$

$$\frac{\partial \mathcal{L}_1}{\partial f_1} = 0, \tag{4.67}$$

yielding

$$\Delta v_r + \frac{2\lambda_1 a_1^2 e_1 \sin f_1}{h_1} = 0 \tag{4.68}$$

$$\Delta v_\theta + \frac{\lambda_1 a_1^2 (1 + e_1 \cos f_1)}{h_1} = 0 \tag{4.69}$$

$$\lambda_1 a_1^2 (\Delta v_r e_1 \cos f_1 - e_1 \sin f_1 \Delta v_\theta) = 0 \tag{4.70}$$

Equations (4.68)–(4.70) together with constraint (4.65) constitute a system of four algebraic equations for the unknowns $\Delta v_r, \Delta v_\theta, \lambda_1, f_1$. There are two possible solutions:

$$\Delta v_r = 0, \Delta v_\theta = \frac{h_1 \Delta a_1}{2a_1^2 (1 + e_1)}, \lambda_1 = -\frac{h_1^2 \Delta a_1}{2a_1^4 (e_1 + 1)^2}, f_1 = 0 \tag{4.71}$$

$$\Delta v_r = 0, \Delta v_\theta = \frac{h_1 \Delta a_1}{2a_1^2 (1 - e_1)}, \lambda_1 = -\frac{h_1^2 \Delta a_1}{2a_1^4 (e_1 - 1)^2}, f_1 = \pi \tag{4.72}$$

Evaluation of the bordered Hessian shows that Eq. (4.71) provides the minimum solution. It implies that the optimal impulsive maneuver should be performed at periapsis ($f_1 = 0$):

$$\Delta \mathbf{v}_1^* = [0, \Delta v_\theta (f_1 = 0), 0]^T \tag{4.73}$$

Hence, we have obtained an indirect indication of the optimal maneuver initiation time, t_i, discussed in the preceding section. The magnitude of the optimal maneuver equals the magnitude of Δv_θ.

Note that the magnitude of the velocity correction in the deputy's frame should be equal to the magnitude of the velocity correction as viewed in the chief's frame, as magnitude of vectors are invariant to a particular selection of reference frames.

4.5 CIRCULAR CHIEF ORBIT

In Section 4.1 we derived the general nonlinear equations of relative motion for arbitrary chief orbits.[1] A simpler, autonomous, form of the relative motion

---

[1] Generally, the chief's orbit can be hyperbolic, and not necessarily elliptic; but this is of little practical value.

equations can be derived, however, if we assume that the chief follows a circular orbit. In many practical cases this is a realistic assumption.

In the circular chief orbit case, $\dot{\theta}_0 = n_0 = \text{const.}, \ddot{\theta}_0 = 0$ and $r_0 = a_0 = \text{const.}$ Substituting into Eqs. (4.14)–(4.16) results in

$$\ddot{x} - 2n_0\dot{y} - n_0^2 x = -\frac{\mu(a_0 + x)}{\left[(a_0 + x)^2 + y^2 + z^2\right]^{\frac{3}{2}}} + \frac{\mu}{a_0^2} \tag{4.74}$$

$$\ddot{y} + 2n_0\dot{x} - n_0^2 y = -\frac{\mu y}{\left[(a_0 + x)^2 + y^2 + z^2\right]^{\frac{3}{2}}} \tag{4.75}$$

$$\ddot{z} = -\frac{\mu z}{\left[(a_0 + x)^2 + y^2 + z^2\right]^{\frac{3}{2}}} \tag{4.76}$$

These equations admit an *equilibria continuum* ($\dot{x} = \ddot{x} = \dot{y} = \ddot{y} = \dot{z} = \ddot{z} = 0$) given by

$$z = 0, (x + a_0)^2 + y^2 = a_0^2 \tag{4.77}$$

It can be straightforwardly shown that, not unexpectedly, the equilibria (4.77) conform to the energy matching condition.

Equation (4.77) defines a circle that coincides with the chief's orbit: It is centered at ($x = -a_0$, $y = 0$), which are the coordinates of the primary in frame $\mathscr{L}$. This result reflects the trivial physical observation that the deputy spacecraft will appear stationary in a chief-fixed frame if the deputy is co-located on the circular orbit of the chief. This type of in-line relative motion is referred to as *co-orbital motion*. From the dynamical systems perspective, we expect that there exist small perturbations near equilibria that will generate periodic orbits about the equilibria. The resulting periodic motion is called *libration*. Libration orbits can be found from the energy matching condition and hence can be viewed as a subset of the 1:1 commensurability periodic orbits.

A similar terminology is used for describing the relative motion between celestial bodies. The dynamic structure of the problem in this case is much more involved, as it includes the mutual gravitation of the bodies [93].

Equations (4.74)–(4.76) can be normalized, so that the distances are measured in units of a_0 and the angular velocities are written in units of $\sqrt{\mu/a_0^3}$. In this case, Eq. (4.77) defines a *unit circle*.[2] A libration about a point moving in a circular orbit is a form of *epicyclic motion*.[3] We will also discuss epicyclic motion in Chapter 6. To illustrate epicyclic motion, we consider the following example.

---

[2]In this context, we should also mention that the triangular equilibrium points in the restricted three body problem, known as Lagrange's points L_4 and L_5, are also equilibria associated to motion in the vicinity of the unit circle.

[3]In Ptolemaic astronomy, a small circle, the center of which moves on the circumference of a larger circle at whose center is the Earth and the circumference of which describes the orbit of one of the planets around Earth.

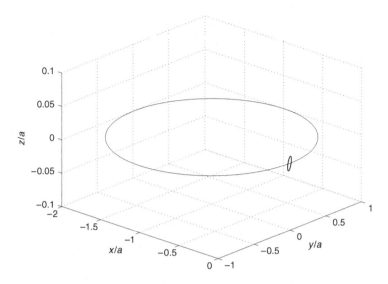

FIGURE 4.7 Libration orbits about a unit circle constitute epicyclic motion.

Example 4.3. *Simulate the libration orbits obtained for the normalized initial conditions*

$$\bar{x}(0) = -0.005019, \ \bar{y}(0) = 0.01, \ \bar{z}(0) = 0.01$$
$$\bar{x}'(0) = 0.01, \ \bar{y}'(0) = 0.01, \ \bar{z}'(0) = 0 \tag{4.78}$$

The energy matching condition with a circular chief orbit, using normalized units, simplifies into (cf. Eq. (4.34)):

$$\bar{\mathcal{E}}_1 = \frac{1}{2}\{(\bar{x}' - \bar{y})^2 + [\bar{y}' + (\bar{x} + 1)]^2 + (\bar{z}')^2\} - \frac{1}{\sqrt{(1 + \bar{x})^2 + \bar{y}^2 + \bar{z}^2}}$$
$$= -\frac{1}{2} \tag{4.79}$$

It can be readily verified that initial conditions (4.78) satisfy this condition. Utilizing these initial conditions, we simulate Eqs. (4.74)–(4.76) for a single orbital period. The results of the simulation are depicted in Fig. 4.7, showing the libration orbit (thick line) with respect to the unit circle (thin line). As previously explained, the unit circle constitutes a continuum of relative equilibria representing the chief's orbit. A libration orbit represents a small oscillation about each fixed point.

4.6 LAGRANGIAN AND HAMILTONIAN DERIVATIONS

We will now show how to obtain the relative motion equations using Lagrangian and Hamiltonian formalisms, following a treatment similar to the one proposed by Kasdin *et al.* [94]. This will be used as a stepping stone for the derivation

of linear equations of motion in the next chapter, and for incorporating perturbations (Chapter 8). For simplicity, we treat the case of a relative motion with
respect to a circular reference orbit. We treat this case because of its simplicity,
allowing us to focus attention on the details of the method.

The first step is to develop the Lagrangian of relative motion in the rotating
frame $\mathscr{L}$. In the circular reference orbit case, the velocity of the deputy in $\mathscr{L}$ is
derived from Eq. (4.30):

$$\mathbf{v}_1 = {}^{\mathscr{I}}\boldsymbol{\omega}^{\mathscr{L}} \times \mathbf{r}_0 + \frac{d^{\mathscr{L}}}{dt}\boldsymbol{\rho} + {}^{\mathscr{I}}\boldsymbol{\omega}^{\mathscr{L}} \times \boldsymbol{\rho} \tag{4.80}$$

where, as before, $\mathbf{r}_0 \in \mathbb{R}^3$ is the inertial position vector of the chief spacecraft
along the reference orbit, $\boldsymbol{\rho} = [x, y, z]^T \in \mathbb{R}^3$ is the relative position vector in
the rotating frame, and ${}^{\mathscr{I}}\boldsymbol{\omega}^{\mathscr{L}} = [0, 0, n]^T$ is the angular velocity of the rotating
frame $\mathscr{L}$ with respect to the inertial frame $\mathscr{I}$. Assuming a circular reference
orbit, denoting $\|\mathbf{r}_0\| = a$ and substituting into Eq. (4.80), we can write the
velocity in a component-wise notation:

$$\mathbf{v}_1 = \begin{bmatrix} \dot{x} - n_0 y \\ \dot{y} + n_0 x + n_0 a_0 \\ \dot{z} \end{bmatrix} \tag{4.81}$$

The kinetic energy per unit mass is given by

$$\mathcal{K} = \frac{1}{2}\|\mathbf{v}_1\|^2 \tag{4.82}$$

The potential energy (for a spherical attracting body) of the deputy, whose
position vector is $\mathbf{r}_1$, is the usual gravitational potential written in terms of
$\rho = \|\boldsymbol{\rho}\|$:

$$\mathcal{U} = -\frac{\mu}{\|\mathbf{r}_1\|} = -\frac{\mu}{\|\mathbf{r}_0 + \boldsymbol{\rho}\|} = -\frac{\mu}{a_0\left[1 + 2\frac{\mathbf{r}_0 \cdot \boldsymbol{\rho}}{a_0^2} + \left(\frac{\rho}{a_0}\right)^2\right]^{1/2}} \tag{4.83}$$

The last term in Eq. (4.83) can be expanded by using Legendre polynomials,
P_k, which we introduced in Subsection 2.5.3:

$$\mathcal{U} = -\frac{\mu}{a_0}\sum_{k=0}^{\infty} P_k(\cos\vartheta)\left(\frac{\rho}{a_0}\right)^k \tag{4.84}$$

where ϑ is the angle between the reference orbit radius vector and the relative
position vector (Fig. 4.1), so that

$$\cos\vartheta = -\frac{\boldsymbol{\rho} \cdot \mathbf{r}_0}{a_0\rho} = \frac{-x}{\sqrt{x^2 + y^2 + z^2}} \tag{4.85}$$

As per the definition appearing in Eq. (3.1), the Lagrangian $\mathcal{L} : \mathbb{R}^3 \times \mathbb{R}^3 \to \mathbb{R}$ is now found by subtracting the potential energy from the kinetic energy:

$$\mathcal{L} = \frac{1}{2} \left\{ (\dot{x} - n_0 y)^2 + (\dot{y} + n_0 x + n_0 a_0)^2 + \dot{z}^2 \right\}$$
$$+ n_0^2 a_0^2 \sum_{k=0}^{\infty} P_k(\cos \vartheta) \left(\frac{\rho}{a_0} \right)^k \tag{4.86}$$

The equations of relative motion, Eqs. (4.14)–(4.16), can now be obtained by applying the Euler–Lagrange equations (3.3). However, due to the use of the Legendre polynomials, one can obtain arbitrary-order approximations to the full nonlinear equations of relative motion by truncating the Legendre polynomials at some desired degree. This exercise is performed in Section 5.3.

We can now proceed with the Hamiltonian formalism. Finding the Hamiltonian for the Cartesian system is straightforward. First, the canonical momenta are found from the definition (3.4):

$$p_x = \frac{\partial \mathcal{L}}{\partial \dot{x}} = \dot{x} - n_0 y$$
$$p_y = \frac{\partial \mathcal{L}}{\partial \dot{y}} = \dot{y} + n_0 x + n_0 a_0 \tag{4.87}$$
$$p_z = \frac{\partial \mathcal{L}}{\partial \dot{z}} = \dot{z}$$

and then, using the Legendre transformation $\mathcal{H} = \mathbf{p}^T \dot{\mathbf{q}} - \mathcal{L}$, as given in Eq. (3.5), the Hamiltonian for relative motion is found:

$$\mathcal{H} = \frac{1}{2} (p_x^2 + p_y^2 + p_z^2) + n_0 y p_x - (n_0 x + n_0 a_0) p_y$$
$$- n_0^2 a_0^2 \sum_{k=0}^{\infty} P_k(\cos \vartheta) \left(\frac{\rho}{a_0} \right)^k \tag{4.88}$$

It is worthwhile to note that the Hamiltonian is a constant of motion. We will use expressions (4.86) and (4.88) in Chapter 5 to derive linear equations of relative motion. A Hamiltonian description of relative motion between satellites in similar Keplerian orbits, satisfying the principles of conservation of angular momentum and the Hamiltonian, has been provided by Palmer and İmre [95], who derive their results in a perigee-fixed coordinate system of the chief, not the $\mathscr{L}$ frame.

4.7 EQUATIONS OF RELATIVE MOTION UNDER THE INFLUENCE OF J_2

The equations of motion of a satellite under the influence of gravitational and thrust perturbations can be written in frame $\mathscr{I}$ as (see our discussion of orbital perturbations in Section 2.5)

$$\ddot{\mathbf{r}} = -\nabla_{\mathbf{r}} \mathcal{V} + \mathbf{u} \tag{4.89}$$

where $\mathbf{r}$ is the position vector, $\mathcal{V} = \mathcal{U} - \mathcal{R}$ is the *total gravitational potential* (the nominal potential plus the J_2 potential), and $\nabla_{\mathbf{r}} \mathcal{V}$ is its gradient. The thrust acceleration is denoted by $\mathbf{u}$. The gravitational potential, including the contribution of J_2, is given by (see Eq. (2.98) and Ref. [96]):

$$\mathcal{V} = -\frac{\mu}{r} \left\{ 1 - \frac{J_2 R_e^2}{2 r^2} \left[\frac{3}{r^2} (\mathbf{r} \cdot \hat{\mathbf{K}})^2 - 1 \right] \right\} \tag{4.90}$$

where r is the magnitude of $\mathbf{r}$ and $\hat{\mathbf{K}}$ is the unit vector along the polar axis of the $\mathscr{I}$ frame; it can be expressed in any desired frame of reference, $\mathscr{I}$ or $\mathscr{L}$. In the $\mathscr{L}$ frame we have:

$$\hat{\mathbf{K}} = \sin \theta_0 \sin i_0 \hat{\mathbf{i}} + \cos \theta_0 \sin i_0 \hat{\mathbf{j}} + \cos i_0 \hat{\mathbf{k}} \tag{4.91}$$

where $\hat{\mathbf{i}}, \hat{\mathbf{j}}$, and $\hat{\mathbf{k}}$, are respectively, the unit vectors along the radial, along-track, and cross-track directions of the $\mathscr{L}$ frame of the satellite. Note that the term $(\mathbf{r} \cdot \hat{\mathbf{K}})$ in Eq. (4.90) is nothing but Z, the polar component of the satellite's position vector, $\mathbf{r}$. The gravitational acceleration, $\mathbf{F}_g$, obtained from Eq. (4.90) is

$$\mathbf{F}_g = -\nabla_{\mathbf{r}} \mathcal{V} = -\frac{\mu}{r^3} \mathbf{r} - \frac{1}{2r^5} \mu J_2 R_e^2 \left\{ 6(\mathbf{r} \cdot \hat{\mathbf{K}}) \hat{\mathbf{K}} \right.$$
$$\left. + \left[3 - \frac{15}{r^2} (\mathbf{r} \cdot \hat{\mathbf{K}})^2 \right] \mathbf{r} \right\} \tag{4.92}$$

Equations (4.91) and (4.92) can be substituted into Eq. (4.89) to obtain the equations of motion. A very convenient $\mathscr{I}$ frame representation of the perturbed motion, including the effects of $J_2 - J_6$, can be found in Ref. [44]; see also our discussion of zonal perturbations in Subsection 2.5.3. These equations are presented below for the special case of J_2-perturbed motion:

$$\ddot{X} = -\frac{\mu X}{r^3} \left[1 - \frac{3}{2} J_2 \left(\frac{R_e}{r} \right)^2 \left(5 \frac{Z^2}{r^2} - 1 \right) \right] \tag{4.93a}$$

$$\ddot{Y} = -\frac{\mu Y}{r^3} \left[1 - \frac{3}{2} J_2 \left(\frac{R_e}{r} \right)^2 \left(5 \frac{Z^2}{r^2} - 1 \right) \right] \tag{4.93b}$$

$$\ddot{Z} = -\frac{\mu Z}{r^3} \left[1 - \frac{3}{2} J_2 \left(\frac{R_e}{r} \right)^2 \left(5 \frac{Z^2}{r^2} - 3 \right) \right] \tag{4.93c}$$

where X, Y, and Z are the components of $\mathbf{r}$ expressed in the $\mathscr{I}$ frame. We note that the Hamiltonian, which, in this case, is identical to the total energy, is a constant of motion,

$$\mathcal{H} = \mathcal{K} + \mathcal{V} = \frac{\dot{X}^2 + \dot{Y}^2 + \dot{Z}^2}{2} - \frac{\mu}{r} \left\{ 1 - \frac{J_2}{2} \frac{R_e^2}{r^2} \left[\frac{3Z^2}{r^2} - 1 \right] \right\}$$

$$= \text{const.} \tag{4.94}$$

and so is the polar component of the angular momentum,[4]

$$\mathbf{h} \cdot \hat{\mathbf{K}} = X\dot{Y} - Y\dot{X} = \text{const.} \tag{4.95}$$

The relative displacement and velocity vectors, expressed in the $\mathscr{I}$ frame, are defined as

$$\delta\mathbf{r} = \mathbf{r}_1 - \mathbf{r}_0 \tag{4.96a}$$

$$\delta\mathbf{v} = \mathbf{v}_1 - \mathbf{v}_0 \tag{4.96b}$$

where $\mathbf{v}$ denotes the inertial velocity vector of a satellite. We remind the reader that variables relating to the chief are indicated with the subscript 0 and those connected with the deputy are denoted by the subscript 1.

A nonlinear simulation for multiple satellites is carried out by numerically integrating copies of Eqs. (4.93a)–(4.93c), one for each satellite, with different sets of initial conditions. The results thus obtained can be transformed into the relative motion states given by Eqs. (4.96a)–(4.96b). For many applications, the relative motion states are typically expressed in the $\mathscr{L}$ frame of the chief. Transformation of the results of Eqs. (4.96a)–(4.96b) into the relative motion variables in the $\mathscr{L}$ frame is discussed next.

4.7.1 Relative motion states in the $\mathscr{L}$ frame

The relative motion between two satellites in general elliptic orbits, written using a slightly different representation compared to previous chapters, can be expressed in the chief's $\mathscr{L}$ frame [99]:

$$x = \frac{\delta\mathbf{r}^T \mathbf{r}_0}{r_0} \tag{4.97a}$$

$$y = \frac{\delta\mathbf{r}^T (\mathbf{h}_0 \times \mathbf{r}_0)}{\|\mathbf{h}_0 \times \mathbf{r}_0\|} \tag{4.97b}$$

$$z = \frac{\delta\mathbf{r}^T \mathbf{h}_0}{h_0} \tag{4.97c}$$

where, as before, x, y, and z indicate, respectively, the radial, along-track,

---

[4]The total energy and the polar component of the angular momentum are the *only* two integrals of the J_2-perturbed problem [97,98]. Since there are three degrees-of-freedom and only two integrals, the J_2-perturbed problem constitutes a *non-integrable Hamiltonian system*, implying that there are phase-space regions in which chaotic motion is likely to exist. However, these regions are extremely small for actual Earth orbits.

and cross-track displacements; and h_0 is the magnitude of $\mathbf{h}_0$, the angular momentum vector of the chief.

The relative velocities in the $\mathscr{L}$ frame can be obtained by differentiating the expressions (4.97a)–(4.97c) with respect to time:

$$\dot{x} = \frac{\delta \mathbf{v}^T \mathbf{r}_0 + \delta \mathbf{r}^T \mathbf{v}_0}{r_0} - \frac{(\delta \mathbf{r}^T \mathbf{r}_0)(\delta \mathbf{r}_0^T \mathbf{v}_0)}{r_0^3} \tag{4.98a}$$

$$\dot{y} = \frac{\delta \mathbf{v}^T (\mathbf{h}_0 \times \mathbf{r}_0) + \delta \mathbf{r}^T (\mathbf{h}_0 \times \mathbf{r}_0 + \mathbf{h}_0 \times \mathbf{v}_0)}{\|\mathbf{h}_0 \times \mathbf{r}_0\|}$$
$$- \frac{\delta \mathbf{r}^T (\mathbf{h}_0 \times \mathbf{r}_0)(\mathbf{h}_0 \times \mathbf{r}_0)^T (\mathbf{h}_0 \times \mathbf{r}_0 + \mathbf{h}_0 \times \mathbf{v}_0)}{\|\mathbf{h}_0 \times \mathbf{r}_0\|^3} \tag{4.98b}$$

$$\dot{z} = \frac{\delta \mathbf{v}^T \mathbf{h}_0 + \delta \mathbf{r}^T \dot{\mathbf{h}}_0}{h_0} - \frac{\delta \mathbf{r}^T \mathbf{h}_0 (\mathbf{h}_0^T \dot{\mathbf{h}}_0)}{h_0^3} \tag{4.98c}$$

Notice that the relative velocities as computed from Eqs. (4.98a)–(4.98c) depend on $\dot{\mathbf{h}}_0 = \mathbf{r}_0 \times \dot{\mathbf{v}}_0$. Hence, the inertial acceleration vector of the chief, including the contributions due to the perturbations acting upon it, is required in the above transformation. However, this requirement is not problematic if the motion of the chief (real or virtual) is assumed to be known accurately.

4.7.2 Coordinate transformation between the $\mathscr{L}$ and $\mathscr{I}$ frames

Many applications involving formation control may require that a vector be transformed from $\mathscr{L}$ to $\mathscr{I}$ or vice versa. Given the $\mathscr{I}$ frame representation of the position and velocity vectors of the chief, the transformation from $\mathscr{L}$ to $\mathscr{I}$ is given by

$$T_{\mathscr{L}}^{\mathscr{I}} = \begin{bmatrix} \hat{\mathbf{r}}_0 & (\hat{\mathbf{h}}_0 \times \mathbf{r}_0) & \hat{\mathbf{h}}_0 \end{bmatrix} \tag{4.99}$$

where $(\hat{\cdot})$ represents a unit vector. The inverse transformation can easily be obtained by using the matrix transpose operation; this leads to the expression

$$T_{\mathscr{I}}^{\mathscr{L}}(\Omega, i, \theta) = \begin{bmatrix} c_\Omega c_\omega - s_\Omega s_\theta c_i & s_\Omega c_\theta + c_\Omega s_\theta c_i & s_\theta s_i \\ -c_\Omega s_\theta - s_\Omega c_\theta c_i & -s_\Omega s_\theta + c_\Omega c_\theta c_i & c_\theta s_i \\ s_\Omega s_i & -c_\Omega s_i & c_i \end{bmatrix} \tag{4.100}$$

where $\theta = \omega + f$ (cf. also our development in Chapter 2).

4.7.3 Initial conditions

It is convenient for numerical simulations to specify the initial conditions of the satellites in terms of their mean orbital elements. These initial mean elements can be transformed into the respective osculating elements via the use of Brouwer theory [66] (for the classical elements) or the procedure of Gim and Alfriend [75,100], for nonsingular and equinoctial elements. The osculating elements for each satellite can be transformed into the $\mathscr{I}$ frame position and velocity coordinates using the principles of orbital mechanics detailed in Chapter 2. The initial conditions thus obtained, for each satellite, are applicable to Eqs. (4.93a)–(4.93c). A discussion of the process of mean-to-osculating orbit element conversion is presented in Section 3.4.

SUMMARY

This chapter presented a variety of models for simulating relative motion in perturbed or unperturbed elliptic orbits. In the absence of perturbations, bounded relative motion between any two spacecraft in elliptic Keplerian orbits can be straightforwardly found from the energy matching condition. By dealing with the full nonlinear problem, global conditions for bounded motion can be found which naturally accommodate the nonlinearity of the Keplerian relative motion equations as well as the reference orbit eccentricity.

Orbital commensurability, guaranteeing bounded relative motion, can be re-established in spite of initialization errors by applying a single thrust impulse, which can be calculated and optimized using relative state variables and has a simple interpretation in terms of the deputy's orbital elements. This single-impulse maneuver also has considerable operational advantages over multiple impulses.

If the chief's orbit is circular, commensurability conditions can be considerably simplified. The relative motion in this case has a simple geometric interpretation: Epicyclic motion is generated if orbital commensurability is violated, and librations are obtained if the motions are commensurable.

A set of useful transformations for converting the relative motion states between the $\mathscr{L}$ and $\mathscr{I}$ frames was introduced. These transformations have been applied in many instances in the chapters to follow. The direct derivation of the perturbed equations of motion in the $\mathscr{L}$ frame is treated in the next chapter.

Chapter | five

Linear Equations of Relative Motion

It is the mark of an educated mind
to rest satisfied with the degree of precision
which the nature of the subject admits
and not to seek exactness
where only an approximation is possible.

Aristotle (384 BC–322 BC)

In this chapter, we present a variety of linear differential equations for modeling relative motion under the two-body assumptions. The systems of equations for the description of relative motion can be classified according to the coordinate system utilized – Cartesian or curvilinear. There are also the issues of the choices of the independent variable – time or an angle variable and the space in which the linearization is carried out, e.g., physical coordinates vs. a set of orbital elements. The focus in this chapter is on the physical coordinate description of relative motion; motion description by the orbital elements is treated extensively in Chapter 6. This chapter begins with a discussion of the derivation of the linearized equations of relative motion with respect to a circular reference orbit, also known as the Clohessey–Wiltshire (CW) equations [101]. The CW equations have found extensive use for the analysis, design, and control of formations and a brief description of these applications is given in Section 5.2. Section 5.3 presents a derivation of the CW equations from the perspective of analytical mechanics, via the Lagrangian and the Hamiltonian. Section 5.5 presents a comparison and error analysis of the linearized equations obtained by using Cartesian and curvilinear coordinates. This section is followed by the presentation of the Tschauner–Hempel equations [102] for elliptic reference orbits. Brief descriptions of several approaches to the derivation of the two-body relative motion State Transition Matrix (STM) are also included. The chapter ends with a discussion of the application of the elliptic reference orbit STM to obtain initial conditions for preventing secular drift and alternatively, for the computation of impulsive maneuvers for establishing or initializing formations.

83

Spacecraft Formation Flying; ISBN: 9780750685337

5.1 THE CLOHESSY–WILTSHIRE EQUATIONS

In Section 4.5 we developed the nonlinear equations of relative motion assuming that the chief's orbit is circular. This simplification rendered an autonomous set of nonlinear differential equations. If the deputy's orbit in the inertial space is only slightly elliptic and slightly inclined with respect to the chief's orbit, the motion of the deputy will appear very close to the chief in a chief-fixed frame, provided that the initial positions are first-order small. In this case, Eqs. (4.74)–(4.76) may be linearized about the origin of the chief-fixed frame, $\mathscr{L}$, and the resulting motion may be solved in closed-form. The linearized equations of motion are called the *Clohessy–Wiltshire equations* (CW) or the *Hill–Clohessy–Wiltshire equations* (HCW).[1] These equations were developed by CW in the early 1960s to analyze spacecraft rendezvous [101].

There are a number of ways to develop and solve the CW equations. We will discuss some of these methods, as each method contributes insight into the relative motion problem.

A straightforward approach for developing the CW equations is to expand the right-hand side of Eqs. (4.74)–(4.76) into a Taylor series about the origin. Taking only the first-order terms and denoting $n_0 = \sqrt{\mu/a_0^3}$ we get

$$-\frac{\mu(a_0 + x)}{\left[(a_0 + x)^2 + y^2 + z^2\right]^{\frac{3}{2}}} \approx n_0^2(2x - a_0) \tag{5.1}$$

$$-\frac{\mu y}{\left[(a_0 + x)^2 + y^2 + z^2\right]^{\frac{3}{2}}} \approx -n_0^2 y \tag{5.2}$$

$$-\frac{\mu z}{\left[(a_0 + x)^2 + y^2 + z^2\right]^{\frac{3}{2}}} \approx -n_0^2 z \tag{5.3}$$

Rearranging and omitting the subscript 0 (so that $n \equiv n_0$ and $a \equiv a_0$) yields

$$\ddot{x} - 2n\dot{y} - 3n^2 x = 0 \tag{5.4}$$

$$\ddot{y} + 2n\dot{x} = 0 \tag{5.5}$$

$$\ddot{z} + n^2 z = 0 \tag{5.6}$$

Equations (5.4)–(5.6) are the CW equations with no disturbing and/or control accelerations. The nonhomogeneous forms of Eqs. (5.4)–(5.6) are

$$\ddot{x} - 2n\dot{y} - 3n^2 x = d_x + u_x \tag{5.7}$$

$$\ddot{y} + 2n\dot{x} = d_y + u_y \tag{5.8}$$

$$\ddot{z} + n^2 z = d_z + u_z \tag{5.9}$$

---

[1] These equations are sometimes referred to as *Hill's equations*, as Hill was the first to develop linearized relative equations of motion to describe the Moon's orbit in a rotating Earth-centric frame. However, we will avoid using the term "Hill's equations" because in mathematics this term is reserved for the equation $\frac{d^2 y}{dx^2} + \left(\theta_0 + 2\sum_{n=1}^{\infty} \theta_n \cos(2nx)\right) y = 0$, where the θ's are constants.

where $[d_x, d_y, d_z]^T$ and $[u_x, u_y, u_z]^T$ are, respectively, the vectors of environmental perturbations and control accelerations. One often normalizes the relative coordinates by the radius of the reference orbit, a, and the angular velocities by n, so that in normalized form the unforced CW equations (5.4)–(5.6) become

$$\bar{x}'' - 2\bar{y}' - 3\bar{x} = 0 \tag{5.10}$$
$$\bar{y}'' + 2\bar{x}' = 0 \tag{5.11}$$
$$\bar{z}'' + \bar{z} = 0 \tag{5.12}$$

where $(\cdot)'$ denotes differentiation with respect to normalized time. It is convenient to write the linear differential equations (5.4)–(5.6) in state-space form. Choosing the state vector $\mathbf{x} = [x, y, z, \dot{x}, \dot{y}, \dot{z}]^T$, Eqs. (5.4)–(5.6) assume the form

$$\dot{\mathbf{x}}(t) = A\mathbf{x}(t) \tag{5.13}$$

where A is the *system matrix*, given by

$$A = \begin{bmatrix} 0 & 0 & 0 & 1 & 0 & 0 \\ 0 & 0 & 0 & 0 & 1 & 0 \\ 0 & 0 & 0 & 0 & 0 & 1 \\ 3n^2 & 0 & 0 & 0 & 2n & 0 \\ 0 & 0 & 0 & -2n & 0 & 0 \\ 0 & 0 & -n^2 & 0 & 0 & 0 \end{bmatrix}, \tag{5.14}$$

and the initial conditions are $\mathbf{x}(t_0) = [x(t_0), y(t_0), z(t_0), \dot{x}(t_0), \dot{y}(t_0), \dot{z}(t_0)]^T$. The eigenvalues of A are $\{\pm nj, \pm nj, 0, 0\}$, so a secular mode is expected to appear in the solution.

Solving the CW equations is straightforward. The solution may be formulated in terms of the transition matrix, $e^{A(t-t_0)}$,

$$\mathbf{x}(t) = e^{A(t-t_0)}\mathbf{x}(t_0) \tag{5.15}$$

Following the usual steps for computing the transition matrix yields (for $t_0 = 0$),

$$e^{At} = \begin{bmatrix} 4 - 3c_{nt} & 0 & 0 & \dfrac{s_{nt}}{n} & \dfrac{2}{n} - \dfrac{2c_{nt}}{n} & 0 \\ -6nt + 6s_{nt} & 1 & 0 & -\dfrac{2}{n} + \dfrac{2c_{nt}}{n} & \dfrac{4s_{nt}}{n} - 3t & 0 \\ 0 & 0 & c_{nt} & 0 & 0 & \dfrac{s_{nt}}{n} \\ 3ns_{nt} & 0 & 0 & c_{nt} & 2s_{nt} & 0 \\ -6n + 6nc_{nt} & 0 & 0 & -2s_{nt} & -3 + 4c_{nt} & 0 \\ 0 & 0 & -ns_{nt} & 0 & 0 & c_{nt} \end{bmatrix} \tag{5.16}$$

We can now determine, by substituting Eq. (5.16) into Eq. (5.15), the solutions to the relative position and velocity components

$$x(t) = \left[4x(0) + \frac{2\dot{y}(0)}{n}\right] + \frac{\dot{x}(0)}{n}\sin(nt) - \left[3x(0) + \frac{2\dot{y}(0)}{n}\right]\cos(nt) \quad (5.17)$$

$$y(t) = -[6nx(0) + 3\dot{y}(0)]t + \left[y(0) - \frac{2\dot{x}(0)}{n}\right] + \left[6x(0) + \frac{4\dot{y}(0)}{n}\right]\sin(nt)$$
$$+ \frac{2\dot{x}(0)}{n}\cos(nt), \quad (5.18)$$

$$z(t) = \frac{\dot{z}(0)}{n}\sin(nt) + z(0)\cos(nt), \quad (5.19)$$

$$\dot{x}(t) = \dot{x}(0)\cos(nt) + [3x(0)n + 2\dot{y}(0)]\sin(nt) \quad (5.20)$$

$$\dot{y}(t) = -[6nx(0) + 3\dot{y}(0)] + [6x(0)n + 4\dot{y}(0)]\cos(nt) - 2\dot{x}(0)\sin(nt) \quad (5.21)$$

$$\dot{z}(t) = \dot{z}(0)\cos(nt) - z(0)n\sin(nt) \quad (5.22)$$

A number of observations are in order. First, from the dynamical systems perspective, the linear system (5.13) has multiple equilibria $\{x = 0, y = \text{const.}, z = 0\}$, while the original, nonlinear system has a continuum of equilibria (4.77). There is some equivalence between the nonlinear and linear systems: In the linear system, the coordinate y can be written as $y = a\delta\theta$, where $\delta\theta$ is a curvilinear coordinate along the circumference of the reference orbit. Hence, $\delta\theta = \text{const.}$ corresponds to an angular shift of the deputy along the chief's reference orbit.

Second, we note that the linearization has decoupled the out-of-plane, *cross-track*, motion from the in-plane motion. The cross-track motion is a simple harmonic. It can be nullified by selecting the initial conditions: $z(0) = \dot{z}(0) = 0$.

Third, the transverse, *along-track*, component shows the presence of drift, varying linearly with time, implying that the in-plane motion is *unstable*. Nevertheless, we can find a *stable subspace* of the state-space, rendering the motion stable. The initial conditions which span the stable subspace are given by

$$\dot{y}(0) = -2nx(0) \quad (5.23)$$

Choosing initial conditions (5.23) will yield a bounded relative motion *to first order*, i.e., only for the CW model. It is important to note that the concept of stability in the context of the CW equations is *local*. There exist initial conditions which violate Eq. (5.23) but satisfy the energy matching condition, yielding a bounded 1:1 commensurable motion described by Eqs. (4.74)–(4.76). This is because the energy matching condition is a *global* boundedness criterion. By using the CW approximation, we truncate the physical solution of the problem, thus introducing an inherent error. The conclusions from the CW stability analysis must therefore be judiciously handled; it must be understood that the local results apply only in a small neighborhood of the chief, and cannot be relied upon for drawing conclusions as to properties of the relative motion with respect to circular, two-body orbits.

Nevertheless, the stable CW solutions have some interesting geometric properties which are worth mentioning. To gain a geometric insight, note that if Eq. (5.23) is satisfied, Eqs. (5.17)–(5.19) may be written in the magnitude-phase form

$$x(t) = \rho_x \sin(nt + \alpha_x) \tag{5.24}$$
$$y(t) = \rho_y + 2\rho_x \cos(nt + \alpha_x) \tag{5.25}$$
$$z(t) = \rho_z \sin(nt + \alpha_z) \tag{5.26}$$

where

$$\rho_x = \frac{\sqrt{\dot{x}^2(0) + x^2(0)n^2}}{n} \tag{5.27}$$
$$\rho_y = [y(0) - 2\dot{x}(0)/n] \tag{5.28}$$
$$\rho_z = \frac{\sqrt{\dot{z}^2(0) + z^2(0)n^2}}{n} \tag{5.29}$$
$$\alpha_x = \tan^{-1}\frac{nx(0)}{\dot{x}(0)} \tag{5.30}$$
$$\alpha_z = \tan^{-1}\frac{nz(0)}{\dot{z}(0)} \tag{5.31}$$

Equations (5.24)–(5.26) constitute a parametric representation of an *elliptic cylinder*. If

$$\alpha_x = \alpha_z \tag{5.32}$$

then Eqs. (5.24)–(5.26) become a parametric representation of a 3D ellipse, generated by a section of the 3D cylinder, as shown in Fig. 5.1. This ellipse is centered on the y-axis at $(0, \rho_y, 0)$. If

$$\alpha_x = \alpha_z, \quad \rho_z = \sqrt{3}\rho_x \tag{5.33}$$

then the relative orbit is a 3D circle with radius $2\rho_x$. This relative orbit is commonly referred to as the *general circular orbit* (GCO).

In general, the xy-projection of the motion is a parametric representation of a 2:1 ellipse with semimajor axis $2\rho_x$, semiminor axis ρ_x and a constant eccentricity of $\sqrt{1 - \rho_x^2/(4\rho_x^2)} = \sqrt{3}/2$. The line of apsides is rotated $\pi/2$ radians relative to the y-axis, and the center of motion is located at the offset $[0, \rho_y]$.

The xz- and yz-projections are, in general, ovals. There are a few special cases, however, where these projections become elliptic, circular, or linear. An elliptic xz-projection is obtained if

$$\alpha_z = \pi/2 + \alpha_x \tag{5.34}$$

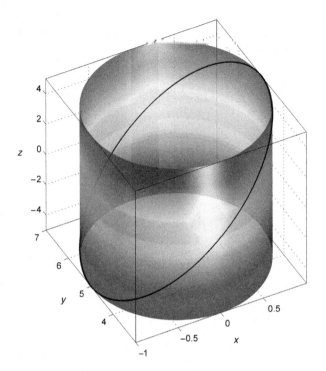

FIGURE 5.1 The bounded relative orbit in the linear approximation lies on a 3D elliptic cylinder. When the phases of the in-plane and out-of-plane motion match, the 3D relative orbit is a section of this cylinder.

and a circular xz-projection is obtained if

$$\alpha_z = \pi/2 + \alpha_x, \quad \rho_x = \rho_z \tag{5.35}$$

A linear xz projection is obtained if Eq. (5.32) holds. Equivalently, an elliptic yz-projection is obtained if Eq. (5.32) holds, and a circular yz-projection is obtained if

$$\alpha_x = \alpha_z, \quad \rho_z = 2\rho_x \tag{5.36}$$

The yz-circular-projection orbit, referred to as the *projected circular orbit* (PCO), has been the subject of a number of studies in the literature due to its potential applications, such as Earth-imaging. We will return to the PCO in Section 5.4. A linear yz-projection is obtained if Eq. (5.34) is satisfied.

For example, setting $x(0) = \dot{z}(0) = 0$ and $\dot{x}(0) = nz(0)$ yields a circular xz-projection and a linear yz-projection; setting $\dot{x}(0) = \dot{z}(0) = 0$ and $z(0) = 2x(0)$ yields a circular yz-projection and a linear xz projection.

Example 5.1. *Two satellites are launched aboard a Delta launcher. The satellites are to fly in a close formation. An initial velocity impulse injects the satellites into a circular LEO of 600 km. An additional velocity impulse applied by*

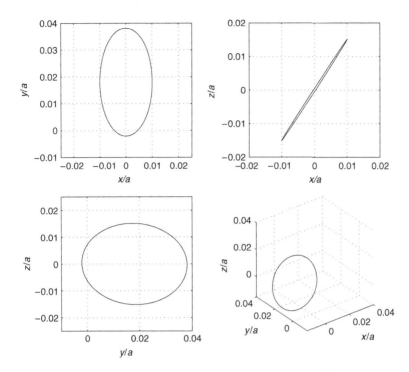

FIGURE 5.2 Geometry of a bounded relative orbit using the CW equations for relative motion. The xy-projection is a 2:1 ellipse.

the deputy de-attaches it from the chief so that the relative position components are

$$x(0) = 69.78 \text{ km}, \quad y(0) = 139.56 \text{ km}, \quad z(0) = 104.67 \text{ km}$$

and the relative velocity components are

$$\dot{x}(0) = 7.5579 \text{ m/s}, \quad \dot{y}(0) = -151.116 \text{ m/s}, \quad \dot{z}(0) = 15.116 \text{ m/s}$$

Determine the geometry of the relative orbit. What initial conditions should be chosen instead of the given initial conditions to yield circular xz- and yz-projections?

The satellite's orbit has $a = 600 + 6378 = 6978$ km, $n = \sqrt{3.986 \cdot 10^5 / 6978^3}$ rad/s. The relative position in normalized coordinates is thus $\bar{x}(0) = x(0)/a = 0.01$, $\bar{y}(0) = y(0)/a = 0.02$, $\bar{z}(0) = z(0)/a = 0.015$ and the normalized velocity is $\bar{x}'(0) = 0.001$, $\bar{y}'(0) = -0.002$, $\bar{z}'(0) = 0.002$. Simulating the CW equations with these initial conditions yields the geometry depicted in Fig. 5.2. The xy-projection of the relative orbit is clearly elliptic. In order to generate a circular xz-projection we need $\bar{x}(0) = 0$, $\bar{x}'(0) = \bar{z}(0) = 0.015$, $\bar{z}'(0) = 0$, and in order to generate a circular yz-projection (PCO), we should set $\bar{z}(0) = 2\bar{x}(0) = 0.02$, $\bar{x}'(0) = \bar{z}'(0) = 0$. The results are shown in Fig. 5.3 and Fig. 5.4, respectively. As expected, the circular xz-projection also

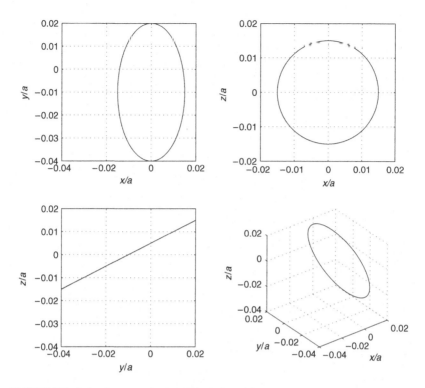

FIGURE 5.3 A circular xz-projection CW relative orbit.

yields a linear yz-projection. Similarly, the circular yz-projection yields a linear xz-projection. In both cases, of course, the projection on the xy plane is a 2:1 ellipse with eccentricity of $\sqrt{3}/2$.

5.2 TWO-IMPULSE LINEAR RENDEZVOUS

As we previously mentioned, the CW equations are linear, admitting an equilibrium at the origin (among other equilibria). Thus, a *ballistic* or *thrust-free* stable linear relative motion is possible if the chief and deputy spacecraft coincide. This is called *rendezvous*, that is, when both the relative position vector and relative velocity vector in the chief-fixed frame are nullified. Rendezvous has many practical applications and has been performed in numerous space missions. For example, docking of the Space Shuttle on the International Space Station requires a rendezvous maneuver.

It is therefore important to find impulsive velocity corrections that guarantee rendezvous. This will be performed in two stages: We will initially find an impulsive maneuver that nullifies the final relative distance between the chief and the deputy. This process is termed *targeting* or *guidance*. At the final time, once the relative distance has been nullified, we will determine the additional impulsive maneuver required to cancel the relative velocity and complete the rendezvous. This scheme is referred to as *two-impulse linear rendezvous*.

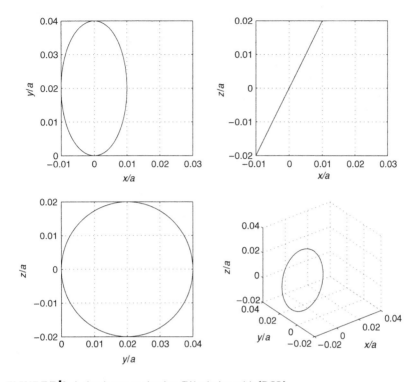

FIGURE 5.4 A circular zy-projection CW relative orbit (PCO).

Designating the moment of the targeting maneuver by $t_0 = 0$, the velocity impulse, $\Delta \mathbf{v}_I$, satisfies

$$\Delta \mathbf{v}_I = \mathbf{v}^+(0) - \mathbf{v}^-(0) \tag{5.37}$$

where $\mathbf{v}^-(0)$ and $\mathbf{v}^+(0)$ are the relative velocity vectors before and after application of the targeting impulse, respectively. The vector $\mathbf{v}^+(0)$, the required initial velocity for targeting, can be found by substituting $x(t_f) = y(t_f) = z(t_f) = 0$ into Eqs. (5.17)–(5.19) and solving for $\dot{x}(0)$, $\dot{y}(0)$, $\dot{z}(0)$, given the initial relative position components $x(0)$, $y(0)$, $z(0)$:

$$
\mathbf{v}^+(0) = \begin{bmatrix} \dot{x}(0) \\ \dot{y}(0) \\ \dot{z}(0) \end{bmatrix}^+
$$

$$
= \begin{bmatrix} -\dfrac{n\left[-4x(0)\,s_{nt_f} + 3x(0)\,nt_f\,c_{nt_f} + 2y(0) - 2y(0)\,c_{nt_f}\right]}{-8 + 8c_{nt_f} + 3nt_f\,s_{nt_f}} \\[2mm] -\dfrac{n\left[-14x(0) + 14x(0)\,c_{nt_f} + 6x(0)\,nt_f\,s_{nt_f} - y(0)\,s_{nt_f}\right]}{-8 + 8c_{nt_f} + 3nt_f\,s_{nt_f}} \\[2mm] -z(0)n\cot(nt_f) \end{bmatrix} \tag{5.38}
$$

$\mathbf{v}^+(0)$ does not exist for all flight times: $\dot{z}^+(0)$ is singular for $nt_f = k\pi$, $k = 0, 1, \ldots$, and the in-plane components $\dot{x}^+(0)$ and $\dot{y}^+(0)$ are singular at $nt_f = 2k\pi$, $k = 0, 1, \ldots$ and at additional points satisfying $8c_{nt_f} + 3nt_f s_{nt_f} = 8$ such as $nt_f = 2.8135\pi$ and $nt_f = 4.8906\pi$.

At t_f, when the deputy satellite – a *chaser* or a *pursuer* – reaches the chief satellite – a *target*, its relative velocity must be nullified to guarantee rendezvous. Therefore, the required velocity impulse at impact must be equal in magnitude and opposite in sign to the final relative velocity:

$$\Delta \mathbf{v}_{II} = -\mathbf{v}(t_f) \tag{5.39}$$

where $\mathbf{v}(t_f)$ is the final relative velocity, obtained by substituting $t = t_f$ into Eqs. (5.20)–(5.22),

$$\mathbf{v}(t_f) = \begin{bmatrix} \dot{x}(t_f) \\ \dot{y}(t_f) \\ \dot{z}(t_f) \end{bmatrix}$$

$$= \begin{bmatrix} \dot{x}^+(0)c_{nt_f} + [3x(0)n + 2\dot{y}^+(0)]s_{nt_f} \\ -[6nx(0) + 3\dot{y}^+(0)] + [6x(0)n + 4\dot{y}^+(0)]c_{nt_f} - 2\dot{x}^+(0)s_{nt_f} \\ \dot{z}^+(0)c_{nt_f} - z(0)ns_{nt_f} \end{bmatrix} \tag{5.40}$$

The total velocity change required for a two-impulse rendezvous is

$$\Delta v = \|\Delta \mathbf{v}_I\| + \|\Delta \mathbf{v}_{II}\| \tag{5.41}$$

In Eq. (5.41), the vector norm depends on the type of thrusters used. If the thrust is applied using a single directional thruster, one must take $\|\cdot\|_2$; if a cluster of body-fixed thrusters are used, providing the required thrust components in each direction, one must substitute $\|\cdot\|_1$ into Eq. (5.41). A good discussion of this topic can be found in Ref. [103].

Example 5.2. *A satellite is in a circular LEO of 600 km. The Space Shuttle is launched to retrieve the satellite. At $t = 0$, a laser ranging system on board the shuttle determines that the relative separation between the Shuttle bay and the satellite is*

$$x(0) = 69.78 \text{ km}, \quad y(0) = 139.56 \text{ km}, \quad z(0) = 104.67 \text{ km}$$

and that the relative velocity is

$$\dot{x}(0) = \dot{y}(0) = \dot{z}(0) = 7.5579 \text{ m/s}$$

Determine the components and the magnitude of the impulsive velocity corrections required for rendezvous after 3.22 hours.

Let us first normalize the given quantities. The satellite's orbit radius is $a = 600+6378 = 6978$ km, and the mean motion is $n = \sqrt{3.986 \cdot 10^5/6978^3}$ rad/s.

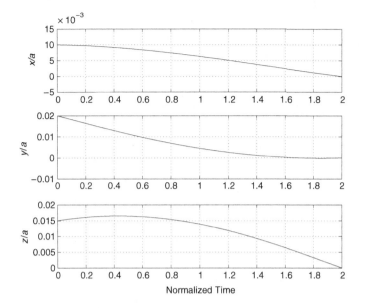

FIGURE 5.5 The relative position components converge to zero following an application of an impulsive targeting maneuver.

The relative position in normalized coordinates is therefore $x(0)/a = 0.01$, $y(0)/a = 0.02$, $z(0)/a = 0.015$ and the normalized velocity is $\bar{x}'(0) = \bar{y}'(0) = \bar{z}'(0) = 7.5579/(na) = 0.001$. The normalized retrieval time is $3.22 \cdot 3600 \cdot n/(2\pi) = 2$. Substituting the data into Eq. (5.38) gives the normalized velocity corrections

$$\Delta\bar{\mathbf{v}}_I = \begin{bmatrix} -0.00178 \\ -0.01927 \\ 0.005865 \end{bmatrix}, \tag{5.42}$$

The normalized final velocity change is computed according to Eq. (5.40),

$$\Delta\bar{\mathbf{v}}_{II} = \begin{bmatrix} -0.00562 \\ 0.00173 \\ -0.0165 \end{bmatrix}, \tag{5.43}$$

Assuming a directional thruster, the total normalized velocity change required is $\Delta\bar{v} = \|\Delta\bar{\mathbf{v}}_I\|_2 + \|\Delta\bar{\mathbf{v}}_{II}\|_2 = 0.03774$. A simulation of the rendezvous has been carried out in MATLAB®. Figure 5.5 depicts the relative position components, showing convergence to zero relative position at the normalized time $\bar{t} = 2$, as expected. Figure 5.6 shows the relative velocity components and the impulsive velocity correction at impact, which nullifies the final values of the relative velocity to yield perfect rendezvous.

From the above discussion, it is obvious that the total velocity change required for rendezvous strongly depends on the time of flight. We can thus ask

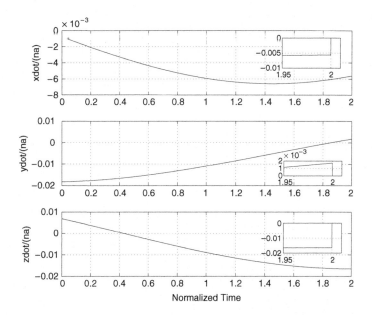

FIGURE 5.6 The relative velocity components are nullified at impact to guarantee rendezvous. A magnified view shows the impulsive velocity correction.

whether there exists a t_f for which the required Δv is minimal given the initial conditions (5.2). We may attempt to answer this query analytically, perform a batch of simulations or simulate the adjoint system to discover the optimum flight time. Adopting the latter approach, we plotted the total Δv vs. t_f and obtained the desired optimum t_f (in normalized units). Figure 5.7 shows a plot of $\Delta \bar{v}$ vs. $\bar{t}_f$. $\Delta \bar{v}$ is maximal near the singularities mentioned above. The minimal velocity change, $\Delta \bar{v} = 0.0375$, is achieved at $\bar{t}_f \approx 4.65$.

We should highlight the fact that the rendezvous methodology implemented in this section relies on the CW equations, and is hence limited to circular unperturbed target orbits. More general solution s for the rendezvous problems have been obtained, see e.g., Ref. [104] (and references therein), in which a bounded solution for the rendezvous problem was derived by neglecting second-order eccentricity terms in the approximation of the chaser's orbit.

To conclude this section, we point out that implementing a certain Δv plan relies upon measuring the specific thrust using, e.g., accelerometers. The challenges associated with this issue are discussed in Chapter 11.

5.3 LAGRANGIAN AND HAMILTONIAN DERIVATIONS OF THE CW EQUATIONS

As in the treatment of Section 5.1, leading to the CW equations, we examine only small deviations from the reference orbit. Thus, we only consider the first three terms of the potential energy (4.84) (here, again, we use the notation

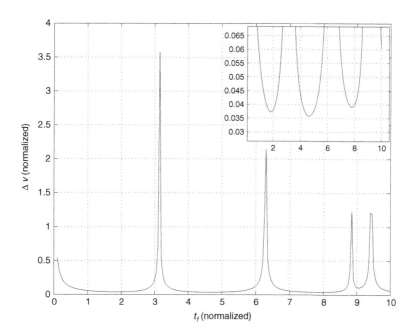

FIGURE 5.7 The total velocity change required for rendezvous varies with the time of flight. The global minimum for the initial conditions given in Example 5.2 is at $t_f \approx 4.65$, as shown in the top right pane.

$n \equiv n_0$ and $a \equiv a_0$)

$$\mathcal{U}^{(0)} = -\frac{\mu}{a} - \frac{\mu}{a^2}\rho\cos\vartheta - \frac{\mu}{a^3}\rho^2\left(\frac{3}{2}\cos^2\vartheta - \frac{1}{2}\right) \tag{5.44}$$

and then use Eq. (4.85) to find the low order Lagrangian,

$$\mathcal{L}^{(0)} = \frac{1}{2}\left(\dot{x}^2 + \dot{y}^2 + \dot{z}^2\right) + n\left(x\dot{y} - y\dot{x} + a\dot{y}\right)$$
$$+ \frac{3}{2}n^2a^2 + \frac{3}{2}n^2x^2 - \frac{n^2}{2}z^2 \tag{5.45}$$

with the perturbed part of the Lagrangian equal to the higher-order terms in the potential $[\mathcal{O}((\rho/a)^3)]$. Omitting the details, it is straightforward to derive the usual CW equations using the Euler–Lagrange equations (3.3):

$$\ddot{x} - 2n\dot{y} - 3n^2x = 0 \tag{5.46}$$
$$\ddot{y} + 2n\dot{x} = 0 \tag{5.47}$$
$$\ddot{z} + n^2z = 0 \tag{5.48}$$

It is helpful before proceeding further to normalize our equations and simplify the notation. Normalizing rates by n (so time is in fractions of an orbit) and relative distances by a, the normalized Lagrangian is given by:

$$\bar{\mathcal{L}}^{(0)} = \frac{1}{2}\left[(\bar{x}')^2 + (\bar{y}')^2 + (\bar{z}')^2\right]$$
$$+ \left[(\bar{x}+1)\bar{y}' - \bar{y}\bar{x}'\right] + \frac{3}{2} + \frac{3}{2}\bar{x}^2 - \frac{1}{2}\bar{z}^2 \qquad (5.49)$$

We can also obtain the equivalent three-degrees-of-freedom normalized Hamiltonian $\bar{\mathcal{H}}^{(0)} : \mathbb{R}^3 \times \mathbb{R}^3 \to \mathbb{R}$ for the linear relative motion dynamics. The canonical momenta are found from the usual definition:

$$p_x = \frac{\partial \bar{\mathcal{L}}^{(0)}}{\partial \bar{x}'} = \bar{x}' - \bar{y}$$
$$p_y = \frac{\partial \bar{\mathcal{L}}^{(0)}}{\partial \bar{y}'} = \bar{y}' + \bar{x} + 1 \qquad (5.50)$$
$$p_z = \frac{\partial \bar{\mathcal{L}}^{(0)}}{\partial \bar{z}'} = \bar{z}'$$

and then, using the Legendre transformation, the normalized Hamiltonian for linear relative motion in Cartesian coordinates is obtained as

$$\bar{\mathcal{H}}^{(0)} = \frac{1}{2}(p_x + \bar{y})^2 + \frac{1}{2}(p_y - \bar{x} - 1)^2 + \frac{1}{2}p_z^2 - \frac{3}{2} - \frac{3}{2}\bar{x}^2 + \frac{1}{2}\bar{z}^2 \quad (5.51)$$

The *linear Hamiltonian equations of relative motion* (in normalized form) are derived from $\bar{\mathcal{H}}^{(0)}$ using Hamilton's equations (3.6):

$$\bar{x}' = p_x + \bar{y} \qquad (5.52)$$
$$\bar{y}' = p_y - \bar{x} - 1 \qquad (5.53)$$
$$\bar{z}' = p_z \qquad (5.54)$$
$$p_x' = p_y + 2\bar{x} - 1 \qquad (5.55)$$
$$p_y' = -p_x - \bar{y} \qquad (5.56)$$
$$p_z' = -\bar{z} \qquad (5.57)$$

It is informative to rewrite $\bar{\mathcal{H}}^{(0)}$ in the tangent space as shown below:

$$\bar{\mathcal{H}}^{(0)} = \frac{1}{2}\left[(\bar{x}')^2 + (\bar{y}')^2 + (\bar{z}')^2\right] - \frac{3}{2} - \frac{3}{2}\bar{x}^2 + \frac{1}{2}\bar{z}^2 \qquad (5.58)$$

For the special case of periodic orbits, satisfying Eq. (5.23), the above equation becomes

$$\bar{\mathcal{H}}^{(0)} = \frac{1}{2}\left[(\bar{x}')^2 + (\bar{z}')^2\right] + \frac{1}{2}(\bar{x}^2 + \bar{z}^2) - \frac{3}{2} \qquad (5.59)$$

Ignoring the constant term $(-3/2)$ in Eq. (5.59), we note that the special case $\bar{\mathcal{H}}^{(0)}$ provides a positive definite function in the reduced configuration space

(y is a dependent function of x); such a function can serve as a Lyapunov function for the CW equations, as long as Eq. (5.23) is satisfied. In general, the Hamiltonian cannot serve as a Lyapunov function for an autonomous dynamical system.

The Lagrangian and Hamiltonian formalisms can be used to obtain corrections for the initial conditions guaranteeing bounded motion in the CW relative motion model. This is performed by taking third-order expressions in the Legendre polynomial expansion of the differential gravitational potential, as shown in the next section.

5.4 ACCOMMODATING SECOND-ORDER NONLINEARITIES

Equations (4.74)–(4.76) can be approximated by either a second-order expansion of the differential gravitational acceleration about a circular orbit, or by taking third-order expressions in the Legendre-polynomial expansion of the Lagrangian, as discussed in Chapter 4. This procedure leads to the following equations, written using normalized coordinates and non-dimensional time:

$$
\begin{aligned}
\bar{x}'' - 2\bar{y}' - 3\bar{x} &= \left[\frac{\bar{y}^2}{2} + \frac{\bar{z}^2}{2} - \bar{x}^2 \right] \\
\bar{y}'' + 2\bar{x}' &= 3\bar{x}\bar{y} \\
\bar{z}'' &= 3\bar{x}\bar{z}
\end{aligned}
\tag{5.60}
$$

It is useful to compare these equations to the normalized CW equations (5.10)–(5.12). Equation (5.60) can be obtained from the normalized Lagrangian [105]

$$
\begin{aligned}
\bar{\mathcal{L}}^{(1)} &= \frac{1}{2} \left[(\bar{x}')^2 + (\bar{y}')^2 + (\bar{z}')^2 \right] + \left[\bar{x}\bar{y}' - \bar{y}\bar{x}' \right] + \frac{1}{2}(3\bar{x}^2 - \bar{z}^2) \\
&\quad - \frac{1}{2}(2\bar{x}^3 - 3\bar{x}\bar{y}^2 - 3\bar{x}\bar{z}^2)
\end{aligned}
\tag{5.61}
$$

A perturbation method for accommodating the effects of second-order nonlinearities in the design of periodic relative orbits was put forth by Vaddi et al. [106]. They obtained a first-order perturbation solution to Eq. (5.60) and showed that the initial conditions provided by the CW solutions can be corrected to achieve bounded relative orbits. Their correction equation, in dimensional form, is

$$
\begin{aligned}
&\dot{y}(0) + 2nx(0) \\
&+ \frac{3n}{2a}\left(2\rho_x{}^2 + 2\rho_y{}^2 + \rho_z{}^2 + 6\rho_x\rho_y\cos\alpha_x + 3\rho_x{}^2\cos 2\alpha_x \right) = 0
\end{aligned}
\tag{5.62}
$$

where ρ_x, ρ_y, and ρ_z are the constants describing the CW solutions, as discussed in Section 5.1 (Eqs. (5.24)–(5.26)). Note that α_z does not affect the above solution. Equation (5.62) can be satisfied by selecting either $x(0)$ or $\dot{y}(0)$. Furthermore, Eq. (5.62) can also be utilized to estimate the effect of nonlinearity on secular along-track growth. If Eq. (5.23) is satisfied, an estimate of the along-track drift rate due to nonlinearity, obtained from Eq. (5.62), is

$$\dot{y}_{\text{drift}} = -\frac{3n}{2a}(2\rho_x^2 + 2\rho_y^2 + \rho_z^2 + 6\rho_x\rho_y \cos\alpha_x + 3\rho_x^2 \cos 2\alpha_x) \quad (5.63)$$

Specializing Eq. (5.63) to the case of a PCO with $2\rho_x = \rho_z \equiv \rho$, $\rho_y = 0$, and $\alpha_x = \alpha_z \equiv \alpha$, a linear estimate of the along-track drift per orbit is

$$y_{\text{drift}} = -\frac{9\pi\rho^2}{4a_0}(2 + \cos 2\alpha) \quad (5.64)$$

Example 5.3. *For a circular orbit with* $a_0 = 7100$ *km,* $\rho = 1$ *km, the drift rate is approximately equal to* 3 *m per orbit, for* $\alpha = 0°$ *and* 1 *m per orbit, for* $\alpha = 90°$.

A perturbation solution based on Eq. (5.60) for the differential energy between two satellites in a formation can be derived from the more general results of Ref. [107] as follows:

$$\delta\bar{E} = (\bar{y}' + 2\bar{x}) + \frac{1}{2}\left[(\bar{x}' - \bar{y})^2 \right.$$
$$\left. + (\bar{y}' + \bar{x})^2 + (\bar{z}')^2 - \left(2\bar{x}^2 - \bar{y}^2 - \bar{z}^2\right)\right] \quad (5.65)$$

An alternate approach, followed by Jiang et al. [108], is to approximate the differential semimajor axis, δa, by a Taylor series expansion of a_1 (calculated from Eq. (4.34)) about the semimajor axis of the chief. For example, a linear approximation valid for circular reference orbits, written in non-dimensional coordinates, is

$$\delta\bar{a}^{(1)} = 2(\bar{y}' + 2\bar{x}) \quad (5.66)$$

A second-order expansion for δa is

$$\delta\bar{a}^{(2)} = \delta\bar{a}^{(1)} + \left[\left(\delta\bar{a}^{(1)}\right)^2 + (\bar{x}' - \bar{y})^2 \right.$$
$$\left. + (\bar{y}' + \bar{x})^2 + \bar{z}^2 - \left(2\bar{x}^2 - \bar{y}^2 - \bar{z}^2\right)\right] \quad (5.67)$$

It can be shown by expanding the normalized form of the energy equation (2.25) that the expressions (5.65) and (5.67) are related as

$$\delta\bar{a}^{(2)} \approx 2\left[\delta\bar{E} + 2\left(\delta\bar{E}\right)^2\right] \quad (5.68)$$

For circular reference orbits, a linear estimate of the non-dimensional along-track drift rate due to second-order nonlinear effects can be obtained from the non-dimensional δa via the relationship

$$\bar{y}' = -\frac{3}{2}\delta\bar{a} \tag{5.69}$$

The presentation in this section shows that expressions for the along-track drift rates due to linearization, or corrections for accommodating the nonlinear effects can be arrived at in several ways. For example, Eq. (5.63) can be obtained from Eqs. (5.69) and (5.67). More general treatments of the $\delta\mathcal{E}$ and δa expressions for eccentric orbits, involving f and e of the chief as parameters, can be found in Refs. [107,108].

5.5 CURVILINEAR VS. CARTESIAN RELATIVE COORDINATES

As we have seen in Section 5.1, the CW equations express the relative motion with respect to a rotating Cartesian coordinate system whose origin coincides with the chief satellite. The assumptions made in their derivation are that (a) the reference of chief orbit is circular, (b) the Earth is spherically symmetric, and (c) the distance between the chief and deputy is small compared to the radius of the orbit so that only the linear terms in the expansion of the differential gravity are retained. In this section, we explore some effects of the linearization assumption and show how to minimize the effect of the neglected nonlinear terms using a change in variables.

In the derivation of the CW equations, a linearization is performed in the expansion of the differential gravity, as shown earlier in Section 5.1. However, the initial conditions are also affected; this happens when obtaining the relative initial conditions from the exact inertial initial conditions of both satellites, or when obtaining the exact initial conditions of the deputy from the chief's initial conditions and the deputy's relative initial conditions.

To illustrate this initial condition impact, consider a simple example of co-orbital motion. The chief is in a circular orbit of radius a_0 and the deputy is in the same orbit, but leading the chief by an angle ς, as shown in Fig. 5.8. Now, obtain the exact relative initial conditions for the deputy. To that end, we first establish an inertial coordinate system, (X, Y, Z), whose axes are aligned at this instant with the LVLH system (x, y, z), but its origin is at the center of the Earth. The initial conditions of the deputy in the $\mathscr{I}$ frame are

$$X(0) = a_0 \cos\varsigma, \quad \dot{X}(0) = -a_0 n \sin\varsigma$$
$$Y(0) = a_0 \sin\varsigma, \quad \dot{Y}(0) = a_0 n \cos\varsigma \tag{5.70}$$
$$Z(0) = 0, \quad \dot{Z}(0) = 0$$

The corresponding initial conditions in the relative frame are

$$x(0) = a_0 (\cos\varsigma - 1), \quad \dot{x}(0) = 0$$
$$y(0) = a_0 \sin\varsigma, \quad \dot{y}(0) = 0 \tag{5.71}$$
$$z(0) = 0, \quad \dot{z}(0) = 0$$

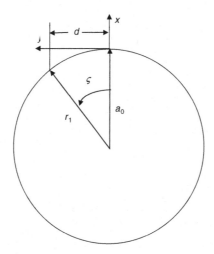

FIGURE 5.8 A simple example of leader–follower co-orbital formation.

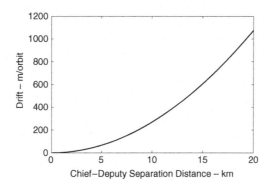

FIGURE 5.9 Drift per orbit caused by the CW equations initial conditions.

The differential semimajor axis resulting from these initial conditions is

$$\delta a = 4x(0) + 2\dot{y}(0)/n_0 = 4a_0(\cos \varsigma - 1) \approx -2a_0\varsigma^2 \qquad (5.72)$$

Thus, the secular term coefficient in Eq. (5.18) is not zero. The secular in-track drift per orbit is

$$3\pi \, \delta a = 12\pi a_0(1 - \cos \varsigma) \approx 6\pi a_0\varsigma^2 \qquad (5.73)$$

Example 5.4. *Evaluate the drift per orbit caused by the CW initial conditions for an initial along-track separation and a circular chief orbit of $a_0 = 7000$ km based on Eq. (5.73).*

Figure 5.9 shows the along-track drift per orbit that occurs when using the relative Cartesian reference frame for selecting the initial conditions. A drift of 269 m/orbit occurs for a 10 km initial separation. This drift would have to be negated by control.

The inverse scenario, that of starting with the initial relative state and obtaining the initial state of the deputy in the $\mathscr{I}$ frame, has a similar problem. With the same problem of a leader–follower formation with an along-track separation $D \approx a_0\varsigma$, as shown in Fig. 5.8, the radius of the deputy's orbit is

$$r_1 = \left(a_0^2 + D^2\right)^{1/2} \approx a_0\left[1 + 0.5\left(D/a_0\right)^2\right] \tag{5.74}$$

Thus

$$\delta r \approx 0.5 a_0 \varsigma^2 \tag{5.75}$$

The deputy velocity is

$$v = \left[(a_0 n_0)^2 + (D n_0)^2\right] \tag{5.76}$$

From the vis-viva equation (2.14),

$$-\frac{\mu}{a} = v^2 - 2\frac{\mu}{r}$$

we can write

$$\frac{\mu}{a_0^2}\delta a = 2v\delta v + 2\frac{\mu}{r_0^2}\delta r \tag{5.77}$$

and hence

$$\delta a = 2a_0 \sin \varsigma^2 \approx 2a_0\varsigma^2 \tag{5.78}$$

Comparing to Eq. (5.72), we see that δa has the same magnitude, but opposite sign. Therefore, the same drift occurs whether or not the deputy's relative initial conditions are obtained from the exact initial conditions of the chief and deputy, or the deputy's exact initial conditions are obtained from the chief's exact and deputy's relative initial conditions. If the initial radius and the corresponding initial relative states are used as the initial conditions for numerical integration or an analytic theory, the two satellites will slowly drift apart, even though they are in the same orbit. Thus, the use of the Cartesian rotating coordinate system results in undesirable errors in the solution due to the nonlinear terms. These nonlinear effects can be reduced if a curvilinear coordinate system is used, as we show now.

To that end, let $(\delta r, \theta_r, \phi_r)$ be the relative coordinates, as shown in Fig. 5.10, where δr is the difference between the radii of the deputy and chief, θ_r is the angle between the projection of the deputy's radius vector in the chief's orbit plane and the chief's radius vector, and ϕ_r is the angle between deputy's radius vector and its projection in the chief's orbit plane. Thus, $(\delta r, a_0\theta_r, a_0\phi_r)$ are analogous to (x, y, z). The radius and velocity vectors of the deputy are

$$\mathbf{r}_1 = r_1\hat{\mathbf{r}}_1 = (a_0 + \delta r)\hat{\mathbf{r}}_1 \tag{5.79}$$

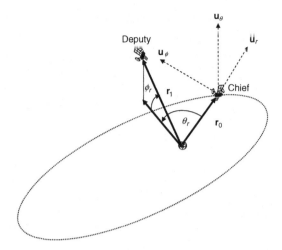

FIGURE 5.10 Relative curvilinear coordinate system.

$$\mathbf{v}_1 = \delta\dot{r}\hat{\mathbf{r}}_1 + (a_0 + \delta r)\,\boldsymbol{\omega}_1 \times \hat{\mathbf{r}}_1 \tag{5.80}$$

where

$$\boldsymbol{\omega}_1 = \boldsymbol{\omega}_0 + \boldsymbol{\omega}_{1/0}$$
$$\boldsymbol{\omega}_0 = n_0\hat{\mathbf{u}}_\phi \tag{5.81}$$
$$\boldsymbol{\omega}_{1/0} = \dot{\theta}_r\hat{\mathbf{u}}_\phi + \dot{\phi}_r\left(\hat{\mathbf{u}}_\theta \cos\theta_r - \hat{\mathbf{u}}_r \sin\theta_r\right)$$
$$\hat{\mathbf{r}}_1 = \hat{\mathbf{u}}_r \cos\phi_r \cos\theta_r - \hat{\mathbf{u}}_\theta \cos\phi_r \sin\theta_r - \hat{\mathbf{u}}_\phi \sin\phi_r \tag{5.82}$$

The acceleration of the deputy is

$$\ddot{\mathbf{r}}_1 = \delta\ddot{r}\hat{\mathbf{r}}_1 + 2\delta\dot{r}\boldsymbol{\omega}_1 \times \hat{\mathbf{r}}_1 + r_1\dot{\boldsymbol{\omega}}_1 \times \hat{\mathbf{r}}_1 + r_1\boldsymbol{\omega}_1 \times \left(\boldsymbol{\omega}_1 \times \hat{\mathbf{r}}_1\right) \tag{5.83}$$

where

$$\dot{\boldsymbol{\omega}}_1 = \ddot{\theta}_r\hat{\mathbf{u}}_r + \ddot{\phi}_r\left(\hat{\mathbf{u}}_\theta \cos\theta_r - \hat{\mathbf{u}}_r \sin\theta_r\right)$$
$$- \dot{\theta}_r\dot{\phi}_r\left(\hat{\mathbf{u}}_r \cos\theta_r + \hat{\mathbf{u}}_\theta \sin\theta_r\right) + \boldsymbol{\omega}_0 \times \boldsymbol{\omega}_{1/0} \tag{5.84}$$

Note that in the curvilinear coordinate formulation there are nonlinear terms in the kinematic equation, whereas in the Cartesian formulation the kinematic equation is linear. Substituting Eqs. (5.81) and (5.84) into Eq. (5.83) and linearizing in δr, θ_r and ϕ_r gives

$$\ddot{\mathbf{r}}_1 = \left(\delta\ddot{r}_1 - 2n_0a_0\dot{\theta}_r - n_0^2\delta r - n_0^2a_0\right)\hat{\mathbf{u}}_r$$
$$+ \left(a_0\ddot{\theta}_r + 2a_0n_0\delta\dot{r} - a_0n_0^2\theta_r\right)\hat{\mathbf{u}}_\theta - a_0\ddot{\phi}_r\hat{\mathbf{u}}_\phi \tag{5.85}$$

The gravitational acceleration is

$$-\frac{\mu}{r_1^2}\hat{\mathbf{r}}_1 = -\frac{\mu}{(a_0+\delta r)^2}\hat{\mathbf{r}}_1 \tag{5.86}$$

Expanding the denominator in a Taylor series in δr, substituting for $\hat{\mathbf{r}}_1$ and linearizing in δr, θ_r and ϕ_r gives

$$-\frac{\mu}{r_1^2}\hat{\mathbf{r}}_1 = -n_0^2\left(a_0\hat{\mathbf{u}}_r - 2\delta r\hat{\mathbf{u}}_r - a_0\theta_r\hat{\mathbf{u}}_\theta - a_0\phi_r\hat{\mathbf{u}}_\phi\right) \tag{5.87}$$

Equating Eqs. (5.85) and (5.87) gives the equations of relative motion in curvilinear coordinates:

$$\begin{aligned} \delta\ddot{r} - 2a_0n_0\dot{\theta}_r - 3n_0^2\delta r &= 0 \\ a_0\ddot{\theta}_r - 2n_0\delta\dot{r} &= 0 \\ \ddot{\phi}_r + n_0^2\phi_r &= 0 \end{aligned} \tag{5.88}$$

Note that these equations are identical to the CW equations with (x, y, z) being replaced by $(\delta r, a_0\theta_r, a_0\phi_r)$. The difference is that the along-track and cross-track distances are measured along arc lengths; they are not linear displacements. This reduces the nonlinear effects in the computation of the initial conditions. Another way to look at this (when using the CW equations) is to let x be the radii difference, y be the arc length displacement along the chief orbit, and z be the arc length displacement in the cross-track direction. Due to the linearization there will still be errors that occur with the use of the curvilinear coordinate equations, but the errors are much smaller. In addition to the effect of the neglected nonlinearities in the initial conditions, there are the neglected nonlinearities that occur from the expansion of the differential gravity. The use of the curvilinear coordinates also reduces these effects.

5.6 ELLIPTIC REFERENCE ORBITS

For circular reference orbits, the choice of the independent variable of either time or true anomaly leads to the same form of the equations of motion. This is not the case for elliptic reference orbits. Time-explicit solutions are attractive since Kepler's equation need not be solved to produce results at specified points of time. However, true-anomaly-based solutions have certain advantages, especially near the perigee, and are more suitable for long-term motion prediction for elliptic orbits.

5.6.1 Time as the independent variable: Melton's STM

Melton [109] has developed expansions of the STM in powers of eccentricity, in both Cartesian as well as cylindrical coordinate systems. A brief outline of his approach is presented here. The linearized equations of relative motion with respect to an unperturbed eccentric reference orbit in a Cartesian coordinate system can be obtained from Eqs. (4.14)–(4.16) as follows (note that the linearization is with respect to the orbit of the chief).

$$\ddot{x} - 2\dot{\theta}\dot{y} - \ddot{\theta}y - \dot{\theta}^2 x = \frac{2\mu x}{r^3} \tag{5.89a}$$

$$\ddot{y} + 2\dot{\theta}\dot{x} + \ddot{\theta}x - \dot{\theta}^2 y = -\frac{\mu y}{r^3} \tag{5.89b}$$

$$\ddot{z} = -\frac{\mu z}{r^3} \tag{5.89c}$$

For unperturbed motion, using the result

$$r^2 = \frac{h}{\dot{\theta}} \tag{5.90}$$

where h is the angular momentum, Eqs. (5.89) can also be written as

$$\ddot{x} = \left[\dot{\theta}^2 + \frac{2\mu}{h^{3/2}}\dot{\theta}^{3/2}\right]x + \ddot{\theta}y + 2\dot{\theta}\dot{y} \tag{5.91a}$$

$$\ddot{y} = -\ddot{\theta}x + \left[\dot{\theta}^2 - \frac{\mu}{h^{3/2}}\dot{\theta}^{3/2}\right]y - 2\dot{\theta}\dot{x} \tag{5.91b}$$

$$\ddot{z} = -\left[\frac{\mu}{h^{3/2}}\dot{\theta}^{3/2}\right]z \tag{5.91c}$$

The equations of motion in the cylindrical coordinate system are obtained from Eq. (5.91) by using non-dimensional in-plane variables $\bar{x} = x/r$ and $\bar{y} = y/r$. Regardless of the coordinate system utilized, the linearized equations can be written in matrix form as follows:

$$\dot{\mathbf{x}}(t) = A(t)\mathbf{x}(t) \tag{5.92}$$

where the state vector is defined as

$$\mathbf{x} = [x, y, z, \dot{x}, \dot{y}, \dot{z}]^T \tag{5.93}$$

The variables r and $\dot{\theta}$ can be expanded in powers of eccentricity, as shown in Battin [33]:

$$r = a[1 - e\cos M - 1/2e^2(\cos 2M - 1) + \cdots] + \mathcal{O}(e^3) \tag{5.94a}$$
$$\dot{\theta} = n[1 + 2e\cos M + 5/2e^2\cos 2M] + \mathcal{O}(e^3) \tag{5.94b}$$

Since $A(t)$ consists of periodic terms, it can also be expanded in powers of e:

$$A(t) = A_0 + eA_1(t) + e^2 A_2(t) + \cdots \tag{5.95}$$

where A_0 is a constant matrix and $A_{1,2,\dots}$ are functions of time. In a similar fashion, the STM can also be expanded as shown below:

$$\Phi(t, t_0) = \Phi_0(t, t_0) + e\Phi_1(t, t_0) + e^2\Phi_2(t, t_0) + \cdots \tag{5.96}$$

where

$$\Phi_0(t, t_0) = \exp[A_0(t - t_0)] \tag{5.97}$$

Since the STM of Eq. (5.92) also satisfies the same differential equation, direct substitution of Eqs. (5.95) and (5.96) yields

$$\dot{\Phi}(t, t_0) = A\Phi(t, t_0)$$
$$= [A_0 + eA_1(t) + \cdots] \times [\Phi_0(t, t_0) + e\Phi_1(t, t_0) + \cdots] \tag{5.98}$$

Collecting the coefficients of like powers of e in the above equation results in

$$\dot{\Phi}_1(t, t_0) = A_0\Phi_1(t, t_0) + A_1(t)\Phi_0(t, t_0)$$
$$\dot{\Phi}_2(t, t_0) = A_0\Phi_2(t, t_0) + A_1(t)\Phi_1(t, t_0) + A_2(t)\Phi_0(t, t_0)$$

$$\cdot$$

$$\tag{5.99}$$

$$\cdot$$

$$\dot{\Phi}_n(t, t_0) = A_0\Phi_n(t, t_0) + A_1(t)\Phi_{n-1}(t, t_0) + \cdots + A_{n-1}(t)\Phi_1(t, t_0)$$
$$+ A_n(t)\Phi_0(t, t_0)$$

Melton presents expressions for the elements of Φ_1 and Φ_2 based on the following integrals:

$$\Phi_1(t, t_0) = \int_{t_0}^{t} \Phi_0(t, s)A_1(s)\Phi_0(s, t_0)ds$$

$$\Phi_2(t, t_0) = \int_{t_0}^{t} \Phi_0(t, s)[A_1(s)\Phi_1(s, t_0) + A_2(s)\Phi_0(s, t_0)]ds$$

$$\tag{5.100}$$

$$\cdot$$

$$\cdot$$

$$\Phi_n(t, t_0) = \int_{t_0}^{t} \Phi_0(t, s)[A_1(s)\Phi_{n-1}(s, t_0) + \cdots A_n(s, t_0)\Phi_0(s, t_0)]ds$$

Melton's solution shows the presence of secular terms for the in-plane motion variables. Secular terms appear in the along-track solution since the periods of the two satellites are, in general, not the same. However, the appearance of secular terms in the radial direction is an artifact of the series approximation utilized. Melton presents numerical results for eccentricities in the range $0 \leq e \leq 0.3$. These results clearly show the successive improvements achieved due to the first and second-order terms in the STM.

5.6.2 Lawden and Tschauner–Hempel Equations

Relative motion equations with true anomaly as the independent variable were derived in Eqs. (4.27)–(4.29). These equations have an interesting history. Lawden [110], during his studies on optimal orbit transfers, developed a set of equations for the so-called *primer vector*, which can be used to describe perturbed motion with respect to elliptic orbits. Later, DeVries [111] and Tschauner and Hempel (TH) [102], independently arrived at the very same equations, while working respectively, on the problems of relative motion of particles and terminal rendezvous of spacecraft. Lawden was able to obtain a closed-form solution to his equations as well, involving an integral that has received considerable attention from subsequent researchers, most notably Carter [112]. The equations presented in this section are referred to as the *Tschauner–Hempel equations*:

$$\bar{x}'' = \frac{3}{k}\bar{x} + 2\bar{y}' \tag{5.101a}$$

$$\bar{y}'' = -2\bar{x}' \tag{5.101b}$$

$$\bar{z}'' = -\bar{z}' \tag{5.101c}$$

where $(\cdot)'$ and $(\cdot)''$ indicate, respectively, the first and second derivatives with respect to f, the true anomaly and $k = 1 + e\cos f$.

As can be seen from the above equations, the in-plane and out-of-plane equations are decoupled from each other. Equation (5.101b) can be integrated once to produce $\bar{y}'$, which, when substituted into Eq. (5.101a), gives

$$\bar{x}'' + \left[4 - \frac{3}{k}\right]\bar{x} = 2d \tag{5.102}$$

where d is a constant of integration. The solution to $\bar{x}$ enables one to obtain $\bar{y}$. Since Eq. (5.101c) describes a simple harmonic motion, its solution can be easily obtained. The solution to Eq. (5.102) developed by Lawden and its subsequent modifications by Carter [112] and Yamanaka and Ankersen [113] for the derivation of the STM are presented next.

5.6.3 Carter's STM

Two linearly-independent solutions to the homogeneous form of Eq. (5.102) (with $d = 0$), obtained by Lawden are

$$\bar{x}_1 = k\sin f \tag{5.103}$$

and

$$\bar{x}_2 = \bar{x}_1 \int \frac{1}{[k(f)\sin f]^2}df \tag{5.104}$$

Carter noted that the above integral is singular when f is a multiple of π. He eliminated this singularity by employing an alternate integral and provided a modified solution as shown below:

$$\bar{x}_2 = 2e\bar{x}_1 K_1(f) - \frac{\cos f}{k} \tag{5.105}$$

where

$$K_1(f) = \int \frac{\cos f}{k(f)^3} df \qquad (5.106)$$

The above integral has a closed-form solution in terms of the eccentric anomaly [112]:

$$K_1(f) = (1 - e^2)^{-5/2} \left[-\frac{3e}{2} E + (1 + e^2) \sin E - \frac{e}{4} \sin 2E \right] \qquad (5.107)$$

Carter's complete solution of Eq. (5.102) is

$$\bar{x} = c_1 \bar{x}_1 + c_2 \bar{x}_2 + c_3 \bar{x}_3 \qquad (5.108)$$

where

$$\bar{x}_3 = -2k \sin f \, K_1(f) - \cos^2 f / k - \cos^2 f \qquad (5.109)$$

and $c_{1:3}$ are constants of integration. In particular, $c_3 = -d$, the constant appearing in the r.h.s. of Eq. (5.102).

Analytical solutions to the remaining variables can be written as

$$\begin{aligned}
\bar{y} &= -2c_1 S(\bar{x}_1) - 2c_2 S(\bar{x}_2) - c_3 S(2\bar{x}_3 + 1) + c_4 \\
\bar{z} &= c_5 \cos f + c_6 \sin f
\end{aligned} \qquad (5.110)$$

where $c_{4:6}$ are additional constants of integration and $S(.)$ represents integration w.r.t. f.

During the process of developing his STM, Carter encountered another singularity, associated with a term involving $K_1(f)/e$, which he eliminated using a modification to Eq. (5.105) as follows:

$$\bar{x}_2 = 2e\bar{x}_1 \left[\frac{\sin f}{k^3} - 3e K_2(f) \right] - \frac{\cos f}{k} \qquad (5.111)$$

where

$$K_2(f) = \int \frac{\sin^2 f}{k(f)^4} df \qquad (5.112)$$

The resulting modified particular integral of Eq. (5.102) is

$$\bar{x}_3 = \left[6e\bar{x}_1 K_2(f) - 2\frac{\sin^2 f}{k^2} - \frac{\cos^2 f}{k} - \cos^2 f \right] \qquad (5.113)$$

Carter constructed his STM by using the three fundamental solutions given by Eqs. (5.103), (5.111), and (5.113). The integral in Eq. (5.112) can also be evaluated in closed-form, in terms of the eccentric anomaly:

$$K_2(f) = (1 - e^2)^{-5/2} \left[\frac{1}{2} E - \frac{1}{4} \sin 2E - \frac{e}{3} \sin^3 E \right] \quad (5.114)$$

Note that the following ordering of the state variables is used in the presentation of Carter's STM: $\bar{\mathbf{x}} = \left[\bar{x}, \bar{x}', y, \bar{y}', z, \bar{z}' \right]^T$. The first step towards the derivation of the STM is to write the relationship between the state vector and the six constants of integration, $\mathbf{c} = [c_1, c_2, c_3, c_4, c_5, c_6]^T$:

$$\bar{\mathbf{x}} = \phi(f)\mathbf{c} \quad (5.115)$$

The constants can be conveniently evaluated at the initial true anomaly as

$$\mathbf{c} = \phi^{-1}(f(0))\mathbf{x}(f(0)) \quad (5.116)$$

Based on Eqs. (5.115) and (5.116), the STM can be written as

$$\Phi = \phi(f)\phi^{-1}(f(0)) \quad (5.117)$$

where

$$\phi(f) = \begin{bmatrix} \bar{x}_1 & \bar{x}_2 & \bar{x}_3 & 0 & 0 & 0 \\ \bar{x}_1' & \bar{x}_2' & \bar{x}_3' & 0 & 0 & 0 \\ -2S(\bar{x}_1) & -2S(\bar{x}_2) & -S(2\bar{x}_3 + 1) & -1 & 0 & 0 \\ -2\bar{x}_1 & -2\bar{x}_2 & -(2\bar{x}_3 + 1) & 0 & 0 & 0 \\ 0 & 0 & 0 & 0 & \cos f & \sin f \\ 0 & 0 & 0 & 0 & -\sin f & \cos f \end{bmatrix} \quad (5.118)$$

and

$$\phi^{-1}(f) = \begin{bmatrix} 4S(\bar{x}_2) + \bar{x}_2' & -\bar{x}_2 & 0 & 2S(\bar{x}_2) & 0 & 0 \\ -(4S(\bar{x}_1) + \bar{x}_1') & \bar{x}_1 & 0 & -2S(\bar{x}_1) & 0 & 0 \\ -2 & 0 & 0 & -1 & 0 & 0 \\ 2S(2\bar{x}_3 + 1) + \bar{x}_3' & -\bar{x}_3 & -1 & S(2\bar{x}_3 + 1) & 0 & 0 \\ 0 & 0 & 0 & 0 & \cos f & -\sin f \\ 0 & 0 & 0 & 0 & \sin f & \cos f \end{bmatrix} \quad (5.119)$$

where

$$S(\bar{x}_1) = -\cos f \left[1 + \left(\frac{e}{2} \right) \cos f \right]$$

$$S(\bar{x}_2) = 3ek^2 K_2(f) - \frac{\sin f}{k} \quad (5.120)$$

$$S(2\bar{x}_3 + 1) = -6k^2 K_2(f) - \frac{2 + k}{2k} \sin 2f$$

5.6.4 Yamanaka and Ankersen's STM

Yamanaka and Ankerson (YA) [113] proposed a new form for the solution to the homogeneous version of Eq. (5.102) by noting the following identity:

$$I = \int_{f(0)}^{f} \frac{1}{k(f)^2} df = \frac{\mu^2}{h^3}(t - t_0) \tag{5.121}$$

Starting with $\bar{x}_1$, the fundamental solution of Lawden, they obtained a second independent solution as shown below:

$$\bar{x}_2 = 3e^2 I \bar{x}_1 + k \cos f - 2e \tag{5.122}$$

Based on the two linearly independent solutions given above, a particular solution to Eq. (5.102) can be written as follows:

$$\bar{x}_3 = (d/e)k \cos f \tag{5.123}$$

The singularity at $e = 0$ in $\bar{x}_3$ is eliminated when the complete solution is assembled:

$$\bar{x} = c_1 k \sin f + c_2 k \cos f + c_3 (2 - 3ek I \sin f) \tag{5.124a}$$
$$\bar{y} = c_4 + c_1 (1 + 1/k) \cos f - c_2 (1 + 1/k) \sin f - 3c_3 k^2 I \tag{5.124b}$$
$$\bar{z} = c_5 \cos f + c_6 \sin f \tag{5.124c}$$

where $c_{1:6}$ are integration constants. The previously described procedure can be used to obtain the STM by eliminating the integration constants. As before, the state vector is defined as $\bar{\mathbf{x}} = \left[\bar{x}, \bar{x}', y, \bar{y}', z, \bar{z}'\right]^T$. The complete solution to Eq. (5.101) is

$$\bar{\mathbf{x}}(t) = \phi(f)\phi^{-1}(f(0))\bar{\mathbf{x}}(t_0) \tag{5.125}$$

where

$$\phi(f) = \begin{bmatrix} s & c & 2 - 3es\,I & 0 & 0 & 0 \\ s' & c' & -3e\left(s'I + \dfrac{s}{k^2}\right) & 0 & 0 & 0 \\ c\left(1 + \dfrac{1}{k}\right) & -s\left(1 + \dfrac{1}{k}\right) & -3k^2 I & 1 & 0 & 0 \\ -2s & e - 2c & -3(1 - 2es\,I) & 0 & 0 & 0 \\ 0 & 0 & 0 & 0 & \cos f & \sin f \\ 0 & 0 & 0 & 0 & -\sin f & \cos f \end{bmatrix} \tag{5.126}$$

with $c = k \cos f$ and $s = k \sin f$.

Note that the ordering of the state variables in this section is different from that used by Yamanaka and Ankersen in Ref. [113]. The explicit inverse of $\phi(f)$

evaluated at $f = f(0)$, which will prove useful later, has also been provided:

$$\phi^{-1}(f(0)) = \frac{1}{\eta^2}$$

$$\times \begin{bmatrix} -3s\dfrac{k+e^2}{k^2} & c-2e & 0 & -s\dfrac{k+1}{k} & 0 & 0 \\ -3\left(e+\dfrac{c}{k}\right) & -s & 0 & -\left(c\dfrac{k+1}{k}+e\right) & 0 & 0 \\ 3k-\eta^2 & es & 0 & k^2 & 0 & 0 \\ -3es\dfrac{k+1}{k^2} & -2+ec & \eta^2 & -es\dfrac{k+1}{k} & 0 & 0 \\ 0 & 0 & 0 & 0 & \eta^2\cos f & -\eta^2\sin f \\ 0 & 0 & 0 & 0 & \eta^2\sin f & \eta^2\cos f \end{bmatrix}$$

$$(5.127)$$

where $\eta = \sqrt{1-e^2}$. The YA STM is very compact and simple to program.

5.6.5 Broucke's STM

Unlike the developments of Carter and Yamanaka and Ankersen, who used true anomaly as the independent variable to construct an analytical solution to the relative motion equations, Broucke [114] presents a form of the STM by using the analytical solution to the two-body problem. A brief treatment of Broucke's development of the fundamental matrix for the in-plane motion is presented here. The STM for the out-of-plane motion, as shown before, is very simple to write. Broucke's elegant approach to determining independent solutions for relative motion is based on expressing the variations of the chief's radius and velocity with respect to the four orbital elements $\text{œ} = \{a \quad e \quad M_0 \quad \omega\}$, along the radial and transverse directions:

$$x = \sum_{j=1}^{4} \frac{\partial r}{\partial \text{œ}_j}\delta\text{œ}_j \tag{5.128a}$$

$$y = \sum_{j=1}^{4} r\frac{\partial \theta}{\partial \text{œ}_j}\delta\text{œ}_j \tag{5.128b}$$

$$\dot{x} = \sum_{j=1}^{4} \frac{\partial \dot{r}}{\partial \text{œ}_j}\delta\text{œ}_j \tag{5.128c}$$

$$\dot{y} = \sum_{j=1}^{4} \left(\dot{r}\frac{\partial \theta}{\partial \text{œ}_j} + r\frac{\partial \dot{\theta}}{\partial \text{œ}_j}\right)\delta\text{œ}_j \tag{5.128d}$$

where $\delta\text{œ}_j$ represents a small change in the jth element of œ. Since r and θ depend on œ, the variations due to these four orbital elements are required for the construction of the fundamental matrix, ϕ. The four columns of the in-plane fundamental matrix are populated with the four linearly-independent solutions of x, y, $\dot{x}$, and $\dot{y}$ obtained from Eqs. (5.128). For example, the vector corresponding to the variations due to ω is

$$\phi(:, 4) = \begin{bmatrix} 0 & r & 0 & \dot{r} \end{bmatrix}^T \qquad (5.129)$$

A modification of the above scheme is required to eliminate the singularity in the fundamental matrix $e = 0$, since the columns generated by M_0 and ω are the same for $e = 0$. Hence, the column generated by the variations due to M_0 is replaced by the following linear combination to produce the desired fundamental matrix:

$$\phi(:, 3) = \frac{\eta}{aen} \left[\frac{\partial r}{\partial M_0} \quad r\frac{\partial \theta}{\partial M_0} \quad \frac{\partial \dot{r}}{\partial M_0} \quad \dot{r}\frac{\partial \theta}{\partial M_0} + r\frac{\partial \dot{\theta}}{\partial M_0} \right]^T$$
$$- \frac{1}{ep} \left[\frac{\partial r}{\partial \omega} \quad r\frac{\partial \theta}{\partial \omega} \quad \frac{\partial \dot{r}}{\partial \omega} \quad \dot{r}\frac{\partial \theta}{\partial \omega} + r\frac{\partial \dot{\theta}}{\partial \omega} \right]^T \qquad (5.130)$$

The details of the elements of ϕ and ϕ^{-1} can be found in Ref. [114]. The final step of the calculation of the STM is given by Eq. (5.125). Note that Kepler's equation has to be solved at each time step of propagation with the Broucke STM.

5.6.6 STM of Lee, Cochran, and Jo

Lee, Cochran, and Jo (LCJ) [115] exploit the analytical solution to the two-body problem presented by Battin [33]. They use the following definition of the state vector in their work: $\mathbf{x} = [x, y, z, \dot{x}, \dot{y}, \dot{z}]^T$. The state vector at time t can be written as

$$\mathbf{x} = D(t) \begin{bmatrix} \delta r & \delta\theta & \delta\dot{r} & \delta\dot{\theta} & \delta i & \delta\Omega \end{bmatrix}^T \qquad (5.131)$$

where

$$D(t) = \begin{bmatrix}
1 & 0 & 0 & 0 & 0 & 0 \\
0 & r & 0 & 0 & 0 & r\cos i \\
0 & 0 & 0 & 0 & r\sin\theta & -r\sin i\cos\theta \\
0 & 0 & 1 & 0 & 0 & 0 \\
0 & \dot{r} & 0 & r & 0 & \dot{r}\cos i \\
0 & 0 & 0 & 0 & \frac{d}{dt}(r\sin\theta) & -\frac{d}{dt}(r\cos\theta)\sin i
\end{bmatrix} \qquad (5.132)$$

Next, they write

$$\begin{bmatrix} \delta r & \delta\theta & \delta\dot{r} & \delta\dot{\theta} & \delta i & \delta\Omega \end{bmatrix}^T = B(t, t_0) \begin{bmatrix} \delta r_0 & \delta\theta_0 & \delta\dot{r}_0 & \delta\dot{\theta}_0 & \delta i & \delta\Omega \end{bmatrix}^T \quad (5.133)$$

where

$$
B(t, t_0) = \begin{bmatrix}
\dfrac{\partial r(t)}{\partial r_0} & \dfrac{\partial r(t)}{\partial \theta_0} & \dfrac{\partial r(t)}{\partial \dot{r}_0} & \dfrac{\partial r(t)}{\partial \dot{\theta}_0} & 0 & 0 \\[2ex]
\dfrac{\partial \theta(t)}{\partial r_0} & \dfrac{\partial \theta(t)}{\partial \theta_0} & \dfrac{\partial \theta(t)}{\partial \dot{r}_0} & \dfrac{\partial \theta(t)}{\partial \dot{\theta}_0} & 0 & 0 \\[2ex]
\dfrac{\partial \dot{r}(t)}{\partial r_0} & \dfrac{\partial \dot{r}(t)}{\partial \theta_0} & \dfrac{\partial \dot{r}(t)}{\partial \dot{r}_0} & \dfrac{\partial \dot{r}(t)}{\partial \dot{\theta}_0} & 0 & 0 \\[2ex]
\dfrac{\partial \dot{\theta}(t)}{\partial r_0} & \dfrac{\partial \dot{\theta}(t)}{\partial \theta_0} & \dfrac{\partial \dot{\theta}(t)}{\partial \dot{r}_0} & \dfrac{\partial \dot{\theta}(t)}{\partial \dot{\theta}_0} & 0 & 0 \\[2ex]
0 & 0 & 0 & 0 & 1 & 0 \\[1ex]
0 & 0 & 0 & 0 & 0 & 1
\end{bmatrix} \tag{5.134}
$$

The elements of the sensitivity matrix, $B(t, t_0)$ have been developed in Ref. [115]. Equations (5.131) and (5.133) can be combined to arrive at the STM in the following form:

$$
\Phi(t, t_0) = D(t) B D^{-1}(t_0) \tag{5.135}
$$

Even though the nature of the LCJ STM appears to be time-explicit, Kepler's equation has to be solved at each time step.

5.6.7 STM of Nazarenko

An alternative approach, utilizing the two-body STM expressed in the chief-fixed rotating frame (the LVLH frame, $\mathscr{L}$), is presented by Nazarenko [116]. Recall that the STM for the two-body problem [33] is most commonly expressed in the inertial coordinate system. This STM can also be used for the propagation of relative motion if the relative state vector is expressed in the inertial reference frame. In this section the form of the relative state vector in the $\mathscr{L}$ frame is as defined by Eq. (5.93).

The relative position and the relative absolute velocity vectors in the $\mathscr{L}$ frame can be transformed into the $\mathscr{I}$ frame state vector as given below:

$$
[\mathbf{x}(t)]_{\mathscr{I}} = \begin{bmatrix} C^T(t) & 0_{3\times3} \\ 0_{3\times3} & C^T(t) \end{bmatrix} \begin{bmatrix} I_3 & 0_{3\times3} \\ \tilde{\omega}(t) & I_3 \end{bmatrix} \mathbf{x}(t) \tag{5.136}
$$

where $C \equiv T_{\mathscr{I}}^{\mathscr{L}}$ is the DCM transforming from the $\mathscr{I}$ to the $\mathscr{L}$ frame (cf. Eq. (4.100)) and $\tilde{\omega}$ is the angular velocity cross-product matrix, defined by the usual convention that for any vector $\mathbf{v} = [v_1, v_2, v_3]^T$,

$$
\tilde{\mathbf{v}} = \begin{bmatrix} 0 & -v_3 & v_2 \\ v_3 & 0 & -v_1 \\ -v_2 & v_1 & 0 \end{bmatrix} \tag{5.137}
$$

Denoting

$$L(t) = \begin{bmatrix} C^T(t) & 0_{3\times3} \\ 0_{3\times3} & C^T(t) \end{bmatrix} \begin{bmatrix} I_3 & 0_{3\times3} \\ \tilde{\omega}(t) & I_3 \end{bmatrix} \qquad (5.138)$$

it can be shown that

$$L(t)\mathbf{x}(t) = \Phi_{2B_{\mathscr{G}}} L(0)\mathbf{x}(0) \qquad (5.139)$$

where $\Phi_{2B_{\mathscr{G}}}$ is the two-body STM, which can be found in Battin [33] (pp. 457–467). We can also make use of the following useful results for computing $\tilde{\omega}(t)$ in Eq. (5.138):

$$\dot{C}(t) = -\tilde{\omega}(t)C(t) \qquad (5.140a)$$
$$\tilde{\omega}(t) = -\dot{C}(t)C^T(t) \qquad (5.140b)$$

The matrices $C(t)$ and $\dot{C}(t)$ can be easily obtained from the solution of the two-body problem. Based on the above developments, the propagation equation for relative motion is

$$\mathbf{x}(t) = L^{-1}(t)\Phi_{2B_{\mathscr{G}}} L(0)\mathbf{x}(0) \qquad (5.141)$$

Hence, the required STM for relative motion can be obtained as

$$\Phi(t,0) = L^{-1}(t)\Phi_{2B_{\mathscr{G}}} L(0) \qquad (5.142)$$

Details of the elements of the STM can be obtained from Ref. [116].

5.6.8 Initial conditions to prevent secular drift

The CW and TH equations and their associated STMs provide convenient means for describing linearized relative motion. They can be used to determine initial conditions for establishing a formation such that secular drift is prevented. Even though they may not be accurate but for the smallest of formations, they still provide good estimates for the desired initial conditions.

As we have seen in Section 4.2, the exact condition for eliminating secular drift for unperturbed motion is to match the periods of the satellites. In other words, the satellites must all have the same semimajor axes. We have also seen in Section 5.1 that the CW equations provide a condition for zero secular drift:

$$\dot{y} + 2nx = 0 \qquad (5.143)$$

which is an approximation to the zero differential semimajor axis condition, since the effects of eccentricity and nonlinearity are not accounted for.

An improvement to the above condition can be obtained from the TH equations, which account for the effects of eccentricity. Inalhan *et al.* [117] noted that the general solution of Carter, Eqs. (5.110), predominantly involves periodic functions. Secular terms in the in-plane equations appear only through the term K_1, as can be seen from Eq. (5.107). Hence, periodic trajectories of the TH equations can be set up by choosing initial conditions consistent with $c_2 = 0$ in Eq. (5.108). In their derivation of a boundedness condition, Inalhan et al. focused on the establishment of the formation at perigee or apogee of the reference orbit. However, a general condition for linear boundedness can be obtained from Carter's STM, which involves the integral $K_2(f)$. It can be seen from Eqs. (5.108), (5.111), and (5.113) that $K_2(f)$ can be eliminated from the in-plane motion solutions if the following relationship is satisfied:

$$c_3 - ec_2 = 0 \tag{5.144}$$

The coefficients c_2 and c_3 can be obtained from Eqs. (5.116) and (5.119) as

$$
\begin{aligned}
c_2 &= -\left[4S(\bar{x}_1) + \bar{x}_1'\right]\bar{x}(f(0)) + \bar{x}_1\bar{x}'(f(0)) - 2S(\bar{x}_1)\bar{y}'(f(0)) \\
c_3 &= -\left[2\bar{x}(f(0)) + \bar{y}'(f(0))\right]
\end{aligned} \tag{5.145}
$$

Substitution of Eqs. (5.145) and the first of Eq. (5.120) into Eq. (5.144) results in the following condition for boundedness:

$$
\begin{aligned}
&k^2(f(0))\bar{y}'(f(0)) + ek(f(0))\sin f(0)\bar{x}'(f(0)) \\
&+ \left[2 + 3e\cos f(0) + e^2\right]\bar{x}(f(0)) = 0
\end{aligned} \tag{5.146}
$$

The same result can also be obtained from the YA STM by setting $c_3 = 0$ in Eq. (5.124), as shown by Sengupta and Vadali [1]. Jiang et al. also obtain the same condition by expanding the differential semimajor axis in terms of the relative motion variables.

Equation (5.146) can also be written by using the unscaled motion variables:

$$
\begin{aligned}
&k(f(0))y'(f(0)) + e\sin f(0)\left[x'(f(0)) - y(f(0))\right] \\
&+ [2 + e\cos f(0)]x(f(0)) = 0
\end{aligned} \tag{5.147}
$$

and with time as the independent variable as well:

$$
\begin{aligned}
&k(f(0))\left[\dot{y}(t_0) + \dot{f}(0)x(t_0)\right] + e\sin f(0)\left[\dot{x}(t_0) - \dot{f}(0)y(t_0)\right] \\
&+ \dot{f}(0)x(t_0) = 0
\end{aligned} \tag{5.148}
$$

where

$$\dot{f}(0) = \sqrt{\frac{\mu}{p^3}}k(f(0))^2 \tag{5.149}$$

The linear boundedness condition at any epoch is a function of the true anomaly, which can be found from Kepler's equation. It can be enforced in several ways, with a given relative position, since it involves both the radial and along-track relative velocities. However, satisfaction of Eq. (5.146) does

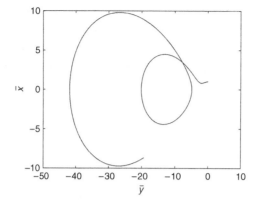

FIGURE 5.11 In-plane motion obtained using the YA STM.

not guarantee that the relative orbit will remain absolutely bounded, since the constraint is only a first-order approximation to $\delta a = 0$.

The use of the STM and the linear boundedness condition are demonstrated via a numerical example.

Example 5.5. *Assume that $e = 0.5$ and $f(0) = \pi/2$. Let the initial state of the relative motion be*

$$\left[\bar{x}, \bar{x}', y, \bar{y}'\right]^T = [\sin f(0), \cos f(0), 2\cos f(0), -2\sin f(0)]^T \quad (5.150)$$

A single-impulse maneuver is to be performed to establish periodic relative motion and center the relative orbit in the along-track direction. Obtain an estimate of the required impulse magnitude and the associated true anomaly of the chief at which the impulse should be performed.

First, the in-plane relative motion over a period of two orbits of the chief, propagated via the YA STM, is shown in Fig. 5.11. The solution obtained from numerical integration of Eqs. (5.101) is virtually indistinguishable from the STM result. It is observed that the motion is not periodic.

Next, an optimization problem is formulated to minimize the velocity correction required to satisfy Eq. (5.148). The velocity correction is sought as a function of f. Hence, the problem can be stated as

$$\text{minimize } k^2\{[(\bar{x}')^+ - (\bar{x}')^-]^2 + [(\bar{y}')^+ - (\bar{y}')^-]^2\}$$

s. t.

$$k^2(f)\bar{y}'(f) + ek(f)\sin f\bar{x}'(f) + \left[2 + 3e\cos f + e^2\right]\bar{x}(f) = 0$$

where $(\cdot)^-$ and $(\cdot)^+$ are respectively, quantities before and after impulse application.

The minimum cost, plotted as a function of f, is shown in Fig. 5.12. As can be seen, the cost is lowest at f values corresponding to multiples of 2π. Figure 5.13 shows the plots of the radial and transverse components of

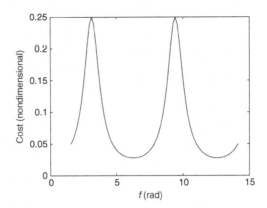

FIGURE 5.12 Cost as a function of f.

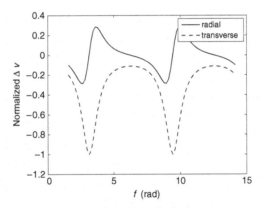

FIGURE 5.13 Normalized radial and transverse velocity increments.

the normalized Δv. Two important observations are: (i) the radial velocity correction is zero when f corresponds to either the periapsis or apoapsis locations; and (ii) the magnitude of the transverse component of the velocity correction has a local minimum at periapsis. These results can also be deduced from Eq. (5.146). It can be seen that when k^2 is at its extremum, $k \sin f = 0$, since $\frac{d}{df}k^2 = 2ke \sin f$. Figure 5.14 shows the result of applying the velocity correction at $f = 2\pi$. The small circle in this figure indicates the location of the maneuver. The normalized, purely transverse velocity correction required is -0.11 units.

There remains the matter of centering the relative orbit with respect to the chief. One of the ways to accomplish this additional objective is to select the initial conditions such that $c_4 = 0$, in either Eq. (5.110) or Eq. (5.124). We see from Eqs. (5.116) and (5.127) that the above condition results in

$$e(k+1) \sin f \bar{y}'(f) + [2 - ek \cos f] \bar{x}'(f)$$
$$+ 3e \frac{k+1}{k} \sin f \bar{x}(f) - \eta^2 \bar{y}(f) = 0 \qquad (5.151)$$

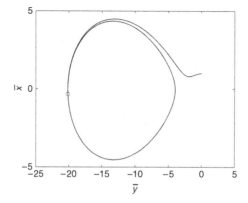

FIGURE 5.14 Result of the orbit establishment maneuver.

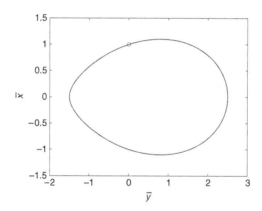

FIGURE 5.15 Single-impulse orbit establishment using two constraints.

Since Eqs. (5.146) and (5.151) represent two constraints involving the two velocity components, except for the choice of the true anomaly, there is no more room for optimization when a single impulse is used. For the initial conditions considered previously, the least cost is obtained when $f = \pi/2$ with a single radial impulse of -0.5 units. The resulting orbit is shown in Fig. 5.15. The small circle in this figure locates the point of application of the impulse.

A linear programming approach for minimum-fuel formation initialization and the analysis of the errors associated with neglecting eccentricity of the reference orbit can be found in Ref. [117]. Several other options of bias removal have been treated in Ref. [1].

5.7 PERIODIC SOLUTIONS TO THE TH EQUATIONS

Sengupta et al. [1] have estimated the relative motion drift rates for elliptic reference orbits due to semimajor axis mismatch by using Eqs. (5.124). By evaluating the expressions (5.124) at the beginning and at the end of an orbit of the chief, it can be shown that

$$\bar{x}_{\text{drift}} = \bar{x}(f(0) + 2\pi) - \bar{x}(f(0)) = -\frac{6k(f(0))\pi c_3}{\eta^3} e \sin f(0)$$

$$\bar{y}_{\text{drift}} = \bar{y}(f(0) + 2\pi) - \bar{y}(f(0)) = -\frac{6k^2\pi c_3}{\eta^3} \qquad (5.152)$$

where c_3 is the constant appearing in the development of the YA STM and it can be obtained from the third row of Eq. (5.127) as

$$\eta^2 c_3 = (3k(f(0)) - \eta^2)\bar{x}(f(0)) + ek(f(0))\sin(f(0))\bar{x}'(f(0))$$
$$+ k(f(0))^2\bar{y}'(f(0)) \qquad (5.153)$$

As noted in Section 5.4, c_3 is also related to the dimensional δa through

$$\frac{\delta a}{a} = 2c_3 \qquad (5.154)$$

It has been shown in Ref. [1], based on Eqs. (5.152), that the radius of the in-plane xy-projection of the relative orbit undergoes a per-orbit drift given by

$$\rho_{\text{drift}} = \left(x_{\text{drift}}^2 + y_{\text{drift}}^2\right)^{1/2} = \frac{3\pi}{\eta}\delta a[1 + e^2 + 2e\cos f(0)]^{1/2} \quad (5.155)$$

Hence, ρ_{drift} is bounded by

$$3\pi\,\delta a\sqrt{\frac{1-e}{1+e}} \le \rho_{\text{drift}} \le 3\pi\,\delta a\sqrt{\frac{1+e}{1-e}} \qquad (5.156)$$

If $c_3 = 0$, then the periodic solutions to the TH equations can be obtained as

$$\bar{x} = \frac{\rho_x}{p}\sin(f + \alpha_x)(1 + e\cos f) \qquad (5.157a)$$

$$\bar{y} = \frac{\rho_x}{p}\cos(f + \alpha_x)(2 + e\cos f) + \frac{\rho_y}{p} \qquad (5.157b)$$

$$\bar{z} = \frac{\rho_z}{p}\sin(f + \alpha_z) \qquad (5.157c)$$

The periodic solutions to the TH equations can also be represented in dimensional form as

$$x = \rho_x\sin(f + \alpha_x) \qquad (5.158a)$$

$$y = 2\rho_x\cos(f + \alpha_x)\frac{(1 + (e/2)\cos f)}{(1 + e\cos f)} + \frac{\rho_y}{(1 + e\cos f)} \qquad (5.158b)$$

$$z = \rho_z\frac{\sin(f + \alpha_z)}{(1 + e\cos f)} \qquad (5.158c)$$

Equations (5.158) show the presence of higher-order harmonics in the motion variables induced due to eccentricity. Fourier series representations of the motion variables with true anomaly and time as the independent variables are provided in Ref. [1]. Only the expressions for y are given here, in truncated forms, to show three different methods for correcting for the along-track bias present in the solution (5.158). The result with f as the independent variable is

$$y(f) = \frac{1}{\eta}\left[-\varepsilon\rho_x \cos\alpha_x + \rho_y\right] + (h.o.t) \tag{5.159}$$

where $\varepsilon = \sqrt{[(1-\eta)/(1+\eta)]}$ and we have, with time as the independent variable,

$$y(t) = \frac{1}{2\eta^2}\left[-e(3+2\eta^2)\rho_x \cos\alpha_x + (3-\eta^2)\rho_y\right] + (h.o.t) \tag{5.160}$$

Equation (5.159) shows that the bias term therein can be removed by enforcing the condition

$$\rho_y = \varepsilon\rho_x \cos\alpha_x \tag{5.161}$$

This correction, labeled *f-correction*, does not have any significance in the time domain. From Eq. (5.160), we get an alternate condition, labeled *t-correction*, for the elimination of the bias term which allows the deputy to spend equal time ahead and behind the chief:

$$\rho_y = [e(3+2\eta^2)/(3-\eta^2)]\rho_x \cos\alpha_x \tag{5.162}$$

A third condition can also be obtained by requiring symmetry of motion given by Eqs. (5.158) at $f = 0$ and $f = \pi$. This correction is termed the *amplitude correction*:

$$\rho_y = e\rho_x \cos\alpha_x \tag{5.163}$$

The results [1] of the application of the three corrections are shown in Fig. 5.16, for $e = 0.6$, $\rho_x = 1/2$ km, $\rho_z = 1$ km and $\alpha_x = \alpha_z = 0$. Of the three corrections, only Eq. (5.163) allows the motion to be symmetrically bounded with respect to the chief.

The modification or distortion of the ideal PCO of the CW equations due to the effect of eccentricity is illustrated in Fig. 5.17 for $2\rho_x = \rho_z = 1$ km and $\alpha_x = \alpha_z = 0$, for $e = 0.2$ and $e = 0.7$. In these figures, the dashed-dotted lines show the ideal PCOs. The application of Eq. (5.148) to determine the corrected $\dot{y}(0)$ provides bounded relative orbits, as shown by the dashed lines. The bias removal is achieved by applying Eq. (5.163), resulting in the solid lines. Note the drastic change produced by the bias correction for $e = 0.7$.

Thus, the relative orbits can be corrected for the effects of nonlinearity as well as eccentricity. Considering the effects of nonlinearity and eccentricity

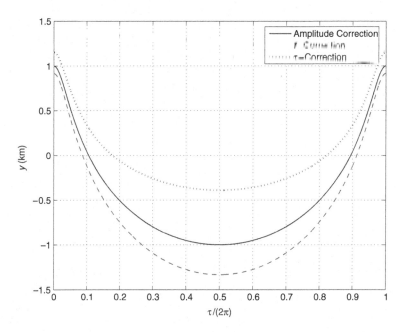

FIGURE 5.16 Effect of the three bias correction formulas, $e = 0.6$ (adapted from Ref. [1]).

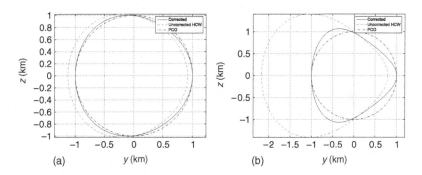

FIGURE 5.17 (a) Corrected PCO relative orbits, $e = 0.2$. (b) Corrected PCO relative orbits, $e = 0.7$ (adapted from Ref. [1]).

separately shows the connections between the CW relative orbit shape-size parameters and the along-track bias and drift rate. However, ultimately, it is the $\delta a = 0$ constraint that has to be satisfied for bounded unperturbed relative motion, and it can be satisfied in several different ways.

SUMMARY

In this chapter, we presented linear relative motion models for circular and elliptic reference orbits. The famous CW equations, which have been widely used for rendezvous and formation design and analysis, were introduced. We also treated the relative motion problem using Lagrangian and Hamiltonian mechanics and derived the higher-order nonlinear versions of the CW equations. The perturbation solution to these equations presented us with a means to estimate the drift rate due to the nonlinear effects associated with a circular reference orbit. The advantages of writing the equations of motion in a curvilinear coordinate system for capturing some of the nonlinear effects were also demonstrated in this chapter. The treatment of the elliptic orbit case lead us to the Tschauner–Hempel equations and the various forms of the two-body STMs for the propagation of relative motion. We also presented a first-order boundedness condition for elliptic orbits and a method for the along-track bias removal to shape the relative orbit geometry.

Chapter six

Modeling Relative Motion Using Orbital Elements

I am a star fallen from the
Blue tent upon the green carpet.
I am the daughter of the elements
With whom Winter conceived.

Khalil Gibran (1883–1931)

In Chapter 4, we discussed the nonlinear, unperturbed, equations of relative motion in a rotating Cartesian coordinate system centered on the reference orbit. In Section 5.1 we developed the CW approximation, which utilized the same rotating coordinate system to derive expressions for the linear relative motion between satellites assuming that the reference orbit is circular. Recognizing some of the limitations of this approach, we presented generalizations of the CW equations for eccentric reference orbit s (the TH equations) in Section 5.6.

An important modification of spacecraft relative motion modeling is the use of orbital elements as constants of unperturbed motion instead of the Cartesian initial conditions. This concept, originally suggested by Hill [118], has been widely used in the analysis of relative spacecraft motion [32,34,119]. One of the main advantages of the orbital-elements approach is the straightforward incorporation of orbital perturbations, yielding Lagrange's planetary equations (LPE) and the Gauss variational equations (GVE), introduced in Chapter 2. Moreover, utilizing orbital elements facilitates the derivation of high-order, nonlinear extensions to the CW solution in Cartesian coordinates.

There have been a few reported efforts to obtain high-order solutions to the relative motion problem. Karlgaard and Lutze proposed formulating the relative motion in spherical coordinates in order to derive second-order expressions [120]. The use of Delaunay elements (cf. Section 3.2) has also been proposed. For instance, Vaddi et al. derived differential equations in order to incorporate **123**

Spacecraft Formation Flying; ISBN: 9780750685337

perturbations and high-order nonlinear effects into the modeling of relative dynamics [106].

In this chapter, we will establish a methodology for obtaining arbitrary high order approximations to the relative motion between spacecraft by utilizing the Cartesian configuration space in conjunction with classical orbital elements [121,122]. In other words, we propose utilizing the known inertial expressions describing vehicles in elliptic orbits in order to obtain, using a Taylor series approximation, a time-series representation of the motion in a rotating frame, where the coefficients of the time series are functions of the orbital elements. We subsequently show that under certain conditions, this time-series becomes a Fourier series. More importantly, in the process of the derivation, there is no need to solve differential equations. This significant merit results directly from utilization of the known inertial configuration space. The high-order approximation we present also provides important insights into boundedness and commensurability of relative formation dynamics. A first-order approximation of the relative motion variables for circular orbits leads to Hill's solution. This solution is utilized to obtain a hybrid set of linear differential equations, explicit in the motion variables and also δa. The hybrid model has the advantage of modeling a significant portion of the nonlinear effects that constitute δa. The geometry of the Hill's solution also provides a means for establishing the initial conditions for PCOs.

6.1 GENERAL SOLUTION TO THE NONLINEAR RELATIVE MOTION EQUATIONS

The unperturbed nonlinear equations of relative motion, Eqs. (4.14)–(4.16), can be solved – in terms of true anomaly – due to the fact that the generating orbits are Keplerian. To see this, we will find an expression for ρ using consecutive Eulerian rotations and a translation.

Let $\mathscr{P}_1$ be a perifocal frame associated with the deputy's orbit (cf. Section 2.1). The initial step is to transform from $\mathscr{P}_1$ to $\mathscr{I}$ using three consecutive clockwise rotations conforming to the common $3 - 1 - 3$ sequence, as we have pointed out in Section 2.3. To that end, we define the line-of-nodes (LON) obtained from the intersection of the deputy's orbital plane and the inertial reference plane. The composite rotation, $T \in SO(3)$, transforming any vector in $\mathscr{P}_1$ into the inertial frame $\mathscr{I}$, is given by [33,44]

$$T(\omega_1, i_1, \Omega_1) = \begin{bmatrix} c_{\Omega_1}c_{\omega_1} - s_{\Omega_1}s_{\omega_1}c_{i_1} & -c_{\Omega_1}s_{\omega_1} - s_{\Omega_1}c_{\omega_1}c_{i_1} & s_{\Omega_1}s_{i_1} \\ s_{\Omega_1}c_{\omega_1} + c_{\Omega_1}s_{\omega_1}c_{i_1} & -s_{\Omega_1}s_{\omega_1} + c_{\Omega_1}c_{\omega_1}c_{i_1} & -c_{\Omega_1}s_{i_1} \\ s_{\omega_1}s_{i_1} & c_{\omega_1}s_{i_1} & c_{i_1} \end{bmatrix} \quad (6.1)$$

where i_1, Ω_1, ω_1 are the deputy's inclination, right ascension of the ascending node (RAAN) and argument of periapsis, respectively, and we have again used the compact notation $s_x = \sin(x), c_x = \cos(x)$. The next step is to transform from $\mathscr{I}$ to the chief's perifocal frame, $\mathscr{P}_0$, using the rotation matrix $T^T(\omega_0, i_0, \Omega_0)$, where i_0, Ω_0, ω_0 are the chief's inclination, RAAN and argument of periapsis, respectively. The transformation of the deputy's position

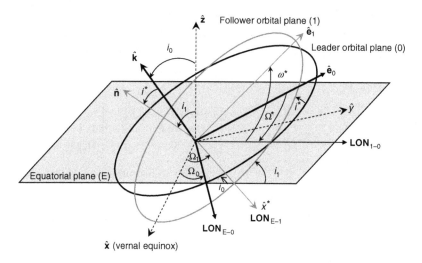

FIGURE 6.1 Relative orbital elements.

vector from $\mathscr{P}_0$ into $\mathscr{L}$, our standard rotating LVLH system centered at the chief, requires an additional rotation,

$$
C(f_0) = \begin{bmatrix} c_{f_0} & s_{f_0} & 0 \\ -s_{f_0} & c_{f_0} & 0 \\ 0 & 0 & 1 \end{bmatrix},
\tag{6.2}
$$

and a translation by $\mathbf{r}_0 = [r_0, 0, 0]^T$, resulting in

$$
\rho = C(f_0) T^T (\omega_0, i_0, \Omega_0) T(\omega_1, i_1, \Omega_1)[\mathbf{r}_1]_{\mathscr{P}_1} - \begin{bmatrix} r_0 \\ 0 \\ 0 \end{bmatrix}
\tag{6.3}
$$

where $[\mathbf{r}_1]_{\mathscr{P}_1}$ is the deputy's position vector in frame $\mathscr{P}_1$, expressed in terms of the deputy's eccentric anomaly, E_1, as (cf. Eq. (2.35))

$$
[\mathbf{r}_1]_{\mathscr{P}_1} = \begin{bmatrix} a_1(\cos E_1 - e_1) \\ b_1 \sin E_1 \\ 0 \end{bmatrix}
\tag{6.4}
$$

and $b_1 = a_1\sqrt{1 - e_1^2}$.

Equation (6.3), written component-wise, is the most general solution (which still entails solution of Kepler's equation, as explained in the sequel) to the relative motion problem, modeled by the differential equations (4.14)–(4.16).

The obtained solution can be simplified if we utilize *relative orbital elements*. These orbital elements describe the orientation of the deputy's orbital plane relative to the chief's orbital plane, where the relative LON is defined by the intersection of these two planes, as shown in Fig. 6.1. Using this LON and some fixed reference line – a good choice would be the eccentricity vector of the deputy's orbit, denoted by $\hat{e}_1$ in Fig. 6.1 – we may define the relative RAAN, Ω^*, the relative argument of periapsis, ω^*, and the relative inclination, i^*. In terms of the relative elements, the expression for the relative position is simplified into

$$\rho = C(f_0)T(\omega^*, i^*, \Omega^*)\begin{bmatrix} a_1(\cos E_1 - e_1) \\ b_1 \sin E_1 \\ 0 \end{bmatrix} - \begin{bmatrix} r_0 \\ 0 \\ 0 \end{bmatrix} \tag{6.5}$$

The matrix $T(\omega^*, i^*, \Omega^*)$ is calculated by substituting $(\omega_1, i_1, \Omega_1) \mapsto (\omega^*, i^*, \Omega^*)$ into Eq. (6.1), and $C(f_0)$ is given in (6.2). Some algebraic manipulations yield [122]

$$x = \frac{1}{2}\left[(k_3 - k_2)s_{f_0 - E_1} + (k_1 + k_4)c_{f_0 - E_1} + (k_3 + k_2)s_{f_0 + E_1} \right. \\ \left. + (k_1 - k_4)c_{f_0 + E_1}\right] - e_1(k_3 s_{f_0} + k_1 c_{f_0}) - r_0 \tag{6.6}$$

$$y = \frac{1}{2}\left[-(k_1 + k_4)s_{f_0 - E_1} + (k_3 - k_2)c_{f_0 - E_1} + (k_4 - k_1)s_{f_0 + E_1} \right. \\ \left. + (k_2 + k_3)c_{f_0 + E_1}\right] + e_1(k_1 s_{f_0} - k_3 c_{f_0}) \tag{6.7}$$

$$z = k_5(c_{E_1} - e_1) + k_6 s_{E_1} \tag{6.8}$$

where

$$k_1 = (c_{\Omega^*}c_{\omega^*} - s_{\Omega^*}s_{\omega^*}c_{i^*})a_1 \tag{6.9}$$

$$k_2 = (-c_{\Omega^*}s_{\omega^*} - s_{\Omega^*}c_{\omega^*}c_{i^*})b_1 \tag{6.10}$$

$$k_3 = (s_{\Omega^*}c_{\omega^*} + c_{\Omega^*}s_{\omega^*}c_{i^*})a_1 \tag{6.11}$$

$$k_4 = (-s_{\Omega^*}s_{\omega^*} + c_{\Omega^*}c_{\omega^*}c_{i^*})b_1 \tag{6.12}$$

$$k_5 = s_{\omega^*}s_{i^*}a_1 \tag{6.13}$$

$$k_6 = c_{\omega^*}s_{i^*}b_1 \tag{6.14}$$

We can simplify Eqs. (6.6)–(6.8) by adopting the magnitude-phase representation

$$x = K_1 \sin(f_0 - E_1 + \Phi_1) + K_2 \sin(f_0 + E_1 + \Phi_2) \\ - K_3 \sin(f_0 + \Phi_3) - r_0 \tag{6.15}$$

$$y = K_1 \sin(f_0 - E_1 - \Phi_1) + K_2 \sin(f_0 + E_1 - \Phi_2) \\ + K_3 \sin(f_0 - \Phi_3) \tag{6.16}$$

$$z = K_4 \sin(E_1 + \Phi_4) - k_5 e_1 \tag{6.17}$$

where

$$K_1 = \frac{1}{2}\sqrt{(k_3 - k_2)^2 + (k_1 + k_4)^2} \tag{6.18}$$

$$K_2 = \frac{1}{2}\sqrt{(k_3 + k_2)^2 + (k_1 - k_4)^2} \tag{6.19}$$

$$K_3 = e_1\sqrt{k_1^2 + k_3^2} \tag{6.20}$$

$$K_4 = \sqrt{k_5^2 + k_6^2} \tag{6.21}$$

$$\Phi_1 = \tan^{-1}\frac{k_1 + k_4}{k_3 - k_2} \tag{6.22}$$

$$\Phi_2 = \tan^{-1}\frac{k_1 - k_4}{k_3 + k_2} \tag{6.23}$$

$$\Phi_3 = \tan^{-1}\frac{k_3}{k_1} \tag{6.24}$$

$$\Phi_4 = \tan^{-1}\frac{k_5}{k_6} \tag{6.25}$$

Thus, we have obtained the *general solution* for the nonlinear differential equations (4.14)–(4.16) modeling the relative motion problem. This general solution lies in the three-dimensional configuration space, comprising the *relative motion invariant manifold*, $\mathfrak{R}$. This manifold is invariant because any solution of the relative motion problem starting on $\mathfrak{R}$ will remain on $\mathfrak{R}$ for all times.

The dynamics on $\mathfrak{R}$ evolves according to Eq. (2.29) and a similar relationship that holds for the deputy's eccentric anomaly, emanating from Kepler's equation,

$$\dot{E}_1 = \frac{n_1}{1 - e_1 \cos E_1} \tag{6.26}$$

where n_1 is the mean motion of the deputy. Thus, the general solution is a function of the chief's orbital elements, œ_0, and the deputy orbital elements, œ_1.

If the mean motions of the chief and deputy are commensurate (e.g. in a 1:1 resonance, i.e. $n_1 = n_0$, as discussed in Section 4.2), then the relative orbit will be a closed smooth curve $\gamma_c(t) \in \mathfrak{R}$ satisfying the periodicity condition with some period T, $\gamma_c(t) = \gamma_c(t + T)$. Otherwise, an open curve $\gamma_o(t) \in \mathfrak{R}$ will be obtained, and the motion will be quasi-periodic. Since the dynamics are always confined to evolve on $\mathfrak{R}$, the relative motion will be always bounded.[1]

Interestingly, in some cases, the manifold $\mathfrak{R}$ can be approximated by parametric representations of familiar geometric shapes. For example, when Ω^* and

---

[1] This observation is trivial, because the relative motion analyzed here is Keplerian. Nevertheless, many of the current works dealing with relative motion tend to distinguish between "bounded" and "unbounded" relative motion, while implicitly referring to 1:1 commensurable and non-commensurable motions, respectively.

ω^* are first-order small, the relative position components x, y, z constitute the parametric equations of an elliptic torus.[2] This issue is discussed in Section 6.3.

6.2 BOUNDS ON MAXIMAL AND MINIMAL DISTANCES

Gurfil and Kholshevnikov [122] developed a number of bounds on the distances between satellites flying on elliptic orbits, using the general solution of the relative motion problem discussed in the previous section. The methodology used in Ref. [122] distinguishes between the commensurable and non-commensurable cases. Some useful results are elaborated in this section.

6.2.1 The non-commensurable case

In the non-commensurable case, assuming circular chief and deputy orbits, the maximal and minimal distances correspond to extremal distances on the mutual LON, so that

$$\max \rho = a_0 + a_1, \qquad \min \rho = |a_0 - a_1|. \tag{6.27}$$

Given that the orbits of the chief and the deputy are coplanar and that the chief's orbit is circular, we distinguish between two cases. If

$$a_1(1 - e_1) \leq a_0 \leq a_1(1 + e_1). \tag{6.28}$$

then

$$\max \rho = a_0 + a_1(1 + e_1), \qquad \min \rho = 0 \tag{6.29}$$

Otherwise,

$$\max \rho = a_0 + a_1(1 + e_1) \tag{6.30}$$

$$\min \rho = \min\{|a_0 - a_1(1 - e_1)|, |a_0 - a_1(1 + e_1)|\} \tag{6.31}$$

6.2.2 The commensurable case

If the orbits of the chief and deputy satellites are circular and the motion is 1:1 commensurable, then for $i^* = 0$

$$\rho = 2a_1 \left| \sin \frac{(M_0)_0 - (M_0)_1}{2} \right| \tag{6.32}$$

where $(M_0)_0$ and $(M_0)_1$ are the respective mean anomalies at epoch of the chief and the deputy. For $i^* > 0$

$$\min \rho^2 = a_1^2(1 + \cos i^*)\{1 - \cos[(M_0)_0 - (M_0)_1]\} \tag{6.33}$$

$$\max \rho^2 = a_1^2\{3 - \cos i^* - (1 + \cos i^*) \cos[(M_0)_0 - (M_0)_1]\} \tag{6.34}$$

---

[2] A surface of revolution which is a generalization of the ring torus. It is produced by rotating an ellipse in the xz-plane about the z-axis.

which constitute particularly simple expressions for evaluating the extremal inter-vehicle distances.

6.3 RELATIVE MOTION APPROXIMATIONS WITH A CIRCULAR-EQUATORIAL REFERENCE ORBIT

In this section, we will use Eq. (6.5) to obtain time-series approximations of the relative motion in terms of classical orbital elements. To avoid unnecessary complications, we will assume that the reference orbit is circular and equatorial. To that end, we re-write Eqs. (4.74)–(4.76) (the differential equations modeling the deputy spacecraft dynamics relative to a circular reference orbit in the chief-fixed rotating LVLH frame $\mathscr{L}$) in vector form:

$$\frac{d}{dt}\begin{bmatrix} \boldsymbol{\rho} \\ \dot{\boldsymbol{\rho}} \end{bmatrix} = \begin{bmatrix} 0_{3\times3} & I_3 \\ A_\rho & A_{\dot{\rho}} \end{bmatrix}\begin{bmatrix} \boldsymbol{\rho} \\ \dot{\boldsymbol{\rho}} \end{bmatrix} + \begin{bmatrix} 0_{3\times3} \\ I_3 \end{bmatrix}\begin{bmatrix} \dfrac{\mu \mathbf{r}_0}{a_0^3} - \dfrac{\mu(\boldsymbol{\rho} + \mathbf{r}_0)}{\|\boldsymbol{\rho} + \mathbf{r}_0\|^3} \end{bmatrix} \qquad (6.35)$$

where

$$A_\rho = \begin{bmatrix} n_0^2 & 0 & 0 \\ 0 & n_0^2 & 0 \\ 0 & 0 & 0 \end{bmatrix}, \qquad A_{\dot{\rho}} = \begin{bmatrix} 0 & 2n_0 & 0 \\ -2n_0 & 0 & 0 \\ 0 & 0 & 0 \end{bmatrix} \qquad (6.36)$$

and $\mathbf{r}_0 = [a_0, 0, 0]^T$, a_0 is the radius of the chief's orbit, μ is the gravitational constant and $n_0 = \sqrt{\mu/a_0^3}$.

Recall that the CW approximate solutions for the configuration space over $\mathscr{L}$ were obtained in Section 5.1 by linearization of Eq. (6.35) about the origin, assuming that ρ/a_0 is first-order small. The resulting linear CW solution for the relative motion, given in Eqs. (5.17)–(5.19), can be equivalently written as

$$\boldsymbol{\rho}[t, \boldsymbol{\rho}(0), \dot{\boldsymbol{\rho}}(0)] = \tilde{\mathbf{C}}(0)[\boldsymbol{\rho}(0), \dot{\boldsymbol{\rho}}(0)] + \tilde{\mathbf{C}}_1[\boldsymbol{\rho}(0), \dot{\boldsymbol{\rho}}(0)]\cos(n_0 t)$$
$$+ \tilde{\mathbf{S}}_1[\boldsymbol{\rho}(0), \dot{\boldsymbol{\rho}}(0)]\sin(n_0 t) + \tilde{\boldsymbol{\rho}}_d[t, \boldsymbol{\rho}(0), \dot{\boldsymbol{\rho}}(0)] \qquad (6.37)$$

where the constant vectors $\tilde{\mathbf{C}}_k$ and $\tilde{\mathbf{S}}_k$ are functions of the initial position and velocity in $\mathscr{L}$, and $\tilde{\boldsymbol{\rho}}_d \in \mathbb{R}^3$ is a non-periodic vector-valued function.

A selection of initial conditions that yield $\tilde{\boldsymbol{\rho}}_d \equiv 0$ renders the periodic terms in Eq. (6.37) a first-order *Fourier series*. Our objective is to generalize the topology of the linearized solution, in order to derive higher-order approximations via higher-order time-series expansions. To this end, we utilize classical orbital elements. This idea also permits a straightforward and natural incorporation of orbital perturbations and control forces via LPE or GVE, as will be discussed in Chapter 7.

To begin, let $\text{œ}_1 = [a_1, e_1, i_1, \Omega_1, \omega_1, (M_0)_1]^T$ be the classical orbital elements (cf. Eq. (2.42)) of a deputy spacecraft. Assuming that the reference orbit is circular and equatorial, i.e., coincides with the fundamental plane of the inertial reference system, we use Eq. (6.5) to get

$$
\rho = \begin{bmatrix} x \\ y \\ z \end{bmatrix}
$$

$$
= \frac{a_1(1 - e_1^2)}{1 + e_1 \cos f_1} \begin{bmatrix} c_{f_1+\omega_1} c_{n_0 t - \Omega_1} + c_{i_1} s_{f_1+\omega_1} s_{n_0 t - \Omega_1} \\ c_{i_1} c_{n_0 t - \Omega_1} s_{f_1+\omega_1} - c_{f_1+\omega_1} s_{n_0 t - \Omega_1} \\ s_{i_1} s_{f_1+\omega_1} \end{bmatrix} - \begin{bmatrix} a_0 \\ 0 \\ 0 \end{bmatrix} \quad (6.38)
$$

Equation (6.38) provides the general, nonlinear, expression for the relative position vector as a function of orbital elements and time in a rotating frame of reference centered on a circular, equatorial orbit. For $\Omega_1, \omega_1 \ll 1$, x, y, z are the parametric equations of an elliptic torus, meaning that Eq. (6.38) can be also written in the form

$$
x = (\kappa_1 + \kappa_2 \cos v) \cos u \qquad (6.39)
$$

$$
y = (\kappa_1 + \kappa_2 \cos v) \sin u \qquad (6.40)
$$

$$
z = \kappa_3 \sin v \qquad (6.41)
$$

where $u, v \in [0, 2\pi]$ are functions of the eccentric anomaly.

Example 6.1. *Assume that the normalized semimajor axes* $a_0 = 1, a_1 = 1.02$, *and* $e_1 = 0.1, i_1 = 15°, \Omega_1 = 5°, \omega_1 = (M_0)_1 = 0°$. *Simulate the deputy's orbit according to Eq. (6.38).*

In order to perform the simulation, we need to integrate a single differential equation (cf. Eq. (2.29)):

$$
\dot{f}_1 = \sqrt{\frac{\mu}{a_1^3 (1 - e_1^2)^3}} (1 + e_1 \cos f_1)^2, \quad f_1(0) = 0
$$

We then plug the time history of f_1 into Eq. (6.38) and plot the results. Figure 6.2 shows the xy-projection of the relative motion using the normalization $a_0 = n_0 = \mu = 1$. As a_0 and a_1 do not commensurate, the energy/period matching condition is violated, so we do not expect a periodic motion. Instead, we see that the motion is quasi-periodic. This is an additional form in which epicyclic motion manifests itself. The deputy spacecraft performs an epicyclic motion along the unit circle in the chief-fixed rotating frame. Thus, although the relative motion is not 1:1 commensurable, it is certainly bounded, as relative motion between elliptic Keplerian orbits will always remain bounded regardless of any particular selection of coordinates or resonance conditions. Figure 6.3 shows the three-dimensional motion and the torus which it lies upon. The motion clearly evolves along the torus. If we change the semimajor axis of the deputy spacecraft to match that of the chief, the drift will stop, and a closed relative orbit will result. The closed orbit lies, again, on the three-dimensional elliptic torus.

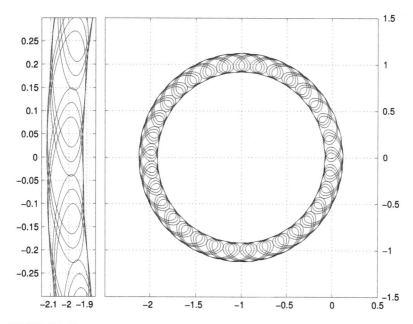

FIGURE 6.2 The xy-projection of an epicyclic motion of a deputy spacecraft in a chief-fixed rotating reference frame. The right pane shows the unit epicycle. The left pane presents a magnified view of an epicycle segment.

Although Eq. (6.38) may be used to investigate the relative dynamics of a spacecraft formation: It provides little insight into the relative dynamics since time dependence is implicit (due to the true anomaly-dependent terms). It is preferable to expand Eq. (6.38) into a time series of the form

$$\boldsymbol{\rho}(t, \boldsymbol{\mathit{œ}}_1, a_0) \approx \mathbf{C}_0(\boldsymbol{\mathit{œ}}_1, a_0) + \sum_{k=1}^{k_{\max}} [\mathbf{C}_k(\boldsymbol{\mathit{œ}}_1, a_0) \cos(k n_0 t)$$
$$+ \mathbf{S}_k(\boldsymbol{\mathit{œ}}_1, a_0) \sin(k n_0 t)] + \boldsymbol{\rho}_d(t, \boldsymbol{\mathit{œ}}_1, a_0) \qquad (6.42)$$

where $\mathbf{C}_k$ and $\mathbf{S}_k$ are vector functions of the (possibly time-varying) orbital elements. Eq. (6.42) constitutes an orbital elements-based generalization of the first-order Cartesian parametrization (6.37). It will be shown shortly that Eq. (6.38) can indeed be represented in the form (6.42). We note that the drift term in Eq. (6.42), $\boldsymbol{\rho}_d$, stems from that fact that the Fourier series expansion of the relative motion is truncated at some $k_{\max} < \infty$. Due to the quasi-periodic nature of the relative orbit, $\boldsymbol{\rho}_d \equiv 0$ for $k_{\max} = \infty$.

6.3.1 High-order time-series approximations

For a circular and equatorial reference orbit, the orbital elements Ω_0 and ω_0 are undefined. We will therefore use instead the degenerate set of orbital elements $\boldsymbol{\mathit{œ}}_0 = [a_0, e_0, i_0, \varepsilon_0]^T$ where $\varepsilon_0 = \Omega_0 + \omega_0 - (M_0)_0$ is the mean

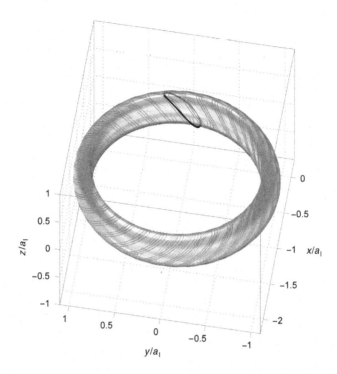

FIGURE 6.3 The relative motion orbits lie on an elliptic torus. Quasi-periodic motion is obtained without commensurability (dashed curve); a closed relative orbit is a result of orbital commensurability (solid curve).

longitude at epoch. In the derivation to follow, we utilize the normalization $a_0 = \mu = n_0 = 1$, resulting in the normalized reference orbital elements

$$\text{\ae}_0 = [1, 0, 0, 0]^T \tag{6.43}$$

The degenerate set of orbital elements for an arbitrary deputy spacecraft in the formation can be thus written as

$$\text{\ae}_1 = \text{\ae}_0 + \delta\text{\ae} \tag{6.44}$$

Therefore, the *orbital element differences* $\delta\text{\ae}$, also termed *differential orbital elements*, can be written in terms of the normalized deputy orbital elements as

$$\delta\text{\ae} = [\bar{a}_1 - 1, e_1, i_1, \varepsilon]^T = [\delta a, e, i, \varepsilon]^T \tag{6.45}$$

The analysis can be completed by expanding Eq. (6.38) into a Taylor series about $\text{\ae}_0$ in powers of $\delta\text{\ae}$. However, in order to obtain a time-series expansion of the form (6.42), we must first relate the deputy's true anomaly to time. This is performed by introducing the deputy's mean anomaly, M, and using the following series expansion solution to Kepler's equation [33]:

$$f = M + 2 \sum_{l=1}^{\infty} \frac{1}{l} \left[\sum_{s=-\infty}^{\infty} J_s(-le) \left(\frac{1 - \sqrt{1 - e^2}}{e} \right)^{|l+s|} \right] \sin(lM) \quad (6.46)$$

where $J_s(\cdot)$ is a Bessel function of the first kind[3] of order s. For illustration, for terms up to order e^2, Eq. (6.46) can be expanded as

$$f = M + 2e \sin M + \frac{5}{4}e^2 \sin 2M + \mathcal{O}(e^3) \quad (6.47)$$

Based on Kepler's equation (2.26), let $\widetilde{M}_0 \triangleq n_0 t_0 - M_0$. Substituting

$$M = n_0 t - \widetilde{M}_0 = \sqrt{\frac{1}{(1 + \delta a)^3}} t - \widetilde{M}_0 \quad (6.48)$$

into Eq. (6.47) yields the desired mapping $f \mapsto t$, which is then substituted into Eq. (6.38). The time-series expansion (6.42) is now available by writing a multi-variable Taylor series for the normalized relative position vector,

$$\rho(t, œ_1) = \rho(œ_0) + \sum_{l=1}^{\infty} \left[\frac{1}{l!} (\delta œ \cdot \nabla)^l \rho(t, œ) \right]_{œ=œ_0} \approx C_0(œ_1)$$

$$+ \sum_{k=1}^{k_{max}} [C_k(œ_1) \cos(kt) + S_k(œ_1) \sin(kt)] + \rho_d(t, œ_1) \quad (6.49)$$

Note that the above expression is written assuming the normalization (6.43).

Based on Eq. (6.49), expressions for the relative motion between deputy spacecraft can be straightforwardly obtained [121].

6.3.2 Second-order approximation

Truncated to include the zero-, first- and second-order terms only, utilizing the identity

$$(\delta œ \cdot \nabla)(\delta œ \cdot \nabla)\rho(t, œ) = \delta œ \cdot \nabla[\delta œ \cdot \nabla(\rho(t, œ))] = \delta œ \cdot [\delta œ \cdot \nabla(\nabla \rho(t, œ))]$$

Eq. (6.49) becomes

$$\rho(t, œ_1) \approx \rho(œ_0) + (\delta œ \cdot \nabla_œ)\rho(t, œ)|_{œ=œ_0}$$

$$+ \frac{1}{2} \delta œ \cdot [\delta œ \cdot \nabla_œ(\nabla_œ \rho(t, œ))]|_{œ=œ_0} \quad (6.50)$$

---

[3]Bessel functions of the first kind, denoted as $J_\alpha(x)$, are solutions of Bessel's differential equation. It is possible to define the function by its Taylor series expansion around $x = 0$: $J_\alpha(x) = \sum_{m=0}^{\infty} \frac{(-1)^m}{m! \Gamma(m+\alpha+1)} \left(\frac{x}{2}\right)^{2m+\alpha}$ where $\Gamma(z)$ is the gamma function.

Performing the symbolic calculation in Eq. (6.50), simplifying, and re-writing into the time-series (6.42) yields

$$\rho(t, \text{œ}_1) \approx C_0(\text{œ}_1) + \sum_{k=1}^{2} [C_k(\text{œ}_1)\cos(kt) + S_k(\text{œ}_1)\sin(kt)]$$
$$+ \rho_d(t, \text{œ}_1) \tag{6.51}$$

where, omitting the subscript 1 for brevity, we have

$$C_0 = \begin{bmatrix} \delta a - \dfrac{1}{2}\varepsilon^2 - \dfrac{1}{4}i^2 - \dfrac{1}{2}e^2 \\ \varepsilon(1 + \delta a) \\ -\dfrac{3}{2}ie\sin(\omega) \end{bmatrix} \tag{6.52}$$

$$C_1 = \begin{bmatrix} 2\varepsilon e \sin(\tilde{M}_0) - (1 + \delta a)e\cos(\tilde{M}_0) \\ -2(1 + \delta a)e\sin(\tilde{M}_0) - \varepsilon e\cos(\tilde{M}_0) \\ i(1 + \delta a)\sin(\omega - \tilde{M}_0) \end{bmatrix} \tag{6.53}$$

$$S_1 = \begin{bmatrix} -(1 + \delta a)e\sin(\tilde{M}_0) - 2\varepsilon e\cos(\tilde{M}_0) \\ 2(1 + \delta a)e\cos(\tilde{M}_0) + \varepsilon e\sin(\tilde{M}_0) \\ i(1 + \delta a)\cos(\omega - \tilde{M}_0) \end{bmatrix} \tag{6.54}$$

$$C_2 = \begin{bmatrix} \dfrac{1}{4}i^2\cos\left[2(\omega - \tilde{M}_0)\right] + \dfrac{1}{2}e^2\cos(2\tilde{M}_0) \\ -\dfrac{1}{4}i^2\sin\left[2(\omega - \tilde{M}_0)\right] - \dfrac{1}{4}e^2\sin(2\tilde{M}_0) \\ \dfrac{1}{2}ie\sin(\omega - 2\tilde{M}_0) \end{bmatrix} \tag{6.55}$$

$$S_2 = \begin{bmatrix} -\dfrac{1}{4}i^2\sin\left[2(\omega - \tilde{M}_0)\right] + \dfrac{1}{2}e^2\sin(2\tilde{M}_0) \\ -\dfrac{1}{4}i^2\cos\left[2(\omega - \tilde{M}_0)\right] + \dfrac{1}{4}e^2\cos(2\tilde{M}_0) \\ \dfrac{1}{2}ie\cos(\omega - 2\tilde{M}_0) \end{bmatrix} \tag{6.56}$$

The non-periodic terms are given by

$$\rho_d = \begin{bmatrix} \dfrac{3}{2}\varepsilon\delta at + \dfrac{3}{2}e\delta a\, t\sin(t - \tilde{M}_0) - \dfrac{9}{8}\delta a^2 t^2 \\ -\dfrac{3}{2}\delta at + \dfrac{3}{2}e\delta a\, t\cos(t - \tilde{M}_0) + \dfrac{3}{8}\delta a^2 t \\ -\dfrac{3}{2}i\delta a\, t\cos(t + \omega - \tilde{M}_0) \end{bmatrix} \tag{6.57}$$

Let us examine expressions (6.52)–(6.57). First, since $C_0 \neq 0$, the center of motion of a deputy spacecraft in the rotating frame is offset relative to the reference orbit.

We can also make a few distinctions regarding the boundedness of the relative motion by examining the terms in ρ_d. An interesting observation is that there exists a second-order secular drift in the out-of-plane direction for $\delta a \neq 0$. This phenomenon is not predicted by the CW approximation. It stems from the converging-diverging nature of the time series approximating the relative motion; if an infinite number of terms is taken in the approximation, we see convergence to the exact expression for the relative position vector (6.38), which is of course bounded if the spacecraft follows an elliptic Keplerian orbit (and periodic, if the orbital rates commensurate). We see here as well that if the energy matching condition is satisfied, i.e. $\delta a = 0$, then the relative motion is bounded.

6.3.3 First-order approximation: Hill's solutions

A first-order, CW-like solution written in terms of orbital elements, also known as *Hill's solution*, can be easily obtained by eliminating the second-order terms in Eqs. (6.52)–(6.57):

$$\mathbf{C}_0 = \begin{bmatrix} \delta a \\ \varepsilon \\ 0 \end{bmatrix} \tag{6.58}$$

$$\mathbf{C}_1 = \begin{bmatrix} -e\cos(\tilde{M}_0) \\ -2e\sin(\tilde{M}_0) \\ i\sin(\omega - \tilde{M}_0) \end{bmatrix} \tag{6.59}$$

$$\mathbf{S}_1 = \begin{bmatrix} -e\sin(\tilde{M}_0) \\ 2e\cos(\tilde{M}_0) \\ i\cos(\omega - \tilde{M}_0) \end{bmatrix} \tag{6.60}$$

$$\rho_d = \begin{bmatrix} 0 \\ -\dfrac{3}{2}\delta at \\ 0 \end{bmatrix} \tag{6.61}$$

Hill's solution may be compactly written using the magnitude-phase representation in dimensional form as

$$x(t) = \delta a - a_0 e\cos(n_0 t - \tilde{M}_0) \tag{6.62}$$

$$y(t) = a_0\left[\varepsilon + 2e\sin(n_0 t - \tilde{M}_0)\right] - \frac{3}{2}\delta a n_0(t - t_0) \tag{6.63}$$

$$z(t) = a_0 i\sin(n_0 t + \omega - \tilde{M}_0) \tag{6.64}$$

It is useful to compare the resulting relative motion components to the CW solution (5.17)–(5.19). Recall that the necessary and sufficient condition for stable linear motion using the CW equations was $\dot{y}(0) = -2nx(0)$ (cf. Eq. (5.23)), which is only a local condition for stability and is not generally related to the global condition for relative motion boundedness known as the energy matching condition. However, in Hill's solution, the necessary and sufficient condition for stability is $\delta a = 0$, which is identical to the energy matching condition. Thus, interestingly, using orbital elements as constants of the motion have yielded a local stability condition which is identical to the global stability condition.

For $\delta a = 0$, Hill's solution becomes

$$x(t) = -a_0 e \cos(n_0 t - \tilde{M}_0) \tag{6.65}$$

$$y(t) = a_0 \left[\varepsilon + 2e \sin(n_0 t - \tilde{M}_0)\right] \tag{6.66}$$

$$z(t) = a_0 i \sin(n_0 t + \omega - \tilde{M}_0) \tag{6.67}$$

Obviously, the xy-projection of the motion is a 2:1 ellipse centered at $(0, \varepsilon)$, having a semiminor axis e, a semimajor axis $2e$, and a constant eccentricity of $\sqrt{1 - e^2/4e^2} = \sqrt{3}/2$. This result should sound familiar; we made the exact same observation in Section 5.1 regarding the CW solution. The advantage of using orbital elements is that the geometric properties of the relative motion are well understood, constituting a much more natural and convenient approach for analyzing both unperturbed and perturbed relative dynamics.

It is also easy to see that if $\delta a = \omega = 0$ and $e = i$, Hill's solution yields a circular xz-projection having radius e. If $i = 2e$ and $\omega = \pi/2$, than the yz-projection is a circle with radius $2e$. Similar observations were made regarding the CW solution.

One should be careful, though, when relating the CW solution to Hill's solution, as the resulting relationships between the Cartesian initial conditions and orbital elements hold only locally. Different relationships will be obtained for higher-order solutions.

Example 6.2. *Simulate the motion of a deputy spacecraft with (the semimajor axis is written in normalized units)*

$$a_1 = 1, \delta a = 0, e = 0.02, i = 15°, \Omega = 5°, \omega = \tilde{M}_0 = 0°$$

relative to a circular equatorial orbit.

Figure 6.4 depicts a comparison of the exact, nonlinear solution (6.38), the second-order approximation (6.52)–(6.57) and Hill's solution (6.62)–(6.64). Notably, the second-order approximation is considerably closer to the exact solution, as Hill's approximation is limited to small relative inclinations and eccentricities.

6.4 ESTABLISHING THE PCO INITIAL CONDITIONS

Equations (6.62)–(6.64) can also be written in terms of the differential nonsingular elements [75] as

$$x = \delta a - a_0 \left[(\cos\theta)\delta q_1 + (\sin\theta)\delta q_2\right] \tag{6.68a}$$

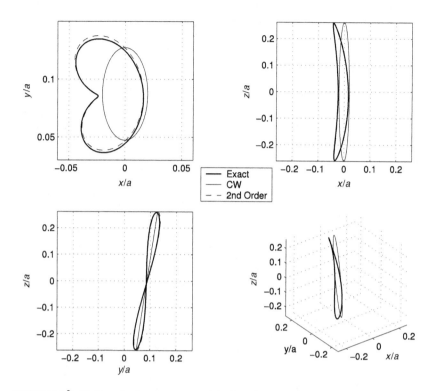

FIGURE 6.4 Motion of a deputy spacecraft in the reference-orbit centered rotating frame: Comparison of exact, first-, and second-order approximations.

$$y = -\frac{3}{2}n_0(t - t_0)\delta a + a_0\left[(\delta\lambda + \delta\Omega\cos i) + 2(\sin\theta)\delta q_1\right.$$
$$\left. - 2(\cos\theta)\delta q_2\right] \tag{6.68b}$$
$$z = a_0\left[(\sin\theta)\delta i - (\cos\theta\sin i)\delta\Omega\right] \tag{6.68c}$$

where δq_1, δq_2, $\delta\lambda$, $\delta\Omega$ and δi are the initial values of the respective differential orbital elements, and they remain constant for the two-body problem.

It can be shown that the differential nonsingular element initial conditions for obtaining a PCO of radius ρ and phase angle α, satisfying Eqs. (5.24)–(5.26) and Eq. (5.36), are [123]

$$\delta a = 0 \tag{6.69}$$

$$\delta q_1 = -\left(\frac{\rho}{2a_0}\right)\sin\alpha \tag{6.70}$$

$$\delta q_2 = -\left(\frac{\rho}{2a_0}\right)\cos\alpha \tag{6.71}$$

$$\delta i = \left(\frac{\rho}{a_0}\right)\cos\alpha \tag{6.72}$$

$$\delta\Omega - \left(\frac{\rho}{a_0}\right)\left(\frac{\sin\alpha}{\sin i_0}\right) \tag{6.73}$$

$$\delta\lambda = \left(\frac{\rho}{a_0}\right)\cot i_0 \sin\alpha \tag{6.74}$$

Equations (6.69)–(6.74) constitute extremely useful relations for establishing PCOs and have been used in many of the example problems in this book, with a slight modification to the condition on δa, required to accommodate the effect of J_2. An alternative form of the set of PCO initial conditions, valid for near-circular orbits, is provided in Section 8.5. In the following section a set of approximate, linearized equations are developed for a direct simulation of relative motion in the $\mathscr{L}$ frame. These equations utilize both the orbital elements and the Cartesian relative motion states.

6.5 HYBRID DIFFERENTIAL EQUATIONS WITH NON-LINEARITY COMPENSATION FOR UNPERTURBED CIRCULAR ORBITS

A linear periodicity condition for relative motion with respect to an elliptic orbit was discussed in Chapter 5 in the context of the two-body STM. In this section, a set of *hybrid differential equations* of relative motion are derived for a two-body circular reference orbit. These equations are termed hybrid because they not only involve the LVLH states explicitly, but also δa, the differential semimajor axis as a parameter. Hence, they require a consistency between their initial conditions and δa of the deputy with respect to the chief. The variables x and z in Eqs. (6.68a)–(6.68c) contain periodic terms; they can be conveniently represented by second-order differential equations. The expression for y is handled in a different manner due to the presence of a secular term. Differentiation of Eq. (6.68b) with respect to time results in

$$\dot{y} = -\frac{3}{2}n_0\delta a + 2a_0n_0\left[(\cos\theta)\delta q_1 + (\sin\theta)\delta q_2\right] \tag{6.75}$$

The above equation can be simplified by substituting Eq. (6.68a) to obtain

$$\dot{y} = \frac{1}{2}n_0\delta a - 2n_0x \tag{6.76}$$

The equations for x and y can also be manipulated as shown below:

$$\ddot{x} = -n_0^2(x - \delta a) \tag{6.77}$$

$$\ddot{z} = -n_0^2 z \tag{6.78}$$

Equations (6.76)–(6.78) reproduce the results of Eqs. (6.68a)–(6.68c) for a consistent set of initial conditions. Even though only five differential equations are included in the derived model, information regarding the initial condition on $\dot{y}$ is required to compute δa. An exact analytical expression for δa in terms

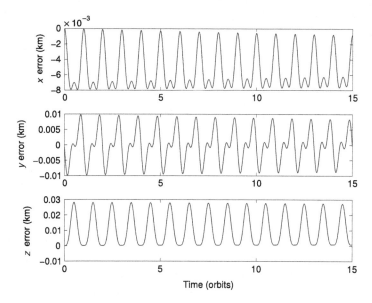

FIGURE 6.5 Errors between the linear and nonlinear propagations ($\rho = 10$ km, $\alpha = 0°$, $\delta a = 0$).

of the relative motion states can be obtained from the equations of two-body motion [90]. Substitution of Eq. (5.66) into Eq. (6.77) results in the radial component of the CW equations. However, the actual δa cannot always be represented accurately by a linear approximation, especially when a long-term propagation is required. The use of the exact or higher-order approximate values of δa incorporates some of the nonlinear effects not captured by the CW equations. Equations (6.76)–(6.78) do not include the effects of perturbations or control inputs, which may cause changes in δa.

Example 6.3. *Compare the results produced by Eqs. (6.76)–(6.78) and a nonlinear simulation, with consistent sets of initial conditions. The following initial orbital elements of the chief are selected:*

$$a_0 = 7100 \text{ km}, \quad \theta_0 = 0, \quad i_0 = 70°$$
$$q_{10} = 0, \quad q_{20} = 0, \quad \Omega_0 = 45°$$

The PCO initial conditions are selected for the deputy. The orbital elements of the chief and deputy are converted into their respective ECI position and velocity states for the nonlinear simulation. A transformation of the relative initial states from the ECI to LVLH frames is used to obtain the required initial conditions for Eqs. (6.76)–(6.78).

Figure 6.5 shows the errors between the results produced by the linear model and a nonlinear simulation, for a PCO with $\rho = 10$ km and $\alpha_x = \alpha_z \equiv \alpha = 0$. Figure 6.6 shows the same for $\rho = 10$ km and $\alpha = 90°$. For both of these examples, $\delta a = 0$. The next set of figures (Figs. 6.7 and 6.8) show the errors

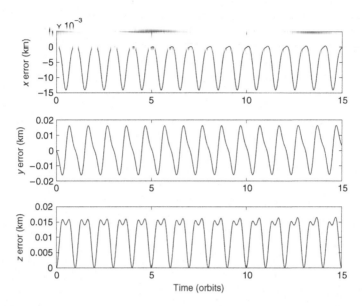

FIGURE 6.6 Errors between the linear and nonlinear propagations ($\rho = 10$ km, $\alpha = 90°, \delta a = 0$).

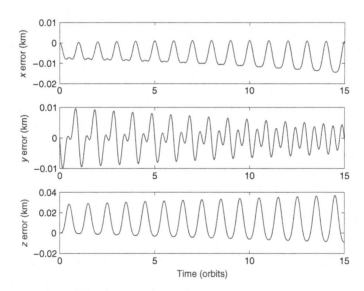

FIGURE 6.7 Errors between the linear and nonlinear propagations ($\rho = 10$ km, $\alpha = 0°$, $\delta a = 50$ m).

for $\delta a = 50$ m. As can be seen from these figures, the use of δa in the linear model compensates for a significant portion of the nonlinear effects, even for large relative displacements.

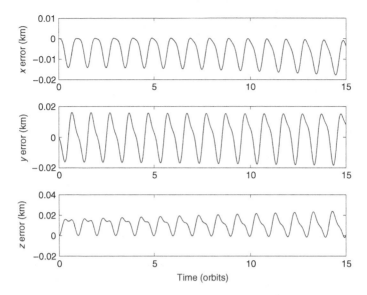

FIGURE 6.8 Errors between the linear and nonlinear propagations ($\rho = 10$ km, $\alpha = 90°, \delta a = 50$ m).

SUMMARY

We have written general expressions modeling the relative spacecraft geometry and have explicitly parameterized the relative motion configuration space using classical orbital elements as constants of the unperturbed Keplerian motion. Based on this geometric insight, we presented analytic expressions for a few bounds on the relative distances, which are important for generating safe and reliable spacecraft formations. These expressions are different for the commensurable and incommensurable cases.

The relative motion geometry evolves on an invariant manifold, which can be easily characterized using relative orbital elements. The motion along this manifold is quasi-periodic in the general case and periodic in the commensurable case. Both types of motion evolve on the relative motion manifold.

We also presented a framework for approximating the relative dynamics of spacecraft formations utilizing the known three-dimensional inertial solutions. The time-series parametrization of the relative position vector lends itself naturally to high orders and hence constitutes a useful analysis and modeling tool, providing much insight into the relative dynamics of spacecraft formations.

Finally, we discussed the process for setting up PCO initial conditions and derived a hybrid differential equation model for relative motion with a circular reference orbit. These equations involve the states in the $\mathscr{L}$ frame and the differential semimajor axis as a parameter.

Chapter seven

Modeling Perturbed Relative Motion Using Orbital Elements

...And situated softly
Upon a pile of wind
Which in a perturbation
Nature had left behind.

Emily Dickinson (1830–1886)

Chapter 6 was devoted to orbital-element-based modeling of relative motion in unperturbed orbits. This chapter is organized into two parts. We begin with the presentation of analytical methods for propagating the perturbed relative motion variables via differential orbital elements and differential Euler parameters. These methods rely on the mean element propagation scheme outlined in Section 3.4 and constitute nonlinear theories. The second part of this chapter is devoted to linear theories including the STM of Gim and Alfriend (GA) and linear differential equation models for J_2-perturbed relative motion about mean circular orbits. A procedure for averaging the short-periodic corrections to the mean elements is discussed, leading to the development of *averaged orbital elements* and *averaged relative motion*. Finally, a more accurate, second-order state transition tensor is developed by incorporating quadratic nonlinearities of the two-body relative motion and the GA STM, which models the linearized J_2 effects.

7.1 THE UNIT-SPHERE APPROACH

The *unit-sphere approach*, as the name implies, projects the motion of the satellites in a formation onto a sphere of unit radius, allowing for application of the rules of spherical trigonometry. Such a projection uncouples from the motion of a satellite, the periodic variations of its radius due to the **143** effects of eccentricity and other perturbations. The unit-sphere approach provides the projected relative motion coordinates and velocities without invoking any small-angle approximations. The solution, presented in terms of differential orbital elements, is exact and reveals the complex structure of the relative motion. The true or unscaled relative motion variables are obtained via re-scaling transformations from their scaled counterparts, involving the radii of the two satellites under consideration.

A satellite's motion can be projected onto a unit sphere by normalizing its Cartesian coordinates by its radius. The projection obtained thus is the sub-satellite point. The relative motion between two sub-satellite points is the focus of this section. Let C_0 and C_1 indicate the direction cosine matrices of the LVLH frames of the two satellites with respect to the ECI frame, $\mathscr{I}$. The relative position vector can be expressed in the rotating frame of the chief, $\mathscr{L}$, as follows (compare to our development in Section 6.1):

$$
\begin{Bmatrix} \bar{x} \\ \bar{y} \\ \bar{z} \end{Bmatrix} = [C_0 C_1^T - I_3] \begin{Bmatrix} 1 \\ 0 \\ 0 \end{Bmatrix} \tag{7.1}
$$

where $\bar{x}$, $\bar{y}$, and $\bar{z}$ are, respectively, the radial, in-track, and cross-track relative positions and I_3 stands for a 3×3 identity matrix. The direction cosine matrices can be parameterized by orbital elements. This allows for explicit solutions for the relative displacement variables in terms of the differential orbital elements as [124]

$$
\begin{aligned}
\bar{x} =\ & -1 + \cos^2 \frac{i_0}{2} \cos^2 \frac{i_1}{2} \cos(\delta\theta + \delta\Omega) + \sin^2 \frac{i_0}{2} \sin^2 \frac{i_1}{2} \cos(\delta\theta - \delta\Omega) \\
& + \sin^2 \frac{i_0}{2} \cos^2 \frac{i_1}{2} \cos(2\theta_0 + \delta\theta + \delta\Omega) \\
& + \cos^2 \frac{i_0}{2} \sin^2 \frac{i_1}{2} \cos(2\theta_0 + \delta\theta - \delta\Omega) \\
& + \frac{1}{2} \sin i_0 \sin i_1 [\cos \delta\theta - \cos(2\theta_0 + \delta\theta)]
\end{aligned} \tag{7.2}
$$

$$
\begin{aligned}
\bar{y} =\ & \cos^2 \frac{i_0}{2} \cos^2 \frac{i_1}{2} \sin(\delta\theta + \delta\Omega) + \sin^2 \frac{i_0}{2} \sin^2 \frac{i_1}{2} \sin(\delta\theta - \delta\Omega) \\
& - \sin^2 \frac{i_0}{2} \cos^2 \frac{i_1}{2} \sin(2\theta_0 + \delta\theta + \delta\Omega) \\
& - \cos^2 \frac{i_0}{2} \sin^2 \frac{i_1}{2} \sin(2\theta_0 + \delta\theta - \delta\Omega) \\
& + \frac{1}{2} \sin i_0 \sin i_1 [\sin \delta\theta + \sin(2\theta_0 + \delta\theta)]
\end{aligned} \tag{7.3}
$$

$$
\bar{z} = -\sin i_0 \sin \delta\Omega \cos \theta_1 - [\sin i_0 \cos i_1 \cos \delta\Omega - \cos i_0 \sin i_1] \sin \theta_1 \tag{7.4}
$$

Recall that the argument of latitude of the chief is $\theta_0 = \omega_0 + f_0$ and thus $\delta\theta$ indicates the differential argument of latitude. The radial and in-track variables involve the sum and difference of θ_0 and θ_1, since $2\theta_0 + \delta\theta = \theta_1 + \theta_0$ and $\delta\theta = \theta_1 - \theta_0$. On the other hand, $\bar{z}$ is an explicit function of θ_1 but not θ_0.

Hence, in general, the fundamental frequencies of the in-plane and cross-track motions are not the same.

Equations (7.2)–(7.4) are exact and valid for large angles. The effect of the J_2-perturbation can be studied by using osculating orbital elements in the above equations. However, one can also substitute expressions for the mean or averaged orbital elements [125], if desired, depending on the usage. The use of mean orbital elements and their secular drift rates, given in Eqs. (2.115), simplifies the process of the analytical propagation of relative motion considerably. However, Kepler's equation has to be solved to obtain a time-explicit solution. Eccentricity expansions for r and θ provide a time-explicit representation of the analytical solution without having to iterate on Kepler's equation. The equations for evaluating the mean drift rates are given below for the jth satellite in a formation:

$$\bar{\Omega}_j = \bar{\Omega}_j(0) + \dot{\bar{\Omega}}_j t \tag{7.5}$$

$$\bar{\omega}_j = \bar{\omega}_j(0) + \dot{\bar{\omega}}_j t \tag{7.6}$$

$$\bar{M}_j = \bar{M}_j(0) + \dot{\bar{M}}_j t \tag{7.7}$$

$$\dot{\bar{\Omega}}_j = -1.5 J_2 (R_e/\bar{p}_j)^2 \bar{n}_j \cos \bar{i}_j \tag{7.8}$$

$$\dot{\bar{\omega}}_j = 0.75 J_2 (R_e/\bar{p}_j)^2 \bar{n}_j \left(5 \cos^2 \bar{i}_j - 1\right) \tag{7.9}$$

$$\dot{\bar{M}}_j = \bar{n}_j \left[1 + 0.75 J_2 \sqrt{1 - \bar{e}_j^2} (R_e/\bar{p}_j)^2 \left(3 \cos^2 \bar{i}_j - 1\right)\right] \tag{7.10}$$

$$\bar{n}_j = \sqrt{\frac{\mu}{\bar{a}_j^3}} \tag{7.11}$$

Note that henceforth M_0 will indicate the mean anomaly of the chief. The mean elements can be converted into the respective osculating elements via the Brouwer transformation [66].

The dimensional relative motion variables are obtained from their unit-sphere counterparts by re-scaling, as shown below:

$$x = r_1(1 + \bar{x}) - r_0 \tag{7.12}$$

$$\begin{Bmatrix} y \\ z \end{Bmatrix} = r_1 \begin{Bmatrix} \bar{y} \\ \bar{z} \end{Bmatrix} \tag{7.13}$$

$$\dot{x} = r_1 \dot{\bar{x}} + \dot{r}_1(1 + \bar{x}) - \dot{r}_0 \tag{7.14}$$

$$\begin{Bmatrix} \dot{y} \\ \dot{z} \end{Bmatrix} = r_1 \begin{Bmatrix} \dot{\bar{y}} \\ \dot{\bar{z}} \end{Bmatrix} + \dot{r}_1 \begin{Bmatrix} \bar{y} \\ \bar{z} \end{Bmatrix} \tag{7.15}$$

Note that the unit-sphere expressions for the in-track and cross-track motion variables are scaled versions of the respective physical coordinates; the scaling factor is the radius of the deputy (not that of the chief). The relative velocity states can be obtained by differentiating Eqs. (7.2)–(7.4) with respect to time.

Equations (7.2)–(7.4) can be simplified for small differential elements via linearization and the resulting solutions for the motion variables can be expressed as

$$x = r_1 - r_0 \tag{7.16}$$

$$y = r_1(\delta\theta + \delta\Omega \cos i_0) \tag{7.17}$$

$$z = r_1(-\sin i_0 \delta\Omega \cos \theta_1 + \delta i \sin \theta_1) \tag{7.18}$$

It is also reasonable for many applications to approximate r_1 and θ_1 in Eqs. (7.17)–(7.18) by r_0 and θ_0, respectively. A higher-order approximation for relative motion with respect to near-circular orbits has been devised by Hill et al. [126], based on the COWPOKE (Cluster Orbits with Perturbation of Keplerian Elements) model, originally developed by Sabol et al. [127]. The differential elements δM, $\delta\Omega$, and $\delta\omega$ are assumed large in this model, and a first-order eccentricity expansion for f is employed. The relative position coordinates are given by

$$x = \frac{(a_0 + \delta a)[1 - (e_0 + \delta e)^2]}{1 + (e_0 + \delta e)\cos(M_0 + 2e_0 \sin M_0 + \delta f)} - \frac{a_0(1 - e_0^2)}{1 + e_0 \cos f} \tag{7.19}$$

$$y = r_0(\delta\theta + \delta\Omega \cos i_0) \tag{7.20}$$

$$z = r_0 \left[-2 \sin i_0 \sin \left(\frac{\delta\Omega}{2} \right) \cos \left(M_0 + \omega_0 + 2e_0 \sin M_0 + \frac{\delta\theta}{2} \right) \right.$$
$$\left. + \delta i \sin(M_0 + \omega_0 + 2e_0 \sin M_0 + \delta\theta) \right] \tag{7.21}$$

where

$$\delta M = \delta M(0) + \left[\sqrt{\frac{\mu}{(a_0 + \delta a)^3}} - \sqrt{\frac{\mu}{a_0^3}} \right] t \tag{7.22}$$

and

$$\delta f = \delta M + 2(e_0 + \delta e)\sin(M_0 + \delta M) - 2e_0 \sin M_0 \tag{7.23}$$

The COWPOKE equations have been found useful for cluster analysis of GEO satellites.

A method based on the numerical integration of the GVE, incorporating the disturbances due to J_2, differential atmospheric drag, and differential solar radiation pressure has been presented by Balaji and Tatnall [128]. The relative motion states are obtained from the perturbed osculating orbital elements of the satellites using an approach similar to the unit-sphere procedure.

If the periodic variations in y are neglected, then we can write

$$y \approx \bar{a}_0(\delta\bar{\lambda} + \delta\bar{\Omega}\cos \bar{i}_0) \tag{7.24}$$

The derivative of the above approximation can be obtained from the secular drift rates of Eqs. (7.8)–(7.10) and thus an *along-track boundedness condition* can be derived:

$$\dot{y} \approx \bar{a}_0(\delta\dot{\bar{\lambda}} + \delta\dot{\bar{\Omega}}\cos\bar{i}_0) = 0 \tag{7.25}$$

This condition results in a constraint on $\delta\bar{a}$, as will be shown in Chapter 8, to mitigate an adverse effect of J_2. An independent derivation of this result, valid for circular orbits, is presented in Eq. (7.140).

7.2 RELATIVE MOTION DESCRIPTION USING QUATERNIONS

The direction cosine matrices in Eq. (7.1) can also be parameterized by two quaternions – sets of Euler parameters – as discussed in Section 2.4 and as shown by Junkins and Turner [129], one each for the chief and the deputy. The applications of this analogy to the determination of relative attitude is discussed in Chapter 9. A closely-related parametrization in terms of the Kustaanheimo–Stiefel variables is given in Ref. [54]. As shown in Section 2.4, the quaternion of a satellite's LVLH frame can be related to its orbital elements as:

$$\begin{aligned}
\beta_0 &= \cos\frac{i}{2}\cos\frac{\Omega+\theta}{2} & \beta_1 &= \sin\frac{i}{2}\cos\frac{\Omega-\theta}{2} \\
\beta_2 &= \sin\frac{i}{2}\sin\frac{\Omega-\theta}{2} & \beta_3 &= \cos\frac{i}{2}\sin\frac{\Omega+\theta}{2}
\end{aligned} \tag{7.26}$$

Equations (7.26), adapted from Ref. [54], can be used to obtain the quaternion via the mean element propagation scheme. The quaternion description is nonsingular. For example, i and Ω can be set to zero for an equatorial orbit.

The direction cosine matrix representing the rotation from ECI to the LVLH frame associated with a satellite can be parameterized in terms of its quaternion as (cf. also the discussion in Section 2.4)

$$C = \begin{bmatrix} \beta_0^2 + \beta_1^2 - \beta_2^2 - \beta_3^2 & 2(\beta_1\beta_2 + \beta_0\beta_3) & 2(\beta_1\beta_3 - \beta_0\beta_2) \\ 2(\beta_1\beta_2 - \beta_0\beta_3) & \beta_0^2 - \beta_1^2 + \beta_2^2 - \beta_3^2 & 2(\beta_2\beta_3 + \beta_0\beta_1) \\ 2(\beta_1\beta_3 + \beta_0\beta_2) & 2(\beta_2\beta_3 - \beta_0\beta_1) & \beta_0^2 - \beta_1^2 - \beta_2^2 + \beta_3^2 \end{bmatrix} \tag{7.27}$$

Given the quaternion for the orbit of a satellite at any instant, its unit radius vector can be obtained from Eq. (7.27) as

$$\hat{\mathbf{r}} = \begin{bmatrix} \beta_0^2 + \beta_1^2 - \beta_2^2 - \beta_3^2 & 2(\beta_1\beta_2 + \beta_0\beta_3) & 2(\beta_1\beta_3 - \beta_0\beta_2) \end{bmatrix}^T \tag{7.28}$$

The quaternion propagation equation is

$$\begin{Bmatrix} \dot{\beta}_0 \\ \dot{\beta}_1 \\ \dot{\beta}_2 \\ \dot{\beta}_3 \end{Bmatrix} = \frac{1}{2} \begin{bmatrix} 0 & -\omega_x & -\omega_y & -\omega_z \\ \omega_x & 0 & \omega_z & -\omega_y \\ \omega_y & -\omega_z & 0 & \omega_x \\ \omega_z & \omega_y & -\omega_x & 0 \end{bmatrix} \begin{Bmatrix} \beta_0 \\ \beta_1 \\ \beta_2 \\ \beta_3 \end{Bmatrix} \tag{7.29}$$

It is well known (cf. Eqs. (7.143b) and (7.150b)) that the osculation constraint results in $\omega_{r_v} - 0$

Relative orientation of the LVLH frame of the deputy with respect to that of the chief can be expressed by a relative quaternion using a finite rotation [29]:

$$\begin{Bmatrix} \delta\beta_0 \\ \delta\beta_1 \\ \delta\beta_2 \\ \delta\beta_3 \end{Bmatrix} = \begin{bmatrix} \beta_{01} & \beta_{11} & \beta_{21} & \beta_{31} \\ \beta_{11} & -\beta_{01} & -\beta_{31} & \beta_{21} \\ \beta_{21} & \beta_{31} & -\beta_{01} & -\beta_{11} \\ \beta_{31} & -\beta_{21} & \beta_{11} & -\beta_{01} \end{bmatrix} \begin{Bmatrix} \beta_{00} \\ \beta_{10} \\ \beta_{20} \\ \beta_{30} \end{Bmatrix} \tag{7.30}$$

where the second subscript on the quaternions refers to a particular satellite, either the chief or deputy. The elements of the relative quaternion vector can be explicitly obtained by substituting Eqs. (7.26) for the chief and deputy into Eq. (7.30) as

$$\delta\beta_0 = \cos\frac{i_0}{2} \cos\frac{i_1}{2} \cos\left(\frac{\delta\theta + \delta\Omega}{2}\right) + \sin\frac{i_0}{2} \sin\frac{i_1}{2} \cos\left(\frac{\delta\theta - \delta\Omega}{2}\right)$$

$$\delta\beta_1 = -\sin\frac{i_0}{2} \cos\frac{i_1}{2} \cos\left(\frac{\delta\theta + \delta\Omega}{2}\right) + \cos\frac{i_0}{2} \sin\frac{i_1}{2} \cos\left(\frac{\delta\theta - \delta\Omega}{2} + \theta_0\right)$$

$$\tag{7.31a}$$

$$\delta\beta_2 = \sin\frac{i_0}{2} \cos\frac{i_1}{2} \sin\left(\frac{\delta\theta + \delta\Omega}{2} + \theta_0\right) - \cos\frac{i_0}{2} \sin\frac{i_1}{2} \sin\left(\frac{\delta\theta - \delta\Omega}{2} + \theta_0\right)$$

$$\delta\beta_3 = \cos\frac{i_0}{2} \cos\frac{i_1}{2} \sin\left(\frac{\delta\theta + \delta\Omega}{2}\right) + \sin\frac{i_0}{2} \sin\frac{i_1}{2} \sin\left(\frac{\delta\theta - \delta\Omega}{2}\right)$$

$$\tag{7.31b}$$

The relative position of the deputy with respect to the chief can be written very simply by noting that $C_0 C_1^T$ in Eq. (7.1) is nothing but the relative direction cosine matrix for a rotation from the deputy's LVLH frame onto that of the chief. Hence, it follows from Eqs. (7.27) and (7.28) that

$$\begin{Bmatrix} \bar{x} \\ \bar{y} \\ \bar{z} \end{Bmatrix} = \begin{Bmatrix} (\delta\beta_0^2 + \delta\beta_1^2 - \delta\beta_2^2 - \delta\beta_3^2) - 1 \\ 2(\delta\beta_1\delta\beta_2 + \delta\beta_0\delta\beta_3) \\ 2(\delta\beta_1\delta\beta_3 - \delta\beta_0\delta\beta_2) \end{Bmatrix} = 2 \begin{Bmatrix} -(\delta\beta_2^2 + \delta\beta_3^2) \\ \delta\beta_1\delta\beta_2 + \delta\beta_0\delta\beta_3 \\ \delta\beta_1\delta\beta_3 - \delta\beta_0\delta\beta_2 \end{Bmatrix} \tag{7.32}$$

The position variables $(1+\bar{x})$, $\bar{y}$, and $\bar{z}$ can be interpreted as the three direction cosines of the unit radius vector of the deputy with respect to the LVLH frame of the chief. Therefore, the following relationship exists between the three relative position variables:

$$(1 + \bar{x})^2 + \bar{y}^2 + \bar{z}^2 = 1 \tag{7.33}$$

Furthermore, $\bar{x} \le 0$.

Focusing attention on $\bar{y}$ in Eq. (7.32), it is observed that for small formations, the term $\delta\beta_1\delta\beta_2$ is at least an order of magnitude smaller than $\delta\beta_0\delta\beta_3$. Hence, under the mentioned assumption, an excellent approximation for the in-track relative position is given by

$$\overline{y} \approx 2\delta\beta_3 \tag{7.34}$$

Similarly, for small formations

$$\overline{z} \approx -2\delta\beta_2 \tag{7.35}$$

The influence of the differential quaternion on $\overline{x}$ is of second order.

The relative quaternion can be propagated directly from the kinematic relationship

$$
\begin{Bmatrix} \delta\dot{\beta}_0 \\ \delta\dot{\beta}_1 \\ \delta\dot{\beta}_2 \\ \delta\dot{\beta}_3 \end{Bmatrix} = \frac{1}{2}
\begin{bmatrix}
0 & -\delta\omega_x & -\delta\omega_y & -\delta\omega_z \\
\delta\omega_x & 0 & \delta\omega_z & -\delta\omega_y \\
\delta\omega_y & -\delta\omega_z & 0 & \delta\omega_x \\
\delta\omega_z & \delta\omega_y & -\delta\omega_x & 0
\end{bmatrix}
\begin{Bmatrix} \delta\beta_0 \\ \delta\beta_1 \\ \delta\beta_2 \\ \delta\beta_3 \end{Bmatrix} \tag{7.36}
$$

with the differential angular velocity vector obtained from

$$
\begin{Bmatrix} \delta\omega_x \\ \delta\omega_y \\ \delta\omega_z \end{Bmatrix} =
\begin{Bmatrix} \omega_{x1} \\ \omega_{y1} \\ \omega_{z1} \end{Bmatrix} - C_1 C_0^T
\begin{Bmatrix} \omega_{x0} \\ \omega_{y0} \\ \omega_{z0} \end{Bmatrix} \tag{7.37}
$$

The matrix product $C_1 C_0^T$ in Eq. (7.37) can be evaluated as the direction cosine matrix for a rotation from the LVLH frame of the chief onto that of the deputy, i.e., it can be obtained by substituting the elements of the differential quaternion into Eq. (7.27). Furthermore, as noted previously, the osculating orbit constraint results in $\omega_{y0} = \omega_{y1} = 0$.

The remaining angular velocities and the radius of each satellite can be evaluated using the orbital equations of motion written in its own LVLH frame:

$$\ddot{r} - \omega_z^2 r = -\frac{\mu}{r^2} + u_x + d_x \tag{7.38a}$$

$$r\dot{\omega}_z + 2\dot{r}\omega_z = u_y + d_y \tag{7.38b}$$

$$r\omega_x\omega_z = u_z + d_z \tag{7.38c}$$

where the components of the control and J_2 disturbance vectors included are as defined in Eqs. (5.7)–(5.9). The components of the J_2 disturbing acceleration vector can be obtained by evaluating the gradient of $\mathcal{R}$ defined in Section 2.5.3 as

$$
\mathbf{d} = \begin{Bmatrix} d_x \\ d_y \\ d_z \end{Bmatrix} = -\frac{3}{2}\frac{J_2 \mu R_e^2}{r^4}
\begin{Bmatrix} 1 - 3\sin^2 i \sin^2 \theta \\ \sin^2 i \sin 2\theta \\ \sin 2i \sin \theta \end{Bmatrix} \tag{7.39}
$$

Equation (7.39) can be represented in terms of the Euler parameters with the help of the following identities:

$$\sin i \sin \theta = 2 (\beta_1 \beta_3 - \beta_0 \beta_2) \tag{7.40a}$$

$$\cos i = \left(1 - 2\beta_1^2 - 2\beta_2^2\right) \tag{7.40b}$$

$$\sin i \cos \theta = 2 (\beta_0 \beta_1 + \beta_2 \beta_3) \tag{7.40c}$$

Finally, we obtain

$$\mathbf{d} = -\frac{3}{2} \frac{J_2 \mu R_e^2}{r^4} \left\{ \begin{array}{c} 1 - 12 (\beta_1 \beta_3 - \beta_0 \beta_2)^2 \\ 8 (\beta_1 \beta_3 - \beta_0 \beta_2) (\beta_0 \beta_1 + \beta_2 \beta_3) \\ 4 \left(1 - 2\beta_1^2 - 2\beta_2^2\right) (\beta_0 \beta_1 + \beta_2 \beta_3) \end{array} \right\} \tag{7.41}$$

In summary, a complete nonlinear simulation with quaternions can be performed by utilizing Eq. (7.29) and Eqs. (7.38a)–(7.38c) for the chief and deputy. Relative motion variables can be extracted from the results of numerical integration using Eqs. (7.12)–(7.15), (7.30), and (7.32). Alternatively, Eq. (7.36) can be simulated instead of Eq. (7.29) to obtain the relative quaternion of the deputy. The total number of differential equations for this nonsingular approach is 14. The number of equations can be reduced by two with the use of $\beta_0 = \sqrt{(1 - \beta_1^2 - \beta_2^2 - \beta_3^2)}$. Gurfil [50] presents the LPE for the quaternion elements subjected to the J_2-perturbation as an alternative approach to a singularity-free simulation; see also Eqs. (2.116).

We have discussed nonlinear methods for the analytical propagation of the relative motion dynamics. Attention is focused on linear methods of propagation in the subsequent sections of this chapter.

7.3 THE GIM–ALFRIEND GEOMETRIC METHOD

The purpose of this section is to derive the state transition matrices for both osculating and mean elements for the relative motion of two neighboring satellites when the reference satellite (chief) is in an elliptic orbit, and both satellites are subjected to the J_2 perturbation. The linear differential equations of relative motion for this system are formidable. A perturbation solution in powers of J_2 could probably be obtained, but there is an easier approach. Rather than solving the equations of motion, the approach is to use the geometric relationship between the relative state and the orbital elements for the two satellites, similarly to the rationale discussed in Chapter 6. If one has the orbital elements for the two objects, then the relative state can be obtained, and vice versa. This approach was used by Garrison [130] to obtain a solution for the relative motion assuming a spherical Earth. With this approach, solving the differential equations of relative motion can be avoided. Since this approach uses the geometric relationship between the relative state and orbital elements, it is referred to as the *Geometric Method* (GM).

7.3.1 J_2 effects revisited

As we have seen in Sections 2.6, 3.4 and 7.1, the gravitational perturbations create secular, short-periodic and long-periodic variations of the orbital elements, where the long-periodic terms have a frequency equal to the perigee rotation rate. In Chapter 5, when assuming a spherical Earth, we showed that to avoid unnecessary fuel expenditure, the chief orbit's eccentricity should be included in the dynamical model through the Lawden or the Tschauner–Hempel equations (instead of the CW equations).

Rewriting Eqs. (7.8)–(7.10), we obtain the mean angle rates subject to a J_2 perturbation:

$$n = \bar{n} + 0.75 J_2 \bar{n} \left(\frac{R_e}{p}\right)^2 \eta \left(2 - 3\sin^2 i\right)$$

$$\dot{\Omega} = -1.5 J_2 \bar{n} \left(\frac{R_e}{p}\right)^2 \cos i \tag{7.42}$$

$$\dot{\omega} = -0.75 J_2 \bar{n} \left(\frac{R_e}{p}\right)^2 \left(1 - 5\cos^2 i\right)$$

where

$$p = a\eta^2, \quad \eta = \left(1 - e^2\right)^{1/2} \tag{7.43}$$

and $\bar{n}$ is the mean motion for the unperturbed orbit. A difference in the semimajor axis, eccentricity or inclination of the satellites in a formation can cause a difference in the secular rates of the two objects, resulting in secular drift. This is in contrast to the spherical Earth problem, in which only a semimajor axis difference can cause secular drift, and only in the along-track direction. Note that only e^2, not e, appears in the secular rate equations. Thus, it is the difference in e^2 that causes the secular drift. This means that for near-circular orbits, the effect of a change in the eccentricity will be much smaller than the effect of a comparable change in inclination. However, for highly eccentric orbits, such as the NASA MMS mission discussed in Section 1.7, the effect of a change in eccentricity can be comparable to the effect of a change in inclination. The differential secular effects are obtained by a Taylor series expansion about the chief's orbit; for an inclination change they are:

$$\delta n = \delta \dot{l} = - \left[4.5 J_2 \bar{n} \left(\frac{R_e}{p_0}\right)^2 \eta_0 \sin i_0 \cos i_0 \right] \delta i \tag{7.44}$$

$$\delta \dot{\Omega} = \left[1.5 J_2 \bar{n} \left(\frac{R_e}{p_0}\right)^2 \sin i_0 \right] \delta i \tag{7.45}$$

$$\delta \dot{\omega} = - \left[7.5 J_2 \bar{n} \left(\frac{R_e}{p_0}\right)^2 \sin i_0 \cos i_0 \right] \delta i \tag{7.46}$$

Example 7.1. *Determine the differential secular effects over one orbit for a chief circular orbit of 7000 km, an inclination of 70° and a differential inclination that results in an out-of-plane motion of 1 km, that is, $\delta i = 1/7000$ rad.*

The differential effects are

$$\delta l = -\left[9 J_2 2\pi \left(\frac{R_e}{a_0} \right)^2 \sin i_0 \cos i_0 \right] \left(\frac{1}{a_0} \right) \tag{7.47}$$

$$\delta \Omega = \left[3 J_2 2\pi \left(\frac{R_e}{a_0} \right)^2 \sin i_0 \right] \left(\frac{1}{a_0} \right) \tag{7.48}$$

$$\delta \omega = -\left[15 J_2 2\pi \left(\frac{R_e}{a_0} \right)^2 \sin i_0 \cos i_0 \right] \left(\frac{1}{a_0} \right) \tag{7.49}$$

The along-track drift is $a_0(\delta l + \delta \omega + \delta \Omega \cos i_0) = -19$ m and the cross-track drift is $a_0 \delta \Omega \sin i_0 = 7.5$ m. Since the along-track drift per orbit is $= -3\pi \delta a$, if the J_2 effects are not modeled, the control system would interpret this drift as a semimajor axis error of approximately 2.02 m and correct for it. However, once the control action is complete, the J_2 effects would start the drift again. Thus, if the secular J_2 effects are not included in the dynamic model, fuel will be wasted. It will be shown later that this along-track drift can be negated with a small change in semimajor axis.

Now, let us consider the case of a leader–follower formation with a mean eccentricity of zero and determine the relative motion. Using the equations in Appendices D, E and F we get

$$\delta \bar{\theta} = \delta \bar{\theta}(t_0) \tag{7.50}$$

$$\delta \theta = \left[1 - J_2 R_e^2 \left(D_{22}^{(sp1)} + D_{22}^{(sp2)} \right) \right] \delta \bar{\theta} \tag{7.51}$$

$$\delta \Omega = -J_2 R_e^2 \left(D_{62}^{(sp1)} + D_{62}^{(sp2)} \right) \delta \bar{\theta} \tag{7.52}$$

$$a_0 = \bar{a}_0 + a^{(sp1)} + a^{(sp2)} \tag{7.53}$$

$$\bar{q}_{10} = \bar{q}_{20} = 0 \tag{7.54}$$

$$q_{10} = \left(q_1^{(sp1)} + q_1^{(sp2)} \right) \tag{7.55}$$

$$q_{20} = \left(q_2^{(sp1)} + q_2^{(sp2)} \right) \tag{7.56}$$

$$r_0 = a_0 / \left(1 + q_{10} \cos \theta_0 + q_{20} \sin \theta_0 \right) \tag{7.57}$$

$$y = \bar{a}_0 \left[1 - \frac{17}{4} J_2 \left(\frac{R_e}{\bar{a}_0} \right)^2 \sin^2 \bar{i}_0 \cos 2\bar{\theta}_0 \right] \delta \bar{\theta} \tag{7.58}$$

$$a = \bar{a}_0 \left[1 + \frac{9}{4} J_2 \left(\frac{R_e}{\bar{a}_0} \right)^2 \left(1 - 3 \cos^2 \bar{i}_0 \right) \right] \tag{7.59}$$

Example 7.2. *Evaluate the along-track drift for a chief satellite orbit with* $\bar{a}_0 = 7000$ km, $\bar{i}_0 = 70°$ *and a separation of* 1 km, *so that* $\delta\bar{\theta} = 1/7000$ rad.

Using Eq. (7.58) we obtain

$$y = \left[1 - 0.0041\cos(2\bar{\theta}_0)\right] \text{ km} \tag{7.60}$$

Thus, there is an along-track oscillation of approximately 4 meters. Whether or not the short-periodic terms should be included in the dynamical model will depend on the preciseness of the required control. If CDGPS is used for the relative navigation, it will provide a relative position error of several centimeters (cf. Chapter 12), so this oscillatory motion will certainly be detected. Over a short period of time, it will resemble a secular drift if it is not included in the dynamical model. The type of control system, the size of the control box and the type of formation will determine whether or not the short-periodic terms need to be included in the dynamic model (a high-fidelity formation flying simulation is discussed in Chapter 13). At a minimum, these terms should be evaluated to determine what effect they may have. For example, in the current problem if the control box was set at ± 3 m then the control system would continually be negating the natural short-periodic oscillations. This would not be wise, if it is avoidable.

This analysis has shown that there is a need for a dynamical model that includes the absolute and differential J_2 effects in addition to the chief orbit's eccentricity. In response to this need, the next subsection derives a state transition matrix that is valid for any eccentricity, and includes the first-order absolute and differential J_2 effects.

7.3.2 The geometric method

Let the relative state of the deputy be $\mathbf{x} = (x, \dot{x}, y, \dot{y}, z, \dot{z})^T$, and the chief orbital elements be $\boldsymbol{\alpha} = (a, \theta, i, q_1, q_2, \Omega)^T$, where θ is the argument of latitude, $q_1 = e\cos\omega$, and $q_2 = e\sin\omega$. This nonsingular set, defined in Section 2.4, is used because the true anomaly and the argument of perigee are undefined for a circular orbit. After obtaining the orbital elements of the deputy by a Taylor series expansion about the orbital elements of the chief, the differential orbital elements between them are obtained by $\delta\boldsymbol{\alpha} = \boldsymbol{\alpha}_1 - \boldsymbol{\alpha}_0$.

In the subsequent discussion, we will use the following notation convention: All the orbital elements without subscript are related to the chief, and $\mathbf{x}$ and $\delta\boldsymbol{\alpha}$ are related to the deputy. To obtain more accurate results, a curvilinear coordinate system, $\mathscr{C}$, represented by unit vectors $(\hat{\mathbf{x}}, \hat{\mathbf{y}}, \hat{\mathbf{z}})$ with the origin at the chief, is used instead of the LVLH Cartesian frame $\mathscr{L}$; see our discussion in Section 5.5. Thus, x is the difference in the radii, and y and z are the curvilinear distances along the imaginary circular orbit on the reference orbital plane and perpendicular to the reference orbit, respectively, as shown in Fig. 7.1. The equivalent deputy-fixed curvilinear coordinate system will be denoted by $\mathscr{D}$.

Using the osculating elements for the chief and the deputy under the influence of J_2, and the total angular velocity $\boldsymbol{\varpi} = \dot{\theta}\hat{\mathbf{x}} + \dot{\Omega}\hat{\mathbf{k}} + di/dt\left(\cos\theta\hat{\mathbf{x}} - \sin\theta\hat{\mathbf{y}}\right)$,

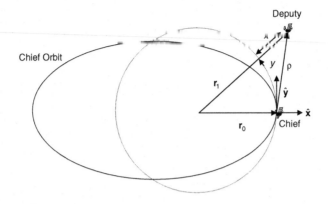

FIGURE 7.1 Curvilinear coordinate system.

where $(\hat{\mathbf{i}}, \hat{\mathbf{j}}, \hat{\mathbf{k}})$ are unit vectors in the ECI reference frame, $\mathscr{I}$, the geometric transformation between $\mathbf{x}(t)$ and $\delta\oe(t)$ is represented by

$$\mathbf{x}(t) = [A(t) + A_2 B(t)]\,\delta\oe(t) \tag{7.61}$$

where $A_2 = 3J_2 R_e^2$, R_e is the equatorial radius of the Earth, and the matrix B contains only the terms perturbed by J_2. Let $\phi_{\oe}$ be the state transition matrix for the relative osculating elements, that is, $\delta\oe(t) = \phi_{\oe}\delta\oe(t_0)$. Therefore, from the solution, $\mathbf{x}(t) = \Phi_{J_2}(t, t_0)\mathbf{x}(t_0)$, the state transition matrix for the relative motion, $\Phi_{J_2}(t, t_0)$, is

$$\Phi_{J_2}(t, t_0) = [A(t) + A_2 B(t)]\,\phi_{\oe}(t, t_0)\,[A(t_0) + A_2 B(t_0)]^{-1} \tag{7.62}$$

Let $\overline{\oe}$ represent the mean elements, and $\oe$ represent the instantaneous, or osculating, elements. Using the transformation matrix $D(t)$, transforming from the relative mean elements to the relative osculating elements (see Appendix F), and the state transition matrix $\bar{\phi}_{\overline{\oe}}(t, t_0)$ for the relative mean elements, we can write

$$\delta\oe_{osc}(t) = D(t)\delta\overline{\oe}(t) = D(t)\bar{\phi}_{\overline{\oe}}(t, t_0)\delta\overline{\oe}(t_0) \tag{7.63}$$

where

$$D(t) = \frac{\partial\oe}{\partial\overline{\oe}} \tag{7.64}$$

$\Phi_{J_2}(t, t_0)$ becomes

$$\Phi_{J_2}(t, t_0) = [A(t) + A_2 B(t)]\,D(t)\bar{\phi}_{\overline{\oe}}(t, t_0)D^{-1}(t_0)$$
$$\times [A(t_0) + A_2 B(t_0)]^{-1} \tag{7.65}$$

The state transition matrix for the mean elements is

$$\bar{\Phi}_{J_2}(t, t_0) = \left[\bar{A}(t) + A_2\bar{B}(t)\right] \bar{\phi}_{\bar{\alpha}}(t, t_0) \left[\bar{A}(t_0) + A_2\bar{B}(t_0)\right]^{-1} \quad (7.66)$$

Since the mean and osculating elements are equal when $J_2 = 0$, $\bar{A}(t) = A(t)$. The angular velocities for the mean and osculating elements are different, consequently, $\bar{B}(t) \neq B(t)$.

THE TRANSFORMATION MATRIX $\Sigma(t) = [A(t) + A_2 B(t)]$

The approach used here is to express the position and velocity of the deputy in terms of the relative state and differential orbital elements and equate the two. The position and velocity of the chief and deputy are

$$\mathbf{r}_0 = r_0\hat{\mathbf{x}} \quad (7.67)$$

$$\mathbf{v}_0 = \dot{r}_0\hat{\mathbf{x}} + r_0\varpi_{n0}\hat{\mathbf{y}} - r_0\varpi_{t0}\hat{\mathbf{z}} \equiv v_{r0J_2}\hat{\mathbf{x}} + v_{t0J_2}\hat{\mathbf{y}} + v_{n0J_2}\hat{\mathbf{z}} \quad (7.68)$$

$$\mathbf{r}_1 = \mathbf{r}_0 + \boldsymbol{\rho} = (r_0 + x)\hat{\mathbf{x}} + y\hat{\mathbf{y}} + z\hat{\mathbf{z}} \quad (7.69)$$

$$\mathbf{v}_1 = (v_{r0J_2} + \dot{x} - y\varpi_{n0} + z\varpi_{t0})\hat{\mathbf{x}} + \left(v_{t0J_2} + \dot{y} + x\varpi_{n0} - z\varpi_{r0}\right)\hat{\mathbf{y}} \quad (7.70)$$

$$+ \left(v_{n0J_2} + \dot{z} - x\varpi_{t0} + y\varpi_{r0}\right)\hat{\mathbf{z}} \quad (7.71)$$

where the subscript J_2 means that the velocity components include the effects of J_2, and

$$\boldsymbol{\varpi} \equiv \varpi_r\hat{\mathbf{x}} + \varpi_t\hat{\mathbf{y}} + \varpi_n\hat{\mathbf{z}} \quad (7.72)$$

The position and velocity vectors of the deputy can also be obtained in curvilinear coordinates by using a Taylor series expansion about the chief and the geometric transformation. The deputy's position in the chief orbit reference frame is given by

$$[\mathbf{r}_1]_{\mathscr{C}} = T^{\mathscr{C}}_{\mathscr{I}} T^{\mathscr{I}}_{\mathscr{D}} [\mathbf{r}_1]_{\mathscr{D}} \quad (7.73)$$

where $T^{\mathscr{C}}_{\mathscr{I}}$ and $T^{\mathscr{I}}_{\mathscr{D}}$ are the coordinate transformations between the chief and inertial frame and the inertial and deputy orbit frames, respectively. The transformation from an orbit frame $\mathscr{O}$, representing either frame $\mathscr{C}$ or $\mathscr{D}$, and the inertial frame $\mathscr{I}$, is given by

$$T^{\mathscr{O}}_{\mathscr{I}} = \begin{pmatrix} c_\theta & s_\theta & 0 \\ -s_\theta & c_\theta & 0 \\ 0 & 0 & 1 \end{pmatrix} \begin{pmatrix} 1 & 0 & 0 \\ 0 & c_i & s_i \\ 0 & -s_i & c_i \end{pmatrix} \begin{pmatrix} c_\Omega & s_\Omega & 0 \\ -s_\Omega & c_\Omega & 0 \\ 0 & 0 & 1 \end{pmatrix}$$

$$= \begin{pmatrix} c_\theta c_\Omega - s_\theta c_i s_\Omega & c_\theta s_\Omega + s_\theta c_i c_\Omega & s_\theta s_i \\ -s_\theta c_\Omega - c_\theta c_i s_\Omega & -s_\theta s_\Omega + c_\theta c_i c_\Omega & c_\theta s_i \\ s_i s_\Omega & -s_i c_\Omega & c_i \end{pmatrix} \quad (7.74)$$

where, as before, we used the compact notation $c_x \equiv \cos x$ and $s_x \equiv \sin x$. Now we expand Eq. (7.73) in a Taylor series about the chief satellite's motion:

$$[\mathbf{r}_1]_{\mathscr{C}} = T_{\mathscr{I}}^{\mathscr{C}} \left(T_{\mathscr{C}}^{\mathscr{I}} + \delta T_{\mathscr{C}}^{\mathscr{I}} \right) \begin{pmatrix} r_0 + \delta r \\ 0 \\ 0 \end{pmatrix}$$

$$[\mathbf{r}_1]_{\mathscr{C}} = \begin{pmatrix} r_0 \\ 0 \\ 0 \end{pmatrix} + r_0 T_{\mathscr{I}}^{\mathscr{C}} \begin{pmatrix} \delta T_{\mathscr{I} \, 11}^{\mathscr{C}} \\ \delta T_{\mathscr{I} \, 12}^{\mathscr{C}} \\ \delta T_{\mathscr{I} \, 13}^{\mathscr{C}} \end{pmatrix} \tag{7.75}$$

The δT_{ij} are

$$\delta T_{\mathscr{I} \, 11}^{\mathscr{C}} = T_{\mathscr{I} \, 21}^{\mathscr{C}} \delta \theta - T_{\mathscr{I} \, 12}^{\mathscr{C}} \delta \Omega + \left(T_{\mathscr{I} \, 13}^{\mathscr{C}} \sin \Omega_0 \right) \delta i \tag{7.76}$$

$$\delta T_{\mathscr{I} \, 12}^{\mathscr{C}} = T_{\mathscr{I} \, 22}^{\mathscr{C}} \delta \theta + T_{\mathscr{I} \, 11}^{\mathscr{C}} \delta \Omega - \left(T_{\mathscr{I} \, 13}^{\mathscr{C}} \cos \Omega_0 \right) \delta i \tag{7.77}$$

$$\delta T_{\mathscr{I} \, 13}^{\mathscr{C}} = T_{\mathscr{I} \, 23}^{\mathscr{C}} \delta \theta + (\sin \theta_0 \cos i_0) \, \delta i \tag{7.78}$$

$$\delta T_{\mathscr{I} \, 21}^{\mathscr{C}} = -T_{\mathscr{I} \, 11}^{\mathscr{C}} \delta \theta - T_{\mathscr{I} \, 22}^{\mathscr{C}} \delta \Omega + T_{\mathscr{I} \, 23}^{\mathscr{C}} \sin \Omega_0 \delta i \tag{7.79}$$

$$\delta T_{\mathscr{I} \, 22}^{\mathscr{C}} = -T_{\mathscr{I} \, 12}^{\mathscr{C}} \delta \theta + T_{\mathscr{I} \, 21}^{\mathscr{C}} \delta \Omega - T_{\mathscr{I} \, 23}^{\mathscr{C}} \cos \Omega_0 \delta i \tag{7.80}$$

$$\delta T_{\mathscr{I} \, 23}^{\mathscr{C}} = -T_{\mathscr{I} \, 13}^{\mathscr{C}} \delta \theta + (\cos \theta_0 \cos i_0) \, \delta i \tag{7.81}$$

$$\delta T_{\mathscr{I} \, 31}^{\mathscr{C}} = -T_{\mathscr{I} \, 32}^{\mathscr{C}} \delta \Omega + (\cos i_0 \sin \Omega_0) \, \delta i \tag{7.82}$$

$$\delta T_{\mathscr{I} \, 32}^{\mathscr{C}} = T_{\mathscr{I} \, 31}^{\mathscr{C}} \delta \Omega - (\cos i_0 \cos \Omega_0) \, \delta i \tag{7.83}$$

$$\delta T_{\mathscr{I} \, 33}^{\mathscr{C}} = - \sin i_0 \delta i \tag{7.84}$$

Substituting Eq. (7.76) into Eq. (7.75) and equating to Eq. (7.69) gives

$$x = \delta r \tag{7.85}$$

$$y = r_0 \left(\delta \theta + \delta \Omega \cos i_0 \right) \tag{7.86}$$

$$z = r_0 \left(\delta i \sin \theta_0 - \delta \Omega \sin i_0 \cos \theta_0 \right) \tag{7.87}$$

where

$$r_0 = \frac{a_0(1 - e_0^2)}{1 + e_0 \cos f_0} = \frac{a_0 \left(1 - q_{10}^2 - q_{20}^2 \right)}{1 + q_{10} \cos \theta_0 + q_{20} \sin \theta_0} \tag{7.88}$$

$$\delta r = \left(\frac{r_0}{a_0} \right) \delta a - 2r_0 \left(\frac{a_0}{p_0} \right) (q_{10} \delta q_1 + q_{20} \delta q_2)$$
$$- \frac{r_0^2}{p_0} [\delta q_1 \cos \theta_0 + \delta q_2 \sin \theta_0 + (q_{20} \cos \theta_0 - q_{10} \sin \theta_0) \, \delta \theta] \tag{7.89}$$

Similarly,

$$[\mathbf{v}_1]_{\mathscr{C}} = T^{\mathscr{C}}_{\mathscr{J}} T^{\mathscr{J}}_{\mathscr{D}} \begin{pmatrix} v_{r1} \\ v_{t1} \\ v_{n1} \end{pmatrix}$$

$$[\mathbf{v}_1]_{\mathscr{C}} = T^{\mathscr{C}}_{\mathscr{J}} \left(T^{\mathscr{J}}_{\mathscr{C}} + \delta T^{\mathscr{J}}_{\mathscr{C}} \right) \begin{pmatrix} v_{r0J_2} + \delta v_r \\ v_{t0J_2} + \delta v_t \\ v_{n0J_2} + \delta v_n \end{pmatrix}$$

$$[\mathbf{v}_1]_{\mathscr{C}} = \begin{pmatrix} v_{r0J_2} + \delta v_r \\ v_{t0J_2} + \delta v_t \\ v_{n0J_2} + \delta v_n \end{pmatrix} + v_{r0J_2} \begin{pmatrix} 0 \\ \delta\theta + \delta\Omega_0 \cos i_0 \\ \delta i \sin\theta_0 - \delta\Omega \cos\theta_0 \sin i_0 \end{pmatrix} \quad (7.90)$$

$$+ v_{t0J_2} \begin{pmatrix} -\delta\theta - \delta\Omega \cos i_0 \\ 0 \\ \delta i \cos\theta_0 + \delta\Omega \sin\theta_0 \sin i_0 \end{pmatrix}$$

$$+ v_{n0J_2} \begin{pmatrix} -\delta i \sin\theta_0 + \delta\Omega_0 \cos\theta_0 \sin i_0 \\ -\delta i \cos\theta_0 - \delta\Omega \sin\theta_0 \sin i_0 \\ 0 \end{pmatrix}$$

The velocity components are now divided into two parts: Those not depending directly on J_2 and those depending on J_2. The first part is expressed in terms of the orbital elements, and has the same form as the unperturbed motion. The second part, denoted by Δ, is the variation caused by only J_2:

$$v_{jJ_2} = v_j + \Delta v_j, \quad \delta v_{jJ_2} = \delta v_j + \delta\Delta v_j, \quad j = r, t, n \quad (7.91)$$

From these equations, the relationship between the state $\mathbf{x}(t)$ and $\delta\boldsymbol{\alpha}(t)$ is

$$\begin{pmatrix} x \\ y \\ z \end{pmatrix} = \begin{pmatrix} \left(\dfrac{r_0}{a_0}\right)\delta a + \left(\dfrac{r_0^2}{p_0}\right)(q_{10}\sin\theta_0 - q_{20}\cos\theta_0)\,\delta\theta \\ -\left(\dfrac{r_0}{p_0}\right)[(2a_0 q_{10} + r_0\cos\theta_0)\,\delta q_1 - (2a_0 q_{20} + r_0\sin\theta_0)\,\delta q_2] \\ r_0\,(\delta\theta + \delta\Omega\cos i_0) \\ r_0\,(\delta i \sin\theta_0 - \delta\Omega \sin i_0 \cos\theta_0) \end{pmatrix} \quad (7.92)$$

$$\begin{pmatrix} \dot{x} \\ \dot{y} \\ \dot{z} \end{pmatrix} = \begin{pmatrix} \delta v_r \\ \delta v_t \\ \delta v_n \end{pmatrix} + v_{r0} \begin{pmatrix} 0 \\ \delta\theta + \delta\Omega \cos i_0 \\ \delta i \sin\theta_0 - \delta\Omega \sin i_0 \cos\theta_0 \end{pmatrix}$$

$$+ v_{t0} \begin{pmatrix} -\delta\theta - \delta\Omega \cos i_0 \\ 0 \\ \delta i \cos\theta_0 + \delta\Omega \sin i_0 \sin\theta_0 \end{pmatrix}$$

$$+ v_{n0} \begin{pmatrix} -\delta i \sin\theta_0 + \delta\Omega \sin i_0 \cos\theta_0 \\ -\delta i \cos\theta_0 - \delta\Omega \sin i_0 \sin\theta_0 \\ 0 \end{pmatrix}$$

$$+ \begin{pmatrix} \delta\Delta v_r \\ \delta\Delta v_t \\ \delta\Delta v_n \end{pmatrix} + \varpi_{n0} \begin{pmatrix} y \\ -x \\ 0 \end{pmatrix} + \varpi_{r0} \begin{pmatrix} 0 \\ z \\ -y \end{pmatrix} + \varpi_{t0} \begin{pmatrix} -z \\ 0 \\ x \end{pmatrix}$$

$$+ \Delta v_{\theta 0} \begin{pmatrix} 0 \\ \delta\theta + \delta\Omega \cos i_0 \\ \delta i \sin\theta_0 - \delta\Omega \sin i_0 \cos\theta_0 \end{pmatrix}$$

$$+ \Delta v_{r0} \begin{pmatrix} 0 \\ \delta\theta + \delta\Omega \cos i_0 \\ \delta i \sin\theta_0 - \delta\Omega \sin i_0 \cos\theta_0 \end{pmatrix}$$

$$+ \delta v_{t0} \begin{pmatrix} -\delta\theta - \delta\Omega \cos i_0 \\ 0 \\ \delta i \cos\theta_0 + \delta\Omega \sin i_0 \sin\theta_0 \end{pmatrix} \tag{7.93}$$

The angular velocity for osculating elements is

$$\varpi = \left(\dot\Omega_0 \sin\theta_0 \sin i_0 + (di_0/dt)\cos\theta_0\right)\hat{\mathbf{x}}$$
$$+ \left(\dot\Omega_0 \cos\theta_0 \sin i_0 - (di_0/dt)\sin\theta_0\right)\hat{\mathbf{y}} + \left(\dot\theta_0 + \dot\Omega_0 \cos i_0\right)\hat{\mathbf{z}} \tag{7.94}$$

With osculating elements, the orbit plane is defined by the position and velocity, so the velocity of the chief must be in the reference plane; therefore, the $\hat{\mathbf{z}}$-component of the osculating angular velocity must be zero. This yields

$$\dot\Omega_0 \cos\theta_0 \sin i_0 = \frac{di_0}{dt}\sin\theta_0 \tag{7.95}$$

and the osculating angular velocity becomes

$$\varpi = \left(\dot\Omega_0 \sin i_0 / \sin\theta_0\right)\hat{\mathbf{x}} + \left(\dot\theta_0 + \dot\Omega_0 \cos i_0\right)\hat{\mathbf{z}} \tag{7.96}$$

The angle rates are needed to obtain the $\Sigma(t)$ matrix. They are obtained from the Hamiltonian, Eqs. (3.10) and (3.18), and the GVE of Section 2.5.4:

$$\frac{di_0}{dt} = -A_2\left(\frac{n_0}{4a_0^2 n_0^7}\right)(1 + q_{10}\cos\theta_0 + q_{20}\sin\theta_0)\sin(2i_0)\sin(2\theta_0) \tag{7.97}$$

$$\dot\Omega_0 = -A_2\left(\frac{n_0}{4a_0^2 n_0^7}\right)(1 + q_{10}\cos\theta_0 + q_{20}\sin\theta_0)\cos i_0 \sin^2\theta_0 \tag{7.98}$$

The angular momentum perpendicular to the orbit plane provides the last angular rate:

$$h_0 = r_0^2 \varpi_{n0} = r_0^2\left(\dot\theta_0 + \dot\Omega_0 \cos i_0\right) = \sqrt{\mu p_0}$$

$$\dot\theta_0 = \sqrt{\frac{\mu}{p_0^3}}(1 + q_{10}\cos\theta_0 + q_{20}\sin\theta_0)^2 - \dot\Omega_0 \cos i_0 \tag{7.99}$$

The velocity components are

$$
v_{r0} = \dot{r}_0 = \sqrt{\mu/p_0} \, (q_{10} \sin\theta_0 - q_{20} \cos\theta_0)
$$
$$
v_{t0} = r_0 \varpi_{n0} = \sqrt{\mu/p_0} \, (1 + q_{10} \cos\theta_0 + q_{20} \sin\theta_0)
$$
$$
v_{n0} = 0
$$
$$
\Delta v_{j0} = \delta \Delta v_{j0} = 0, \quad j = r, t, n
$$

(7.100)

The last quantities needed are the variations of the velocity components, which are

$$
\delta v_r = -\left[\left(\frac{n_0}{2\eta_0^3} \right) \delta a - \left(\frac{3 n_0 a_0}{\eta_0^4} \right) \delta\eta \right] (q_{10} \sin\theta_0 - q_{20} \cos\theta_0)
$$
$$
+ \left(\frac{a_0 n_0}{\eta_0^3} \right) [\delta q_1 \sin\theta_0 - \delta q_2 \cos\theta_0 + (q_{10} \cos\theta_0 + q_{20} \sin\theta_0) \delta\theta] \quad (7.101)
$$
$$
\delta v_t = -\left[\left(\frac{n_0}{2\eta_0^3} \right) \delta a - \left(\frac{3 n_0 a_0}{\eta_0^4} \right) \delta\eta \right] (1 + q_{10} \cos\theta_0 + q_{20} \sin\theta_0)
$$
$$
+ \left(\frac{a_0 n_0}{\eta_0^3} \right) [\delta q_1 \cos\theta_0 + \delta q_2 \sin\theta_0 + (-q_{10} \sin\theta_0 + q_{20} \cos\theta_0) \delta\theta] \quad (7.102)
$$

Substituting these quantities into Eqs. (7.92) and (7.93) yields the matrices $\Sigma(t)$ and $\Sigma^{-1}(t)$, the elements of which are given in Appendices A and B, respectively.

THE TRANSFORMATION MATRIX $\bar{\Sigma}(t) = [\bar{A}(t) + A_2 \bar{B}(t)]$

Since the matrix $A(t)$ is a function of the osculating elements without J_2 effects, it is functionally equal to $\bar{A}(t)$ with mean elements being substituted for osculating elements, that is, $\bar{A} = A(\bar{\alpha})$. In addition to mean elements being used, the constraint that the velocity be in the orbit plane no longer applies. Thus, the angular velocity is different and the velocity components affected by J_2 are different. From Eqs. (7.42), the secular angle rates, denoted by the superscript $(\cdot)^{(s)}$, are

$$
\frac{di_0^{(s)}}{dt} = 0
$$

(7.103)

$$
\dot{\Omega}_0^{(s)} = -0.5 A_2 \left(\frac{n_0}{a_0^2 \eta_0^4} \right) \cos i_0
$$

(7.104)

$$
\dot{\omega}_0^{(s)} = -0.25 A_2 \left(\frac{n_0}{a_0^2 \eta_0^4} \right) (1 - 5 \cos^2 i_0)
$$

(7.105)

$$
\dot{M}_0^{(s)} = n_0 + 0.25 A_2 \left(\frac{n_0}{a_0^2 \eta_0^3} \right) (2 - 3 \sin^2 i_0)
$$

(7.106)

The mean argument of latitude rate is given by Eq. (7.99), but using $\dot{\Omega}_0^{(s)}$ instead of the osculating rate. The mean angular velocity is

$$\overline{\varpi} = \left(\dot{\Omega}_0^{(s)} \sin \theta_0 \sin i_0 \right) \hat{x} + \left(\dot{\Omega}_0^{(s)} \cos \theta_0 \sin i_0 \right) \hat{y}$$
$$+ \left(\dot{\theta}_0^{(s)} + \dot{\Omega}_0^{(s)} \cos i_0 \right) \hat{z} \tag{7.107}$$

The variations of the velocity components due to J_2 are not zero, and are given by

$$\Delta v_r = - \left(\frac{r_0^2}{p_0} \right) (q_{10} \sin \theta_o - q_{20} \cos \theta_0) \dot{\omega}_0^{(s)}$$
$$\Delta v_t = r_0 \dot{\Omega}_0^{(s)} \cos i_0 \tag{7.108}$$
$$\Delta v_n = -r_0 \dot{\Omega}_0^{(s)} \sin i_0 \cos \theta_0$$

Substituting Eqs. (7.108) and their variation into Eqs. (7.92) and (7.93) gives the matrix $\bar{B}(t)$, whose elements are given in Appendix C.

THE MEAN ELEMENT STATE TRANSITION MATRIX $\bar{\phi}_{\overline{\alpha}}$

With only J_2 perturbations, the mean elements are

$$a = a(0), \quad i = i(0), \quad e = e(0)$$
$$q_1 = e \cos \omega = q_1(0) \cos \left[\dot{\omega}^{(s)} (t - t_0) \right] + q_{20} \sin \left[\dot{\omega}^{(s)} (t - t_0) \right]$$
$$q_2 = e \sin \omega = q_1(0) \sin \left[\dot{\omega}^{(s)} (t - t_0) \right] - q_{20} \cos \left[\dot{\omega}^{(s)} (t - t_0) \right] \tag{7.109}$$
$$\Omega = \Omega(0) + \dot{\Omega}^{(s)} (t - t_0)$$

Now, define the mean argument of latitude, λ, and the eccentric argument of latitude, F:

$$\lambda = M + \omega, \quad F = E + \omega \tag{7.110}$$

From Kepler's equation, $M = E - e \sin E$ (Eq. (2.26)), the *modified Kepler equation* for λ and F reads

$$\lambda = F - q_1 \sin F + q_2 \cos F \tag{7.111}$$

From Ref. [75], the relationship between the true and eccentric arguments of latitude is given by

$$\tan F = \frac{r \left(1 + \beta q_1^2 \right) \sin \theta - \beta q_1 q_2 \cos \theta + a q_1}{r \left(1 + \beta q_2^2 \right) \cos \theta - \beta q_1 q_2 \sin \theta + a q_2} \tag{7.112}$$

where

$$\beta = \frac{1}{\left[(1 - e^2) + \sqrt{1 - e^2}\right]} = \frac{1}{\eta(1 + \eta)} \tag{7.113}$$

The relationship between λ and θ becomes

$$\lambda = F - \frac{\sqrt{1 - q_1^2 - q_2^2}\,(q_1 \sin \theta - q_2 \cos\theta)}{(1 + q_1 \cos \theta + q_2 \sin \theta)} \tag{7.114}$$

with F given by Eq. (7.112). Now, define

$$G \equiv \lambda - \lambda(t_0) = M(0) + \omega(0) + \left(\dot{M}^{(s)} + \dot{\omega}^{(s)}\right)(t - t_0) \tag{7.115}$$

We now have all the elements of $\overline{\text{œ}}$ as a function of time. The state transition matrix, $\phi_{\overline{\text{œ}}}(t, t_0)$ of $\delta\overline{\text{œ}}(t)$, is obtained by expanding these equations in a Taylor series about the chief elements, $\overline{\text{œ}}_0(t)$. The elements of $\phi_{\overline{\text{œ}}}(t, t_0)$ are given in Appendix D.

MEAN-TO-OSCULATING TRANSFORMATION

As we have seen in Section 3.4, the first-order mean-to-osculating transformation is given by

$$\text{œ} = \overline{\text{œ}} + \epsilon\,\{\overline{\text{œ}}, W_1\} \tag{7.116}$$

where $\{\overline{\text{œ}}, W_1\}$ is the Poisson bracket. Usually, the mean-to-osculating transformation is performed by two transformations, mean-to-long-periodic using $W_1^{(lp)}$, and then long-periodic-to-osculating using $W_1^{(sp)}$. However, since we are only performing the first-order transformation, it can be performed as a single transformation, and the differences are of $\mathcal{O}(J_2^2)$. For convenience, the short-periodic generating function is separated into two parts: $W_1^{(sp1)}$, which is a function of only the true anomaly, and $W_1^{(sp2)}$, which is a function of the true anomaly and the argument of perigee. From Section 3.4, the generating functions are

$$W_1^{(lp)} = -\left(\frac{1}{32G^3}\right)\left(1 - \frac{G^2}{L^2}\right)\left(1 - 5\frac{H^2}{G^2}\right)^{-1}$$
$$\times \left(1 - 16\frac{H^2}{G^2} + 15\frac{H^4}{G^4}\right)\sin 2g \tag{7.117}$$

and

$$W_1^{(sp1)} = -\frac{1}{4G^3}\left(1 - 3\frac{H^2}{G^2}\right)(f - l + e\sin f) \tag{7.118}$$

$$W_1^{(sp2)} = \frac{3}{8G^3}\left(1 - \frac{H^2}{G^2}\right)$$
$$\times \left[\sin(2f + 2g) + e\sin(f + 2g) + \frac{e}{3}\sin(3f + 2g)\right] \quad (7.119)$$

We define

$$\text{œ} = \overline{\text{œ}} - J_2 R_e^2 \left(\text{œ}^{(lp)} + \text{œ}^{(sp1)} + \text{œ}^{(sp2)}\right) \quad (7.120)$$

where

$$\text{œ}^{(lp)} = \frac{1}{R_e^2}\{W_1^{(lp)}, \overline{\text{œ}}\} \quad (7.121)$$

$$\text{œ}^{(sp1)} = \frac{1}{R_e^2}\{W_1^{(sp1)}, \overline{\text{œ}}\} \quad (7.122)$$

$$\text{œ}^{(sp2)} = \frac{1}{R_e^2}\{W_1^{(sp2)}, \overline{\text{œ}}\} \quad (7.123)$$

The generating functions, Eqs. (7.117)–(7.119), depend upon normalized variables. To perform the above Poisson bracket operations, they need to be transformed back to regular variables. The mean-to-osculating transformations for the elements of œ are given in Appendix E.

The long-periodic terms appear as secular terms of $\mathcal{O}(J_2^2)$ for periods of time much less than the argument of perigee rotation period, T_{pr}. Since the second-order secular terms have not been considered, the long-periodic terms could probably be dropped in the mean-to-osculating transformation if the prediction time is small compared to T_{pr}, say $0.1T_{pr}$. In addition, note the term $(1 - G^2/L^2) = e^2$ in $W^{(lp)}$. This means that all the long-periodic terms in the mean-to-osculating transformation are multiplied by e or e^2. Therefore, for near-circular orbits, the long-periodic terms in the mean-to-osculating transformation could be ignored.

THE D MATRIX

The D matrix is given by

$$D = \frac{\partial \text{œ}}{\partial \overline{\text{œ}}} \quad (7.124)$$

Using Eq. (7.116) we get

$$D = I + \epsilon \frac{\partial\{W_1^{(lp)} + W_1^{(sp1)} + W_1^{(sp2)}, \overline{\text{œ}}\}}{\partial \overline{\text{œ}}} \quad (7.125)$$

Define

$$D = I - J_2 R_e^2 \left(D^{(lp)} + D^{(sp1)} + D^{(sp2)}\right) \quad (7.126)$$

where

$$D^{(lp)} = \frac{\partial\{W_1^{(lp)}, \overline{\text{œ}}\}}{\partial\overline{\text{œ}}} \tag{7.127}$$

$$D^{(sp1)} = \frac{\partial\{W_1^{(sp1)}, \overline{\text{œ}}\}}{\partial\overline{\text{œ}}} \tag{7.128}$$

$$D^{(sp2)} = \frac{\partial\{W_1^{(sp2)}, \overline{\text{œ}}\}}{\partial\overline{\text{œ}}} \tag{7.129}$$

The elements of $D^{(lp)}$, $D^{(sp1)}$ and $D^{(sp2)}$ are given in Appendix F.

NUMERICAL EVALUATION

To evaluate the Geometric Method, the predicted relative motion is compared with the results obtained by a numerical integration, performed according to the following stages: (i) integrating the equations of motion for both satellites in an ECI frame with a $J_2 - J_5$ gravity field; (ii) transforming the position and velocity obtained in the ECI frame to the chief LVLH frame; (iii) transforming the position and velocity into the curvilinear frame; and (iv) differencing the position and velocity to obtain the relative state. The initial conditions are chosen such that the projection of the relative orbit in the horizontal plane is a 0.5-km radius circle when the chief orbit is circular with the same semimajor axis. This means that the CW equations will predict a PCO with a radius of 500 m. The out-of-plane motion is created by a differential inclination in order to create secular out-of-plane drift. The eccentricity is chosen to be 0.1. Tables 7.1 and 7.2 give the respective initial conditions of the chief and the deputy, which were selected for the numerical evaluation.

Figure 7.2 shows the three-dimensional relative motion predicted by the CW equations, numerical integration and the Geometric Method for one day. Note

Table 7.1 Chief initial conditions

Element	a (km)	θ (deg)	i (deg)	q_1	q_2	Ω (deg)
Osculating	8500	170	70	9.397×10^{-2}	3.420×10^{-2}	45
Mean	8494.549	170.003	69.9929	9.420×10^{-2}	3.407×10^{-2}	45.006

Table 7.2 Deputy osculating initial conditions

δa (m)	$\delta\theta$ (deg)	δi (deg)	δq_1	δq_2	$\delta\Omega$ (deg)
-103.624	-1.104×10^{-3}	7.076×10^{-4}	4.262×10^{-5}	-9.708×10^{-6}	3.227×10^{-3}

x (m)	y (m)	z (m)	$\dot{x}$ (m/s)	$\dot{y}$ (m/s)	$\dot{z}$ (m/s)
250	0	500	0	-0.403	0

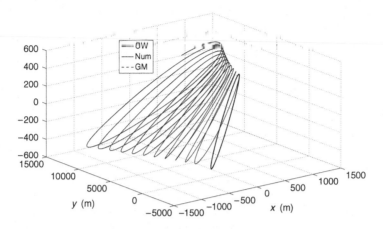

FIGURE 7.2 3D relative trajectory.

that the CW equations predict a periodic trajectory, but the actual motion is much different. To the scale of the figure, the numerical and GM trajectories are the same. Figure 7.3 shows the along-track, radial and out-of-plane time histories. The CW time histories are periodic, because the initial conditions were set up to result in a PCO. As with the 3D plot, the difference in the numerical integration and the GM is not evident with the scale in the plot. Those differences are presented in the next set of plots. The along-track plot shows the secular growth that occurs when not including J_2 and the chief eccentricity in the model. There is out-of-plane secular growth, but over the time period of one day it is only about 10 m, so it does not show up. Thus, the out-of-plane time histories appear equal. The errors in the osculating relative state in the GM are shown in Figs. 7.4 and 7.5. The position errors are less than 2 m and the velocity errors are less than 2 mm/s after one day. The errors in the mean state are shown in Figs. 7.6 and 7.7. The maximum mean state errors are greater than the osculating state errors, because the mean error is based on the true osculating state; thus, the larger mean error is due to the unmodeled short-periodic variations. However, the errors are still small. More detailed error evaluations can be found in Ref. [131].

The mean to osculating transformation (see Appendix E), contains the term $\Theta = \left(1 - 5\cos^2 i\right)^{-1}$. Thus, there is a singularity at $\cos^2 i = 1/5, i = 63.435°, 116.565°$. These inclinations are called the *critical inclinations*. This means that the transformation is not valid near the critical inclination. Figure 7.8 shows the along-track error as a function of the inclination for various values of the eccentricity. The error increases with eccentricity, because the long-periodic terms are multiplied by e or e^2. One can conclude from this figure that there is no problem until the inclination is within 0.2 degrees of the critical inclination. Details on the performance of the GM near the critical inclination are given in Ref. [131]. Theoretically, a new theory needs to be developed. However, what is often done is to set $(1 - 5\cos^2 i) = \varepsilon\text{sign}(1 - 5\cos^2 i)$ whenever $|1 - 5\cos^2 i| < \varepsilon$, where ε is a small number, e.g., 0.05.

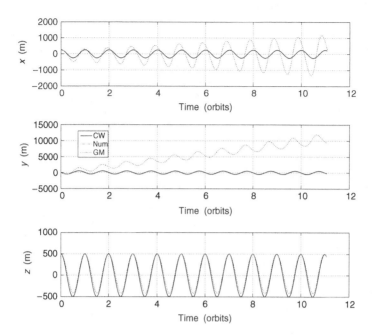

FIGURE 7.3 Position time histories.

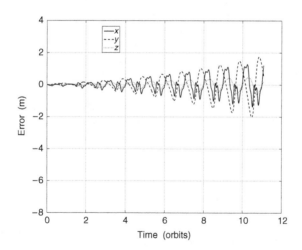

FIGURE 7.4 Geometric method osculating position errors.

7.4 AVERAGED RELATIVE MOTION

The previous developments in this chapter made use of the mean elements extensively. One can, for convenience, define and compute mean motion variables from the differential mean elements instead of the differential osculating elements. However, since the averages of the short-periodic variations in the elements do not necessarily equal zero, errors of $\mathcal{O}(J_2)$ can result due to a

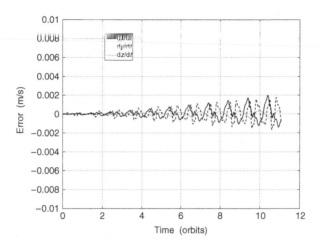

FIGURE 7.5 Geometric method osculating velocity errors.

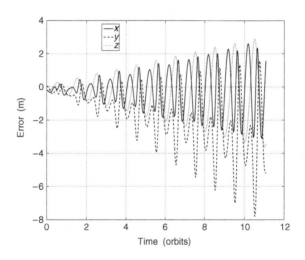

FIGURE 7.6 Geometric method mean position errors.

direct substitution of the differential mean elements into the expressions for the motion variables. Sengupta et al. [125] have obtained corrections to the mean motion variables, resulting in the so-called "averaged" expressions for the relative motion variables. This analytical filtering process of removing the short-periodic oscillations, while simultaneously accounting for their orbit-averaged values, is discussed in this section.

The relationship between the jth osculating element and its short-periodic variation can be written as [132]

$$\mathbf{æ}_j{}^{(sp)} = -J_2 R_e^2 \{\overline{\mathbf{æ}}_j, W^{(sp)}\} \tag{7.130}$$

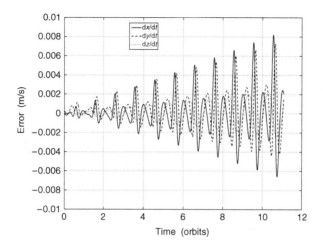

FIGURE 7.7 Geometric method mean velocity errors.

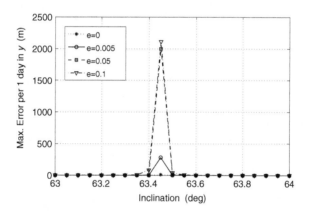

FIGURE 7.8 Along-track errors near the critical inclination.

where $\{\cdot, \cdot\}$ is the Poisson bracket as defined by Eq. (2.93), and $W^{(sp)}$ is the short-periodic generating function (3.22). In a more explicit form, the short-periodic variation of the jth element can be written as

$$
\begin{aligned}
\text{œ}_j^{(sp)} = -J_2 R_e^2 \Bigg[&\left(\frac{\partial \overline{\text{œ}}_j}{\partial l}\right)\left(\frac{\partial W^{(sp)}}{\partial L}\right) + \left(\frac{\partial \overline{\text{œ}}_j}{\partial g}\right)\left(\frac{\partial W^{(sp)}}{\partial G}\right) \\
&+ \left(\frac{\partial \overline{\text{œ}}_j}{\partial \hbar}\right)\left(\frac{\partial W^{(sp)}}{\partial H}\right) - \left(\frac{\partial \overline{\text{œ}}_j}{\partial L}\right)\left(\frac{\partial W^{(sp)}}{\partial l}\right) \\
&- \left(\frac{\partial \overline{\text{œ}}_j}{\partial G}\right)\left(\frac{\partial W^{(sp)}}{\partial g}\right) - \left(\frac{\partial \overline{\text{œ}}_j}{\partial H}\right)\left(\frac{\partial W^{(sp)}}{\partial \hbar}\right) \Bigg]
\end{aligned}
\tag{7.131}
$$

Sengupta et al. [125] show that Eq. (7.130) can also be used for functions of the mean elements. We present samples of these results here:

$$r^{(sp)} = \bar{a}\,\bar{\eta}^2 \left[(1 - 3\cos^2 \bar{i}) \left(\frac{\bar{\eta} + \bar{\zeta}}{4(1 + \bar{\eta})} + \frac{\bar{\eta}}{2\bar{\zeta}} \right) + \frac{1}{4}\sin^2 \bar{i}\,\cos 2\bar{\theta} \right] \quad (7.132a)$$

$$\begin{aligned}
\theta^{(sp)} &= \frac{\bar{\xi}}{2} - \frac{3}{4}(1 - 5\cos^2 \bar{i})(\bar{\theta} - \bar{\lambda}) - \frac{1}{4}(1 - 3\cos^2 \bar{i})\frac{\bar{\xi}}{(1 + \bar{\eta})}(\bar{\zeta} + 4\bar{\eta} + 5) \\
&\quad - \frac{1}{4}\cos^2 \bar{i}\,(\bar{q}_1 \sin 3\bar{\theta} - \bar{q}_2 \cos 3\bar{\theta}) + \frac{1}{8}(1 - 7\cos^2 \bar{i})\sin 2\bar{\theta} \\
&\quad + \frac{1}{4}(2 - 5\cos^2 \bar{i})(\bar{q}_1 \sin \bar{\theta} + \bar{q}_2 \cos \bar{\theta}) \quad (7.132b)
\end{aligned}$$

where the two new variables ζ and ξ are functions of q_1 and q_2 and are defined as

$$\zeta + j\xi = 1 + (q_1 - j\,q_2)\exp(j\theta) \quad (7.133)$$

and $j = \sqrt{-1}$. Short-periodic corrections are applied to the expressions obtained directly from the mean elements, as shown for the case of r:

$$\bar{r} = \frac{\bar{a}\bar{\eta}^2}{\bar{\zeta}} \quad (7.134a)$$

$$r = \bar{r} + Jr^{(sp)} \quad (7.134b)$$

where

$$J = J_2 \left(\frac{R_e}{\bar{a}\bar{\eta}^2} \right)^2$$

The averaged orbital-elements are obtained by adding the orbit averaged short-periodic variations to the respective mean elements, as follows:

$$\widehat{\text{œ}}_j = \overline{\text{œ}}_j + \frac{1}{2\pi} \int_0^{2\pi} \text{œ}_j^{(sp)}\, dM \quad (7.135)$$

where the averaging is performed with respect to the mean anomaly. Equation (7.135) can also be used for any function of the orbital elements. For example, the averaged perturbed radius of a satellite in a mean circular orbit can be obtained from Eqs. (7.132a) and (7.135) as

$$\widehat{r}\,|_{\bar{q}_1 = \bar{q}_2 = 0} = \bar{a}\left[1 + \frac{3}{4}J(1 - 3\cos^2 \bar{i}) \right] \quad (7.136)$$

In a similar manner, starting with the mean differential element descriptions of the relative motion, Sengupta et al. obtained the following expressions by averaging the short-periodic effects:

$$\frac{\widehat{x}(t)}{\bar{a}_0} = \left[1 - \frac{3}{4}J\left(1 - 3\cos^2\bar{i}_0\right)\right]\frac{\delta\bar{a}}{\bar{a}_0} - \cos(\bar{\lambda}_0 - \dot{\bar{\omega}}_0 t)\delta\bar{q}_1(0)$$
$$- \sin(\bar{\lambda}_0 - \dot{\bar{\omega}}_0 t)\delta\bar{q}_2(0) + \frac{9}{4}J\sin 2\bar{i}_0\delta\bar{i} \qquad (7.137)$$

$$\frac{\widehat{y}(t)}{\bar{a}_0} = \left[-\frac{3}{2} + \frac{33}{8}J\left(1 - 3\cos^2\bar{i}_0\right)\right](\bar{n}_0 t)\frac{\delta\bar{a}}{\bar{a}_0}$$
$$+ \left[1 + \frac{3}{4}J(1 - 3\cos^2\bar{i}_0)\right]\delta\bar{\lambda}(0)$$
$$- \frac{21}{4}J\sin 2\bar{i}_0(\bar{n}_0 t)\delta\bar{i} + 2\sin(\bar{\lambda}_0 - \dot{\bar{\omega}}_0 t)\delta\bar{q}_1(0)$$
$$- 2\cos(\bar{\lambda}_0 - \dot{\bar{\omega}}_0 t)\delta\bar{q}_2(0)$$
$$+ \left[1 + \frac{3}{4}J\left(1 - 3\cos^2\bar{i}_0\right)\right]\cos\bar{i}_0\delta\bar{\Omega}(0) \qquad (7.138)$$

$$\frac{\widehat{z}(t)}{\bar{a}_0} = -\frac{21}{8}J\sin 2\bar{i}_0\cos\bar{\lambda}_0(\bar{n}_0 t)\frac{\delta\bar{a}}{\bar{a}_0}$$
$$+ \left[\sin\bar{\lambda}_0 - \frac{3}{2}J\sin^2\bar{i}_0\cos\bar{\lambda}_0(\bar{n}_0 t)\right]\delta\bar{i}$$
$$- \cos\bar{\lambda}_0\sin\bar{i}_0\delta\bar{\Omega}(0) - \frac{3}{4}J\sin 2\bar{i}_0(\delta\bar{q}_1(0) - \delta\bar{q}_2(0)) \quad (7.139)$$

The differences between the averaged and mean-element-based expressions may be small, but are very important nonetheless for long-term *formation maintenance*. An alternative along-track boundedness condition can be derived from Eq. (7.138) for mean circular orbits by setting the secular growth terms therein to zero:

$$\frac{\delta\bar{a}}{\bar{a}_0} = -\frac{7}{2}J_2\left(\frac{R_e}{\bar{a}_0}\right)^2\sin 2\bar{i}_0\delta\bar{i} + \mathcal{O}(J_2^2) \qquad (7.140)$$

Equation (7.140) shows that, in general, $\delta a \neq 0$ for locally bounded along-track motion.

7.5 LINEARIZED J_2-DIFFERENTIAL EQUATIONS FOR CIRCULAR ORBITS

In this section, a set of linearized equations for perturbed motion are presented. They are derived by using the mean orbital elements, their mean drift rates due to J_2, and the contributions from the short-periodic effects. A set of non-linear differential equations for evolving perturbed relative motion dynamics directly in the LVLH frame of the chief has been derived by Kechichian [133]. Kechichian's equations utilize a dragging and precessing reference frame to account for the effects of drag and J_2, and they are quite complex.

The GA STM provides a means for propagating the states (position and velocity vectors) of relative motion for elliptic reference orbits in a curvilinear coordinate system, However, the expressions for building the STM involve both the short- and long-periodic effects of J_2. The long-periodic terms can be ignored for simplifying the STM, especially for prediction times less than the perigee rotation period and for an $\mathcal{O}(J_2)$ approximation. For mean circular orbits, this is not an approximation, since there are no long-periodic terms in the osculating elements. Hamel and Lafontaine [134] consider such an alternative for propagating the differential orbital elements, but include the short-period effects only for the semimajor axis; they ignore the short-period effects on the other elements. However, since they deal with the classical elements, their fundamental matrix suffers from a singularity for zero mean eccentricity of the reference orbit. Short-periodic corrections to the mean orbital elements for near-circular orbits can be found in Born et al. [135]. Sengupta et al. [125] derive such corrections for elliptic orbits using nonsingular elements. These results have been applied in Ref. [136] for the development of a set of linearized relative motion equations with respect to a mean circular reference orbit. The structure of the model developed is similar to that of Ref. [137], the differences are in the short-periodic approximations utilized. The details of the linear model are presented next.

7.5.1 Development of the model

The relative position vector of a deputy defined in the chief's LVLH frame is denoted by

$$\rho = [x \; y \; z]^T \tag{7.141}$$

and the angular velocity vector of the LVLH frame is

$$\omega = \left[\omega_x \; \omega_y \; \omega_z\right]^T \tag{7.142}$$

with

$$\omega_x = \dot{\Omega}_0 \sin i_0 \sin \theta_0 + \dot{i}_0 \cos \theta_0 \tag{7.143a}$$
$$\omega_y = \dot{\Omega}_0 \sin i_0 \cos \theta_0 - \dot{i}_0 \sin \theta_0 = 0 \tag{7.143b}$$
$$\omega_z = \dot{\Omega}_0 \cos i_0 + \dot{\theta}_0 \tag{7.143c}$$

where Ω_0 is the longitude of the ascending node, θ_0 is the argument of latitude, and i_0 is the inclination of the reference orbit. Equations (7.143a)–(7.143c) involve the osculating elements. The osculating orbit constraint (7.143b) can be verified by substituting the components of **d** from Eq. (7.39) into Eqs. (2.107).

The equations of relative motion can be written compactly [137] as

$$\ddot{\rho} = -2\omega \times \dot{\rho} - \omega \times (\omega \times \rho) - \dot{\omega} \times \rho + \nabla F_{g2B} + \nabla F_{gJ_2} \tag{7.144}$$

where ∇F_{g2B} is the gravity gradient acceleration due to the two-body gravity field and ∇F_{gJ_2} is that due to the J_2 potential. The two-body gravity gradient acceleration, expressed in the LVLH frame (see Eqs. (4.14)–(4.16) and

Ref. [106]), is

$$\nabla F_{g2B} = -\mu \begin{bmatrix} \dfrac{(r_0 + x)}{[(r_0+x)^2 + y^2 + z^2]^{\frac{3}{2}}} - \dfrac{1}{r_0^2} \\[2ex] \dfrac{y}{[(r_0+x)^2 + y^2 + z^2]^{\frac{3}{2}}} \\[2ex] \dfrac{z}{[(r_0+x)^2 + y^2 + z^2]^{\frac{3}{2}}} \end{bmatrix} \tag{7.145}$$

where μ is the gravitational parameter and r_0 is the radius of the chief's orbit. Equation (7.145) can be linearized about r_0 to obtain

$$\nabla F_{g2B} \approx -\frac{\mu}{r_0^3} \begin{bmatrix} -2x \\ y \\ z \end{bmatrix} \tag{7.146}$$

The linearized J_2 differential acceleration vector, obtained from Eq. (7.39), is [137]

$$\nabla F_{gJ_2} = \Upsilon \begin{bmatrix} 1 - 3s_{i_0}^2 s_{\theta_0}^2 & s_{i_0}^2 s_{2\theta_0} & s_{2i_0} s_{\theta_0} \\[1ex] s_{i_0}^2 s_{2\theta_0} & s_{i_0}^2 \left(\frac{7}{4}s_{\theta_0}^2 - \frac{1}{2}\right) - \frac{1}{4} & -\frac{1}{4} s_{2i_0} c_{\theta_0} \\[1ex] s_{2i_0} s_{\theta_0} & -\frac{1}{4} s_{2i_0} c_{\theta_0} & s_{i_0}^2 \left(\frac{5}{4}s_{\theta_0}^2 + \frac{1}{2}\right) - \frac{3}{4} \end{bmatrix} \rho \tag{7.147}$$

where s and c stand for sin and cos, respectively and

$$\Upsilon = 6J_2 \left(\frac{\mu R_e^2}{r_0^5} \right)$$

7.5.2 Short-periodic effects

The relationships between the osculating and mean elements for mean circular orbits can be obtained from the detailed expressions given in Ref. [132]. These expressions are also available in Ref. [125] in a convenient form. The relevant variables for the case of circular orbits, derived from those given in Ref. [125], are

$$r_0 = \bar{a}_0 \left[1 + J \left\{ \frac{3}{4}(1 - 3\cos^2 \bar{i}_0) + \frac{1}{4} \sin^2 \bar{i}_0 \, \cos 2\bar{\theta}_0 \right\} \right] \tag{7.148a}$$

$$\theta_0 = \bar{\theta}_0(0) + \dot{\bar{\theta}}_0 t + \frac{1}{8} J(1 - 7\cos^2 \bar{i}_0 \,) \sin 2\bar{\theta}_0 \tag{7.148b}$$

$$i_0 = \bar{i}_0 + \frac{3}{8} J \sin 2\bar{i}_0 \cos 2\bar{\theta}_0 \tag{7.148c}$$

$$\Omega_0 = \bar{\Omega}_0(0) + \dot{\bar{\Omega}}_0 t + \frac{3}{4} J \cos \bar{i}_0 \sin 2\bar{\theta}_0 \tag{7.148d}$$

where $\bar{\theta}_0 = \bar{\theta}_0(0) + \dot{\bar{\theta}}_0 t$, $J = J_2(\frac{R_e}{\bar{a}_0})^2$, and

$$\dot{\bar{\theta}}_0 = n_0 \left[1 - \frac{3}{2} J(1 - 4\cos^2 \bar{i}_0) \right] \tag{7.149a}$$

$$\dot{\bar{\Omega}}_0 = -\frac{3}{2} J n_0 \cos \bar{i}_0 \tag{7.149b}$$

Substitution of Eqs. (7.148b)–(7.149b) into Eqs. (7.143a)–(7.143c) and the approximation $\theta_0 \approx \bar{\theta}_0$ in the evaluation of the trigonometric functions therein, results in the following expressions for the angular velocities:

$$\omega_x = 2\dot{\bar{\Omega}}_0 \sin \bar{i}_0 \sin \bar{\theta}_0 \tag{7.150a}$$

$$\omega_y = 0 \tag{7.150b}$$

$$\omega_z = \dot{\bar{\Omega}}_0 \cos \bar{i}_0 + \dot{\bar{\theta}}_0 + \frac{1}{4} J n_0 \cos 2\bar{\theta}_0 \sin^2 \bar{i}_0 \tag{7.150c}$$

The mean element approximation is performed by setting $\dot{\Omega}_0 = \dot{\bar{\Omega}}_0$, $\dot{\theta}_0 = \dot{\bar{\theta}}_0$, and $\dot{i}_0 = 0$ in Eqs. (7.143a)–(7.143c). Note that the ω_x expressions, with and without accounting for the short-periodic variations, differ by a factor of two. Equation (7.150b) shows that the approximations developed in this section satisfy the osculating orbit condition Eq. (7.143b). The short-periodic correction to the mean ω_z is obvious from Eq. (7.150c).

7.5.3 The linear model

The relative motion equations are assembled in a matrix form by substituting the relevant expressions derived above into Eq. (7.144), including the linear approximation to ∇F_{g2B} given by Eq. (7.146). The linear model [138] is

$$\begin{bmatrix} \dot{\rho} \\ \ddot{\rho} \end{bmatrix} = \begin{bmatrix} 0 & 0 & 0 & 1 & 0 & 0 \\ 0 & 0 & 0 & 0 & 1 & 0 \\ 0 & 0 & 0 & 0 & 0 & 1 \\ a_{41} & a_{42} & a_{43} & 0 & 2\omega_z & 0 \\ a_{51} & a_{52} & a_{53} & -2\omega_z & 0 & 2\omega_x \\ a_{61} & a_{62} & a_{63} & 0 & -2\omega_x & 0 \end{bmatrix} \begin{bmatrix} \rho \\ \dot{\rho} \end{bmatrix} \tag{7.151}$$

where

$$a_{41} = \omega_z{}^2 + 2\frac{\mu}{r_0{}^3} + \Upsilon(1 - 3\sin^2 \bar{i}_0 \sin^2 \bar{\theta}_0) \tag{7.152a}$$

$$a_{42} = \dot{\omega}_z + \Upsilon(\sin^2 \bar{i}_0 \sin 2\bar{\theta}_0) \tag{7.152b}$$

$$a_{43} = -\omega_x \omega_z + \Upsilon(\sin 2\bar{i}_0 \sin \bar{\theta}_0) \tag{7.152c}$$

$$a_{51} = -\dot{\omega}_z + \Upsilon(\sin^2 \bar{i}_0 \sin 2\bar{\theta}_0) \tag{7.152d}$$

$$a_{52} = \omega_x{}^2 + \omega_z{}^2 - \frac{\mu}{r_0{}^3} + \Upsilon\left[-\frac{1}{4} + \sin^2 \bar{i}_0 \left(\frac{7}{4} \sin^2 \bar{\theta}_0 - \frac{1}{2} \right) \right] \tag{7.152e}$$

$$a_{53} = \dot{\omega}_x + \Upsilon \left(-\frac{1}{4} \sin 2\bar{i}_0 \cos \bar{\theta}_0 \right) \tag{7.152f}$$

$$a_{61} = -\omega_x \omega_z + \Upsilon (\sin 2\bar{i}_0 \sin \bar{\theta}_0) \tag{7.152g}$$

$$a_{62} = -\dot{\omega}_x + \Upsilon \left(-\frac{1}{4} \sin 2\bar{i}_0 \cos \bar{\theta}_0 \right) \approx 0 \tag{7.152h}$$

$$a_{63} = \omega_x{}^2 - \frac{\mu}{r_0{}^3} + \Upsilon \left[-\frac{3}{4} + \sin^2 \bar{i}_0 \left(\frac{5}{4} \sin^2 \bar{\theta}_0 + \frac{1}{2} \right) \right] \tag{7.152i}$$

Equations (7.152) can be simplified further by making several approximations. For example, it can be shown that $a_{62} \approx 0$ to $\mathcal{O}(J_2)$. The variable y is sensitive to the term a_{53}, especially for initial conditions corresponding to a nonzero nodal difference between the two satellites, i.e., $\delta\Omega(0) \neq 0$. This shows the existence of a coupling between cross-track and in-track variables for certain initial conditions. Another observation is that the terms involving $\mu/r_0{}^3$ in Eq. (7.152) can be reduced by using a binomial expansion of $\mathcal{O}(J_2)$. However, this simplification increases the motion propagation errors over the long run, leading to the conclusion that where possible, terms of $\mathcal{O}(J_2{}^2)$ and higher should be retained.

We shall now illustrate the fidelity of the linear model by presenting several examples. Let the reference orbit have the following initial mean nonsingular orbital elements:

$$\bar{a}_0 = 7100 \text{ km}, \quad \bar{\theta}_0 = 0, \quad \bar{i}_0 = 70°$$
$$\bar{q}_{10} = 0, \quad \bar{q}_{20} = 0, \quad \bar{\Omega}_0 = 45°$$

The deputy is set up in a PCO in the examples considered in this section. The initial conditions for setting up such a relative orbit, parameterized by ρ and α, are obtained from Eqs. (6.69) and (7.140) (see also Refs. [1,139]).

Example 7.3. *The following initial conditions on the state variables are obtained for a PCO with $\rho = 0.5$ km and $\alpha = 0$:*

$$\rho(0) = \begin{bmatrix} -0.000288947081 & 0.500033326318 & 0.000175666681 \end{bmatrix}^T \text{ km}$$
$$\dot{\rho}(0) = \begin{bmatrix} 0.000263388377 & 0.000000272412 & 0.000527371445 \end{bmatrix}^T \text{ km/s}$$

The corresponding initial conditions of the chief and deputy in the ECI frame are

$$\mathbf{r}_0(0) = \begin{bmatrix} 5023.558528005 & 5023.558528005 & 0 \end{bmatrix}^T \text{ km}$$
$$\mathbf{v}_0(0) = \begin{bmatrix} -1.810956397226 & 1.810956397226 & 7.041120373157 \end{bmatrix}^T \text{ km/s}$$
$$\mathbf{r}_1(0) = \begin{bmatrix} 5023.437579954 & 5023.679067423 & 0.469973680 \end{bmatrix}^T \text{ km}$$
$$\mathbf{v}_1(0) = \begin{bmatrix} -1.810792589537 & 1.810419297938 & 7.041300610075 \end{bmatrix}^T \text{ km/s}$$

The errors, in the three components of ρ, between the linear and the nonlinear simulation results, are shown in Fig. 7.9. The radial and cross-track errors are slightly biased but are predominantly periodic, unlike the in-track error. The in-track drift rate, $\dot{y}$, estimated from Fig. 7.9, is -11 m in 15 orbits of the chief. The drift predicted by Eq. (5.64) for the data of this example is -11.2 m in 15 orbits. Hence, the in-track drift seen from Fig. 7.9 is predominantly due to the linearization of the two-body relative gravitational acceleration.

Example 7.4. *The initial conditions for this example are for $\rho = 0.5$ km and $\alpha = 90^o$:*

$$\rho(0) = \begin{bmatrix} 0.250014418391 & 0.000198338483 & 0.500288022195 \end{bmatrix}^T \text{ km}$$

$$\dot{\rho}(0) = \begin{bmatrix} -0.000000124335 & -0.000527557529 & -0.000000019840 \end{bmatrix}^T \text{ km/s}$$

The corresponding initial conditions of the deputy in the ECI frame are

$$r_1(0) = \begin{bmatrix} 5024.067715322 & 5023.402914470 & 0.171195964 \end{bmatrix}^T \text{ km}$$

$$v_1(0) = \begin{bmatrix} -1.810892863426 & 1.810892391776 & 7.040872374521 \end{bmatrix}^T \text{ km/s}$$

Figure 7.10 shows that except for a change in the in-track drift rate, the error profiles are quite similar to those seen for the previous example. The radial error is oscillatory with a small superimposed linear growth rate. The in-track drift rate estimated from Fig. 7.10 is -3.5 m in 15 orbits, compared to -3.73 m during the same period, predicted by Eq. (5.64).

Example 7.5. *A PCO with $\rho = 1$ km is considered to show the effects of the size of the relative orbit on the model errors.*

Figures 7.11 and 7.12, respectively, show the errors between the linear and nonlinear simulations for $\alpha = 0°$ and $\alpha = 90°$. The errors increase with the size of the relative orbit. The amplitudes of the radial and cross-track errors, as well as the in-track growth rate, show quadratic relationships with ρ for $\alpha = 0°$, as is to be expected from Eq. (5.64). The quadratic relationships of the error amplitudes with ρ are not well satisfied for $\alpha = 90°$, showing a more dominant J_2-effect for this case.

Example 7.6. *The extent of the errors, solely due to the approximation of the J_2 disturbance, can be ascertained by replacing the linear two-body differential gravitational acceleration of Eq. (7.146) by the difference of the two-body accelerations for the deputy and the chief given by Eq. (7.145). The effect of this change is evaluated for the 1 km PCO considered previously.*

Figure 7.13 shows the errors in the states with respect to the nonlinear simulation results for $\alpha = 0°$ and Fig. 7.14 shows the same for $\alpha = 90°$. The in-track error due to the approximation of the J_2-effect is more for $\alpha = 90°$ as compared to the case of $\alpha = 0$, corroborating the conclusion drawn previously. However, the secular effect due to the J_2-approximation is

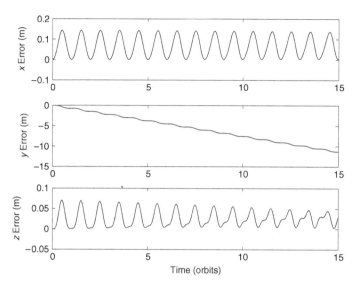

FIGURE 7.9 Errors between the linear and nonlinear models ($\rho = 0.5\,\mathrm{km}$, $\alpha = 0°$).

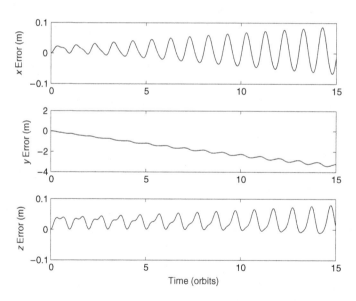

FIGURE 7.10 Errors between the linear and nonlinear models ($\rho = 0.5\,\mathrm{km}$, $\alpha = 90°$).

very small compared to that due to linearization of the two-body differential gravitational acceleration.

The examples considered in this section show that the linear model is accurate to $\mathcal{O}(J_2)$ and is also consistent with respect to the errors expected from linearization.

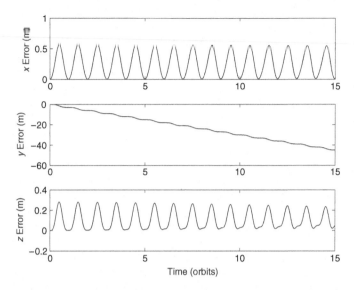

FIGURE 7.11 Errors between the linear and nonlinear models ($\rho = 1$ km, $\alpha = 0°$).

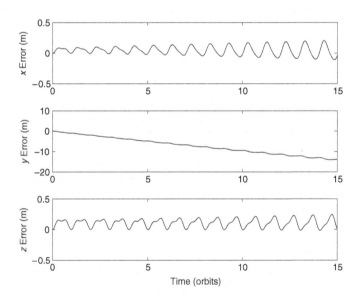

FIGURE 7.12 Errors between the linear and nonlinear models ($\rho = 1$ km, $\alpha = 90°$).

7.6 DIFFERENTIAL EQUATIONS FROM THE GIM–ALFRIEND STM

In principle, a set of linear differential equations can be derived for a dynamical system from its STM. However, the process can be complicated, especially for perturbed eccentric orbits. Simplifications can be achieved by neglecting the long-periodic effects [125,134]. There are no long-periodic effects of the

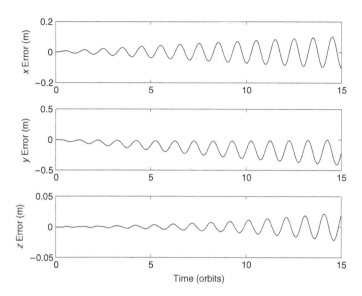

FIGURE 7.13 Errors between the linear-J_2-nonlinear two-body gravitational acceleration model and nonlinear simulation ($\rho = 1$ km, $\alpha = 0°$).

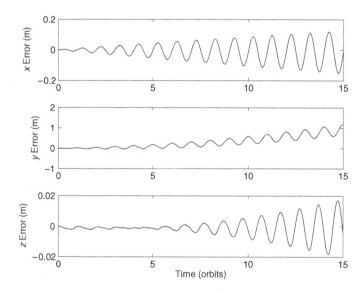

FIGURE 7.14 Errors between the linear-J_2-nonlinear two-body gravitational acceleration model and nonlinear simulation ($\rho = 1$ km, $\alpha = 90°$).

J_2 perturbation on the elements of a mean circular orbit. However, the partial derivatives of the long-periodic correction terms with respect to $\bar{q}_1$ and $\bar{q}_2$ are not zero. These partial derivatives of the long-periodic terms are significant near the critical inclination.

The relationship between the relative state vector and the osculating or mean differential orbital elements at any instant can be written as

$$\mathbf{x}(t) = \Sigma(t)\delta\boldsymbol{\alpha}(t) = \Sigma(t)D\delta\overline{\boldsymbol{\alpha}}(t) = F_1\delta\overline{\boldsymbol{\alpha}}(t) \tag{7.153}$$

where $F_1 = \Sigma(t)D$. The matrix D is the partial derivative matrix of the osculating elements with respect to the mean elements. The elements of $\Sigma(t)$ and D have been derived in Ref. [132] and are also provided in Appendix G for orbits with small eccentricity, but with terms of $\mathcal{O}(eJ_2)$ neglected. The relationship between the current and the initial differential mean elements, either nonsingular or equinoctial, can also be expressed as

$$\delta\overline{\boldsymbol{\alpha}}(t) = \overline{\phi}_{\boldsymbol{\alpha}}(t, t_0)\delta\overline{\boldsymbol{\alpha}}(0) \tag{7.154}$$

where $\overline{\phi}_{\boldsymbol{\alpha}}$ is the state transition matrix for propagating the mean differential elements, which can be obtained from Appendix G. Thus, the relationship between the current relative state and the initial differential mean element vector can be expressed as

$$\mathbf{x}(t) = F_1\overline{\phi}_{\boldsymbol{\alpha}}(t, t_0)\delta\overline{\boldsymbol{\alpha}}(0) = F\delta\overline{\boldsymbol{\alpha}}(0) \tag{7.155}$$

where $F = F_1\overline{\phi}_{\boldsymbol{\alpha}}(t, t_0)$. Since $\delta\overline{\boldsymbol{\alpha}}(0)$ is a constant vector, an approach to obtain a linear state-space model is to differentiate Eq. (7.155) and eliminate $\delta\overline{\boldsymbol{\alpha}}(0)$:

$$\dot{\mathbf{x}}(t) = \dot{F}F^{-1}\mathbf{x}(t) \tag{7.156}$$

Determination of the derivatives of the entries of F is a tedious process for elliptic orbits. The algebra involved is simpler for circular orbits. However, the approach discussed above results in a model that has the same accuracy as that given by Eq. (7.151), since expressing the initial mean differential elements in terms of the current states introduces significant linearization errors. A better approach is outlined next.

Differentiation of Eq. (7.153) and simple variable elimination leads to

$$\dot{\mathbf{x}}(t) = \dot{F_1}\delta\overline{\boldsymbol{\alpha}}(t) + F_1\delta\dot{\overline{\boldsymbol{\alpha}}}(t) = \dot{F_1}F_1^{-1}\mathbf{x}(t) + F_1\dot{\overline{\phi}}_{\overline{e}}(t, t_0)\delta\overline{\boldsymbol{\alpha}}(0) \tag{7.157}$$

Equation (7.157) is nonhomogeneous due to the appearance of a reference input that depends on $\delta\overline{\boldsymbol{\alpha}}(0)$. Note the similarity of structures between Eqs. (6.76)–(6.78) and Eq. (7.157); these are hybrid equations, involving the position and velocity states as well as the initial differential orbital elements.

Note that for circular orbits all the periodic terms in D are functions of $\overline{\theta}_0$ and hence the derivative of D is easily obtained. The singularity associated with the computation of D near the critical inclination can be avoided by using the limiting procedure given in Section 7.3.2.

The state variable propagation errors due to the use of Eq. (7.157) are shown for $\rho = 1$ km in Figs. 7.15 and 7.16, for $\alpha = 0°$ and $\alpha = 90°$, respectively, and for the initial elements of the chief considered in Section 7.5.

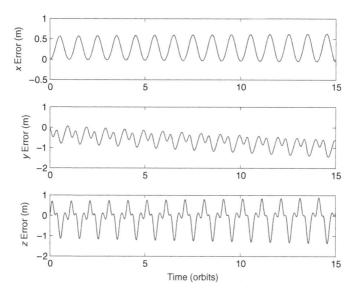

FIGURE 7.15 Errors between the STM-based model and nonlinear propagations ($\rho = 1$ km, $\alpha = 0°$).

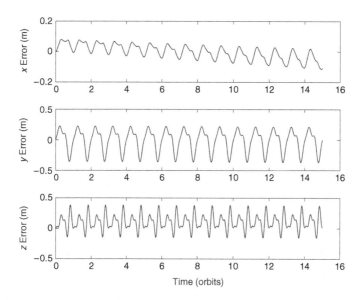

FIGURE 7.16 Errors between the STM-based and nonlinear propagations ($\rho = 1$ km, $\alpha = 90°$).

It is observed by comparing Figs. 7.11 and 7.15 that the secular in-track error has been eliminated considerably by the STM-based hybrid model. There is no appreciable difference between the radial errors from the two models to the scale of the figure. It is remarkable that the cross-track error amplitude from the

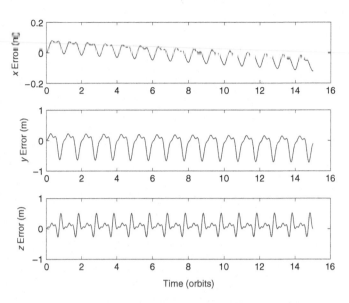

FIGURE 7.17 Errors between the short-periodic-only-STM-based and nonlinear propagations ($\rho = 1$ km, $\alpha = 90°$, and critical inclination).

state-space linearized model is less than that for the STM-based model. Similar observations apply to Figs. 7.12 and 7.16.

There are no appreciable differences in the errors by including or neglecting the long-periodic effects in D, for the examples considered in this section and for the levels of accuracy expected from a first-order theory. In fact, for the critical inclination, neglecting the long-periodic terms results in the errors shown in Fig. 7.17.

7.7 A SECOND-ORDER STATE PROPAGATION MODEL

A second-order state transition approach for propagating perturbed relative motion about eccentric orbits has been proposed by Sengupta et al. [2]. It models the nonlinearities occurring in the unperturbed problem to second order and couples this development with the first-order GA STM, to account for the J_2 perturbation. The main result of this work is the nonlinear propagation equation for the normalized relative motion states with θ as the independent variable:

$$\bar{x}(\theta_2) = \Phi^{(1)}(\theta_2, \theta_1)\bar{x}(\theta_1) + \frac{1}{2}\Phi^{(2)}(\theta_2, \theta_1) \otimes \bar{x}(\theta_1) \otimes \bar{x}(\theta_1) \quad (7.158)$$

where θ_1 and θ_2 define the propagation domain, $\Phi^{(1)}$ is the first-order STM, and $\Phi^{(2)}$ is a third-order tensor; $\otimes$ indicates the tensor product. The relative state vector is indicated by $\bar{x} = \left[\bar{x}, \bar{x}', y, \bar{y}', z, \bar{z}' \right]^T$. The development of Eq. (7.158) is based on a second-order propagation of the mean differential elements and a series reversion process to determine the mean differential elements at any

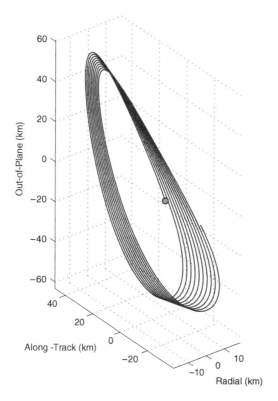

FIGURE 7.18 Relative motion trajectory (adapted from Ref. [2]).

epoch, given the relative state at this time. An example [2] showing the accuracy of the second-order approach is presented next.

Example 7.7. *Let the initial mean elements of the chief be*

$$\bar{a}_0 = 13000 \text{ km}, \quad \bar{\theta}_0 = 0.1 \text{ rad}, \quad \bar{i}_0 = 0.87266 \text{ rad}$$
$$\bar{q}_{10} = 0.29886, \quad \bar{q}_{20} = 0.02615, \quad \bar{\Omega}_0 = 0.34907 \text{ rad}$$

Simulate the evolution of the error in the relative separation distance between the chief and deputy for the initial differential mean elements given below using first- and second-order theories:

$$\delta\bar{a} = 0.2 \text{ km}, \quad \delta\bar{\theta}_0 = -0.15526 \times 10^{-2} \text{ rad}, \quad \delta\bar{i}_0 = 0.005 \text{ rad}$$
$$\delta\bar{q}_{10} = 0.11452 \times 10^{-3}, \quad \delta\bar{q}_{20} = 0.13415 \times 10^{-2}, \quad \delta\bar{\Omega}_0 = 0.2 \times 10^{-3} \text{ rad}$$

Figure 7.18 shows the relative motion trajectory for this example. The errors resulting from the application of the linear and second-order theories are shown in Fig. 7.19 for 10 orbit periods of the chief. The truth model is a nonlinear numerical simulation including J_2. The linear theory results in a final position error of nearly 200 m, whereas the second-order theory reduces the error by one

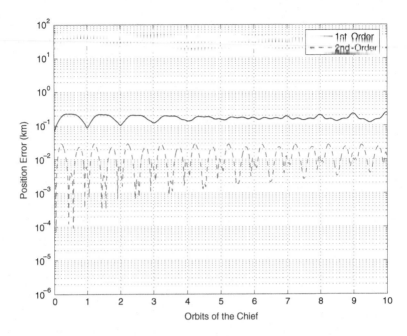

FIGURE 7.19 Position errors resulting from the application of linear and second-order theories (adapted from Ref. [2]).

order of magnitude. For the example considered, the eccentricity and nonlinear effects contribute more significantly to the error than does J_2.

SUMMARY

In this chapter, we discussed the effects of the J_2 perturbation on the orbital elements via the Brouwer Theory. Brouwer's transformation between the mean and osculating orbital elements plays a significant role in the developments in Chapters 8 and 10. We presented the unit-sphere approach to the analytical propagation of relative motion via differential orbital elements and differential Euler parameters.

The GA STM, derived based on the Geometric Method, was developed for nonsingular elements. This STM is valid for any eccentricity and includes the first-order absolute and differential J_2 effects. The relative state used curvilinear coordinates to mitigate the effects of the neglected nonlinear terms that occur when using a relative Cartesian state. The GA STM has been derived for both osculating and mean elements. The mean element version is less accurate, because it does not include the J_2 short-periodic terms, but the secular effects are captured and it is much less complex. The GA STM was developed using the relationship between the relative state in curvilinear coordinates and nonsingular differential orbital elements.

The GA STM presented herein used nonsingular variables for representing the eccentricity. It is not valid near $i = 0$. For the small-inclination case, another

STM had been developed using equinoctial variables. The theory and results can be found in Refs. [100] and [131].

The concept of averaged elements was introduced by accounting for the orbit-averaged short-periodic corrections to the mean elements. The short-periodic corrections to the mean orbital elements were utilized in the development of a consistent set of linearized differential equations for perturbed relative motion. The GA STM can be incorporated into a second-order state transition method to achieve higher levels of accuracy.

A nonlinear relative motion theory due to Yan and Alfriend including J_2^2 terms is discussed in Section 8.2. Several other approaches not discussed in this book account for luni-solar and higher-order geopotential perturbations [140] and use more detailed propagation models such as the Draper Semianalytic Satellite Theory [21] and symplectic integration [141]. Wiesel [142] presents a solution process for relative motion based on the Hamiltonian approach and Floquet theory. The reference orbit for describing relative motion dynamics is a near-circular periodic orbit defined in a nodal frame which regresses about the polar axis of the $\mathscr{I}$ frame.

Chapter eight

Perturbation Mitigation

All space, all time,
(The stars, the terrible perturbations of the suns,
Swelling, collapsing, ending, serving their longer, shorter use,)
Fill'd with eidolons only.

Walt Whitman (1819–1892)

In the first part of this chapter we show how to adjust or constrain some of the initial conditions to reduce the amount of fuel required to counter the undesirable effects resulting from the J_2 perturbation. In contrast to most of the previous developments, which have used the relative position in the rotating reference frame, the current development utilizes Hamiltonian mechanics, discussed in Chapter 3, and is based on Brouwer's theory [66] introduced in Section 3.4. The first-order analysis is extended to the second-order in the Yan–Alfriend theory. A comparison of several linear and nonlinear analytical relative motion propagation theories is presented on the basis of a *nonlinearity error index*, which is a measure of the relative importance of modeling geometric nonlinearity, eccentricity, and J_2 as a function of the formation size.

The second part of the chapter deals with the perturbations in the fundamental in-plane and cross-track frequencies due to J_2. We also present an intelligent scheme for formation maintenance and inter-satellite fuel balancing, which accommodates a significant portion of the perturbations rather than countering them with thrust.

8.1 DYNAMIC CONSTRAINTS FOR J_2 MITIGATION

The attractive solutions of the CW or Tschauner–Hempel–Lawden equations derived in Chapter 5, obtained under the assumption of a spherical Earth, get disturbed in the presence of gravitational perturbations, as shown in Chapter 7. In particular, differences in the mean semimajor axis, eccentricity and inclination cause along-track, radial, and cross-track drift. In addition, perigee rotation causes in-plane and cross-track frequencies to no longer be equal. This means, depending on the desired formation, that control may be required to maintain the formation. For example, as shown in Section 2.6, if the chief is in a 7100 km circular orbit with $i = 70°$, the Δv required to counter the differential nodal

185

precession for a relative motion orbit that has a maximum cross-track displacement of 1 km caused by an inclination difference is approximately 41 m/s/yr. This estimate is based on the assumption that the corrections are impulsive and are performed once per orbit. In addition, the fuel requirements for countering along-track drift and perigee rotation effects can be prohibitive, depending on the reference orbit and size of the formation.

We note that if the momenta of the satellites in a formation are equal then the angle rates are equal and the satellites will not drift apart. If the momenta are not equal then, in general, some of the angle rates will not be equal, and the two satellites will drift apart. The development in this section focuses on how to select the initial momenta to minimize this drift.

The dimensionless averaged Hamiltonian, Eqs. (3.19) and (3.20), provide the dimensionless angle rates

$$\dot{l} = \frac{\partial \bar{\mathcal{H}}}{\partial L} = \frac{1}{L^3} + \epsilon \left(\frac{3}{4L^7} \right) \left(\frac{L}{G} \right)^3 \left(1 - 3\frac{H^2}{G^2} \right) \tag{8.1}$$

$$\dot{g} = \frac{\partial \bar{\mathcal{H}}}{\partial G} = \epsilon \frac{3}{4L^7} \left(\frac{L}{G} \right)^4 \left(1 - 5\frac{H^2}{G^2} \right) \tag{8.2}$$

$$\dot{h} = \frac{\partial \bar{\mathcal{H}}}{\partial H} = \epsilon \frac{3}{2L^7} \left(\frac{L}{G} \right)^4 \left(\frac{H}{G} \right) \tag{8.3}$$

Note that $\varepsilon = -J_2$. It is advantageous now to transform from the generalized momenta (L, G, H) to the orbital elements (a, η, i). Using Eqs. (8.1)–(8.3) the angle rates become

$$\dot{l} = \frac{1}{a^{3/2}} + \epsilon \left(\frac{3}{4a^{7/2}\eta^3} \right) \left(1 - 3\cos^2 i \right) \tag{8.4}$$

$$\dot{g} = \epsilon \left(\frac{3}{4a^{7/2}\eta^4} \right) \left(1 - 5\cos^2 i \right) \tag{8.5}$$

$$\dot{h} = \epsilon \left(\frac{3}{2a^{7/2}\eta^4} \right) \cos i \tag{8.6}$$

It can be shown from Eqs. (8.1) and (8.2) that the mean argument of latitude rate is

$$\dot{\lambda} = \dot{l} + \dot{g} = \frac{1}{a^{3/2}} + \epsilon \left(\frac{3}{4a^{7/2}\eta^4} \right) \left[\eta \left(1 - 3\cos^2 i \right) + \left(1 - 5\cos^2 i \right) \right] \tag{8.7}$$

Any two satellites in a formation will drift apart if the difference in any of their respective angle rates is not zero, i.e., if $\delta\dot{\lambda} = (\dot{\lambda}_2 - \dot{\lambda}_1) \neq 0$, or $\delta\dot{g} = (\dot{g}_2 - \dot{g}_1) \neq 0$, or $\delta\dot{h} = (\dot{h}_2 - \dot{h}_1) \neq 0$. Cross-track drift is determined by $\delta\dot{h} = 0$ and the radial drift is determined primarily by $\delta\dot{g}$. As discussed in Section 7.1 and also in Ref. [143], the in-plane or along-track drift is determined by $(\delta\dot{\lambda} + \delta\dot{h} \cos i)$.

Expanding the three angle rate equations, Eqs. (8.7), (8.5), and (8.6), into a Taylor series about the reference or chief satellite orbit, denoted by the subscript "0", and retaining only first-order terms, the angle rate differences are

$$
\delta\dot\lambda = -\left(\frac{3}{2a_0^{5/2}}\right)\delta a - \epsilon\left(\frac{21}{8a_0^{9/2}n_0^4}\right)\left[\eta_0\left(1 - 3\cos^2 i_0\right) + \left(1 - 5\cos^2 i_0\right)\right]
$$
$$
\times\,\delta a + \epsilon\left(\frac{3}{4a_0^{7/2}n_0^5}\right)\left[\left[(9\eta_0 + 20)\cos^2 i_0 - (3\eta_0 + 4)\right]\delta\eta\right.
$$
$$
\left. + \eta_0\,(3\eta_0 + 5)\,\delta i\,\sin 2i_0\right] \tag{8.8}
$$

$$
\delta\dot g = -\epsilon\left(\frac{21}{8a_0^{9/2}n_0^4}\right)\left(1 - 5\cos^2 i_0\right)\delta a
$$
$$
+ \epsilon\left(\frac{3}{4a_0^{7/2}n_0^5}\right)\left[(5\eta_0\sin 2i_0)\,\delta i - 4\left(1 - 5\cos^2 i_0\right)\delta\eta\right] \tag{8.9}
$$

$$
\delta\dot h = -\epsilon\left(\frac{21}{4a_0^{9/2}n_0^4}\right)\cos i_0\delta a
$$
$$
- \epsilon\left(\frac{3}{2a_0^{7/2}n_0^5}\right)(4\delta\eta\cos i_0 + \eta_0\delta i\sin i_0) \tag{8.10}
$$

Per the discussion above, instead of $\delta\dot\lambda$ we want to use $\left(\delta\dot\lambda + \delta\dot h\cos i_0\right)$, which is given by

$$
\left(\delta\dot\lambda + \delta\dot h\cos i_0\right) = -\left(\frac{3}{2a_0^{5/2}}\right)\delta a
$$
$$
- \epsilon\left(\frac{3}{4a_0^{9/2}n_0^4}\right)(1 + \eta_0)\left(1 - 3\cos^2 i_0\right)\delta a
$$
$$
- \epsilon\left(\frac{3}{4a_0^{7/2}n_0^5}\right)(3\eta_0 + 4)\left[\left(1 - 3\cos^2 i_0\right)\delta\eta - (\eta_0\sin 2i_0)\,\delta i\right] \tag{8.11}
$$

Note that the initial values of the angle differences $(\delta l, \delta g, \delta h)$ have no effect on the drift; only the differences in the momenta $(\delta L, \delta G, \delta H)$ or $(\delta a, \delta\eta, \delta i)$ affect the drift. To prevent drift between the satellites we want $\delta\dot\lambda + \delta\dot h\cos i_0 = \delta\dot g = \delta\dot h = 0$. The drift can be controlled by thrusting but, depending on the desired formation, the amount of fuel required could be prohibitive. What we are investigating here is how to adjust the initial conditions, i.e., $(\delta a, \delta\eta, \delta i)$, to minimize the amount of fuel required to maintain the desired formation.

Equations (8.9), (8.10) and (8.11) are three equations with three unknowns, $(\delta a, \delta\eta, \delta i)$. The only solution is the trivial solution if the determinant of the coefficients is not equal to zero. This singular case will be ignored for the moment. Except for the singular case, setting any of the differential angle rates, $(\delta\lambda, \delta g, \delta\dot{h})$, establishes a constraint on the formation design

The equations can be solved by inverting the matrix of coefficients, but a simpler approach is possible. Note that these equations are infinite series truncated at first-order, i.e., $\mathcal{O}(\epsilon)$. From Eq. (8.11) we see that $\delta a = \mathcal{O}(\epsilon)$. Therefore, we can neglect the δa terms in Eqs. (8.9) and (8.10). This simplifies the solutions and they are just as accurate as the solutions obtained from inverting the 3×3 matrix. The solutions to Eqs. (8.9), (8.10) and (8.11) are

$$\delta a = -\epsilon \left(\frac{1}{2a_0\eta_0^5}\right)(3\eta_0 + 4)\left[\left(1 - 3\cos^2 i_0\right)\delta\eta - (\eta_0 \sin 2i_0)\,\delta i\right] \quad (8.12)$$

$$\delta\eta = \frac{5\eta_0 \sin 2i_0}{4\left(1 - 5\cos^2 i_0\right)}\delta i \quad (8.13)$$

$$\delta\eta = -\frac{\eta_0 \tan i_0}{4}\delta i \quad (8.14)$$

Obviously, the only solution to Eqs. (8.13) and (8.14) is the trivial solution, $\delta\eta = \delta i = 0$. Either or both of these constraints may be incompatible with the desired mission. If only one or neither of these constraints is used, then Eq. (8.12) provides the δa that negates the along-track drift between the satellites. Since negating the along-track drift with this value of δa does not affect or constrain the choice of formations, this constraint should always be used. In terms of dimensional coordinates, it is

$$\delta a = 0.5J_2a_0\left(\frac{R_e}{a_0}\right)^2\left(\frac{3\eta_0 + 4}{\eta_0^5}\right)$$
$$\times\left[\left(1 - 3\cos^2 i_0\right)\delta\eta - (\eta_0 \sin 2i_0)\,\delta i\right] \quad (8.15)$$

The effect of the constraints given by Eqs. (8.12), (8.13) and (8.14) on the formation design and the associated fuel to maintain the formation are now investigated.

8.1.1 Three constraints

As discussed above the three constraints are

$$\delta a = \delta i = \delta\eta = \delta e = 0 \quad (8.16)$$

This means the only freedom in the formation design is in the selection of the initial values of δl, δg, and δh. The only formations possible with these three constraints are a leader–follower with some cross-track motion resulting from the differential right ascension, δh, or $\delta\Omega$. These formations can be characterized as a dog wagging its tail. If the reference or chief orbit is eccentric, the along-track separation will be periodic with a period equal to the anomalistic period. Different types of along-track motion occur depending on the selection of the initial values of δl and δg.

Consider two cases with the chief orbit having a small eccentricity and $(\delta l + \delta g) = K$, a constant. In the first case $\delta l = 0$, $\delta g = K$. This corresponds to two ellipses with different arguments of perigee and both objects are at perigee and apogee at the same time. The result is a constant angular separation of the two satellites and the in-plane separations at perigee and apogee are

$$d_p \approx a_0(1 - e_0)\,K, \quad d_a \approx a_0(1 + e_0)\,K. \tag{8.17}$$

In the second case select $\delta l = K$, $\delta g = 0$. In this case the ellipses coincide and the two satellites are in the same orbit, but just separated by a constant mean anomaly difference. Using $\delta v \approx \delta l(1 + 2e \cos l)$ the separations at perigee and apogee are

$$d_p \approx a_0(1 + e_0)\,K, \quad d_a \approx a_0(1 - e_0)\,K \tag{8.18}$$

This is exactly the opposite to case one, the maximum separation is at perigee and the minimum separation is at apogee. Since there is no drift, no fuel is required to negate the drift. However, if there is cross-track motion, the relative motion orbit will not be periodic. The in-plane and cross-track motions have different periods due to the rotation of perigee of the chief's orbit. The only way to negate this is by thrusting or to be at the critical inclination. There is another inclination [139] at which the difference between the in-plane and cross-track frequencies is minimized. Except for very small eccentricities, the fuel for controlling perigee is prohibitive. The Δv to negate perigee rotation for near circular orbits is (cf. the GVE in Section 2.5.4 and Ref. [144])

$$\Delta v \approx 0.75 J_2 \pi \sqrt{\frac{\mu}{a_0}} \left(\frac{R_e}{a_0}\right)^2 \left(1 - 5\cos^2 i_0\right) e_0 \text{ m/s/orbit} \tag{8.19}$$

For $a = 7500$ km, the annual Δv is

$$\Delta v \approx 65.6 \left(1 - 5\cos^2 i_0\right) e_0 \text{ km/s/yr} \tag{8.20}$$

Thus, unless $e_0 < 0.0001$ or $i_0 \approx 63.4°$ the fuel required to maintain perigee will be prohibitive.

Recall that $\eta^2 = (1 - e^2)$. Consequently, $\delta\eta = -e_0\delta e/\eta_0$. Therefore, for near-circular orbits, the primary effect on δa will be that due to the cross-track motion resulting from δi. Only for highly-elliptic orbits will the differential eccentricity have an effect on the semimajor axis adjustment.

8.1.2 Two constraints

As discussed above, the along-track drift negation constraint given by Eq. (8.12) should always be implemented because its use does not impact the formation design. Therefore, this is one of the two constraints. There are two options for the second constraint: Negating the differential nodal precession or the differential perigee rotation. These constraints are given in Eqs. (8.14) and (8.13), respectively. The effect of the differential perigee rotation is a slow radial drift. There is no along-track drift due to the along-track drift constraint. What is happening here is a slow relative rotation of the two ellipses. For orbits with a small eccentricity, this effect is generally less than the differential nodal precession. However, for highly eccentric orbits, such as the NASA MMS mission, the differential perigee rotation effect can be larger than the differential nodal precession.

DIFFERENTIAL NODAL PRECESSION NEGATION

First, consider low near-circular orbits with the chief orbit semimajor axis $a = 7500$ km and vary the chief eccentricity and inclination. Note that for near-polar orbits, due to the $\tan i$ term in Eq. (8.14), if the cross-track motion is created with a δi, then $\delta \eta$ will have to be large, resulting in a large relative motion orbit; this is likely to be impractical. Conversely, the δi required to counter the effect of a prescribed $\delta \eta$ will be very small for near polar orbits.

Let the relative motion orbit have an average cross-track component of 1 km, consequently, $\delta i = 1/7500$ rad. Figure 8.1 shows the differential eccentricity required to negate the nodal precession as a function of eccentricity for various inclinations. For small eccentricities, the δe required is relatively large, likely too large for the desired formation. For example, for $e_0 = 0.01$ and $i_0 = 65°$ Eq. (8.14) gives, $\delta e = 0.0056$, resulting in a maximum radial difference of approximately 42 km. Note the case with $i = 95°$. For $e < 0.025$ the nodal precession cannot be negated. Figure 8.2 shows the differential semimajor axis to negate the along-track drift. As expected from the above discussion, it is essentially constant with increasing chief eccentricity, and is only a few meters. The relative motion orbits in this case of two constraints are called J_2-invariant relative orbits [32].

Now, consider highly eccentric orbits. Consider a critically inclined 12-hour orbit, $a_0 = 26,610$ km, and vary the eccentricity. The critical inclination was selected because control of perigee rotation is prohibitive for non-critical inclination highly eccentric orbits. The differential eccentricity required to negate the nodal precession is shown in Fig. 8.3 and the differential semimajor axis required to negate the along-track drift is shown in Fig. 8.4. Since the reference orbit is at the critical inclination, the in-plane and cross-track frequencies are equal and periodic relative motion orbits are possible. However, for each eccentricity and δi there is only one such relative motion orbit.

Using the constraint of Eq. (8.14), the approximate Δv per orbit required to negate the differential perigee rotation is

$$\Delta v = 0.75 J_2 \pi v_0 \left(\frac{R_e}{a_0}\right)^2 \left(\frac{e_0}{\eta_0^5}\right) \left(1 + 5\cos^2 i_0\right) \tan i_0 \delta i \text{ m/s/orbit} \quad (8.21)$$

or

$$\Delta v = 3 J_2 \pi v_0 \left(\frac{R_e}{a_0}\right)^2 \left(\frac{e_0}{\eta_0^5}\right) \left(1 + 5\cos^2 i_0\right) \delta \eta \text{ m/s/orbit} \quad (8.22)$$

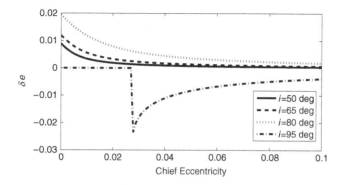

FIGURE 8.1 Differential eccentricity required to negate the differential nodal precession; small eccentricity.

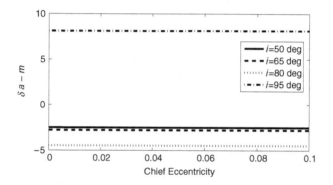

FIGURE 8.2 Differential semimajor axis required to negate the along-track drift; small eccentricity.

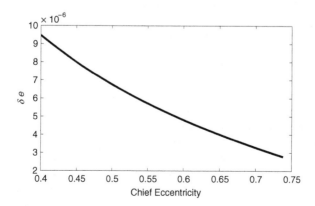

FIGURE 8.3 Differential eccentricity required to negate the differential nodal precession for a 12-hour orbit.

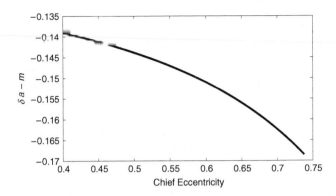

FIGURE 8.4 Differential semimajor axis required to negate the along-track drift for a 12-hour orbit.

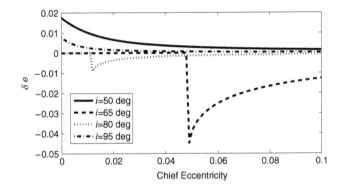

FIGURE 8.5 Differential eccentricity required to negate the differential perigee rotation.

where v_0 is the speed in a circular orbit with a radius equal to the semimajor axis, i.e., $r = a_0$. For a 7000 km near-circular chief orbit ($e_0 \approx 0$) with a 1 km cross-track motion caused by a differential inclination, the annual fuel requirement is $\Delta v = 45.1$ m/s/yr for $i_0 = 50°$. Even for moderate eccentricities, this is a small amount of fuel.

DIFFERENTIAL PERIGEE ROTATION NEGATION

To negate the differential perigee rotation, Eq. (8.13) is used to obtain $\delta\eta$. Figure 8.5 and 8.6 show the differential eccentricity and differential semimajor axis required to negate the differential perigee rotation and the along-track drift for the same low-Earth orbit example with $a_0 = 7500$ km. Figure 8.5 shows that when the chief eccentricity is small, it may not be possible to negate the differential perigee rotation. As with the differential nodal precession negation, this occurs when δe is negative and $|\delta e| > e_0$, which occurs for the cases $i_0 = 65°$ and $i_0 = 80°$. Figure 8.6 shows the same trend of an essentially constant δa of just a few meters, required to negate the along-track drift.

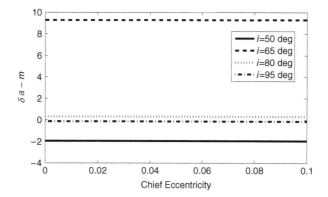

FIGURE 8.6 Differential semimajor axis required to negate the along-track drift.

Using the constraint Eq. (8.14), the approximate Δv per orbit required to negate the differential nodal rotation is

$$\Delta v = 1.2 J_2 \pi v_0 \left(\frac{R_e}{a_0}\right)^2 \left(\frac{\tan i_0}{\eta_0^4}\right) \left(1 + 5 \cos^2 i_0\right) \delta \eta \text{ m/s/orbit} \quad (8.23)$$

or

$$\Delta v = 3 J_2 \pi v_0 \left(\frac{R_e}{a_0}\right)^2 \left(\frac{\sin^2 i_0}{\eta_0^3}\right) \frac{\left(1 + 5 \cos^2 i_0\right)}{\left(1 - 5 \cos^2 i_0\right)} \tan i_0 \, \delta i \text{ m/s/orbit} \quad (8.24)$$

For a 7000 km near circular chief orbit ($e_0 \approx 0$) with a 1 km cross-track motion caused by a differential inclination, the annual fuel requirement is $\Delta v = 83.4$ m/s/yr for $i_0 = 50°$. This supports the earlier statement that for near-circular orbits, the best approach is to negate the differential nodal precession, not the differential perigee rotation.

8.1.3 One constraint

As stated earlier, when there is only one constraint, it should be used for negating the along-track drift. The Δv per orbit for negating the differential nodal precession and differential perigee rotation are, respectively (cf. the GVE in Section 2.5.4),

$$\Delta v = 0.75 J_2 \pi v_0 \left(\frac{R_e}{a_0}\right)^2 \left(\frac{e_0}{\eta_0^6}\right) \left[5 \eta_0 \delta i \sin 2 i_0 + 4 \delta \eta \left(5 \cos^2 i_0 - 1\right)\right]$$

$$(8.25)$$

$$\Delta v = 3 J_2 \pi v_0 \left(\frac{R_e}{a_0}\right)^2 \left(\frac{\sin i_0}{\eta_0^4}\right) (\eta_0 \delta i \sin i_0 + 4 \delta \eta \cos i_0) \quad (8.26)$$

8.1.4 Energy considerations

When there are no perturbations, the condition for periodic relative-motion orbits is that the semimajor axes be equal, i.e., $\delta a = 0$, which means that the energies of the two orbits are equal. This was referred to as the energy matching condition, discussed in Section 4.2. Now, let us revisit the energy matching condition when the J_2 perturbation exists. The mean energy of the orbit as a function of (a, i, η) is just the mean Hamiltonian given by Eqs. (3.19) and (3.20), with a proper substitution of the momenta L, G, H:

$$\bar{\mathcal{H}} = -\frac{1}{2a} - \frac{\epsilon}{4a^3\eta^3}\left(1 - 3\cos^2 i\right) \tag{8.27}$$

For a Keplerian motion, the energy is a function of only the semimajor axis, a. However, when J_2 is included, the energy depends on all three momenta elements (a, e, i). The mean energy difference $\delta\bar{\mathcal{H}}$ between a reference and a neighboring orbit is approximated by

$$\delta\bar{\mathcal{H}} = \bar{\mathcal{H}} - \bar{\mathcal{H}}_0 \approx \frac{\partial\bar{\mathcal{H}}_0}{\partial a}\delta a + \frac{\partial\bar{\mathcal{H}}_0}{\partial\eta}\delta\eta + \frac{\partial\bar{\mathcal{H}}_0}{\partial i}\delta i \tag{8.28}$$

where the subscript 0 denotes the reference orbit. Substituting for the partial derivatives and keeping in mind that $\delta a = \mathcal{O}(\epsilon)$ the energy difference becomes

$$\delta\bar{\mathcal{H}} = \frac{1}{2a_0^2}\delta a + \epsilon\frac{3}{4a_0^3\eta_0^4}\left[\left(1 - 3\cos^2 i_0\right)\delta\eta - (\eta_0\sin 2i_0)\,\delta i\right] \tag{8.29}$$

Substituting Eq. (8.12) for δa gives

$$\delta\bar{\mathcal{H}} = -\frac{\epsilon}{a_0^3\eta_0^5}\left[\left(1 - 3\cos^2 i_0\right)\delta\eta - (\eta_0\sin 2i_0)\,\delta i\right] \tag{8.30}$$

This means that for $\delta\bar{\mathcal{H}} = 0$, one has $\left(1 - 3\cos^2 i_0\right)\delta\eta = (\eta_0\sin 2i_0)\,\delta i$. However, this condition does not negate either the differential nodal precession or differential perigee rotation unless $\delta\eta = \delta i = 0$. Therefore, equal energy orbits result in no along-track drift, but not differential nodal precession or differential perigee rotation. It is important to note that the condition of no along-track drift does not result in equal energy orbits except for specific singular cases such as $\delta i = 0$ and $\cos i_0 = 1/\sqrt{3}$.

Now, consider the energy difference for negation of the differential nodal precession. Substituting Eq. (8.14) into the energy difference equation, Eq. (8.30), gives

$$\delta\bar{\mathcal{H}} = \epsilon\left(\frac{\tan i_0}{4a_0\eta_0^5}\right)\left(1 + 5\cos^2 i_0\right)\delta i \tag{8.31}$$

Thus, $\delta\bar{\mathcal{H}} = 0$ only if $\delta i = \delta\eta = \delta a = 0$. A similar result is obtained for negating the differential perigee rotation.

8.1.5 Numerical results

THREE CONSTRAINTS

The following example is used to demonstrate the three-constraint case. Table 8.1 gives the mean orbital elements of the chief. The deputy trails the chief by a nominal distance of 1 km with a nominal cross-track motion of 1 km. The differential orbital elements are $\delta a = \delta \eta = \delta i = \delta l = 0$ and $\delta g = \delta h = -1/a_0 = -0.0001429$ radians. With these differential elements there should be no drift in any direction. However, the mean-to-osculating transformation is not exact; it is correct only to first order, so some minor drift will occur. In addition, the in-plane and out-of-plane frequencies are different; therefore, the relative motion orbit will not be periodic.

With the chosen orbital element values, the perigee rotation period is approximately 70 days. Hence, it will take 17.5 days to obtain a 90-degree phase shift between the out-of-plane and in-plane motions. The purpose of the simulations is to (a) illustrate the effectiveness of the three constraints $\delta a = \delta \eta = \delta i = 0$ in maintaining a formation in close proximity, i.e., no drift, and (b) to demonstrate the negative effect of neglecting the effect of J_2 in developing the initial conditions. This is implemented by assuming that the mean elements are osculating elements. The truth orbits for both the chief and deputy were determined by numerically integrating the equations of motion with $J_2 - J_5$. The elements given in Table 8.1 are mean elements for the chief satellite. The deputy mean elements are obtained by adding the differential elements to the chief elements. Both are then transformed to osculating elements and the equations of motion are numerically integrated.

Figure 8.7 shows the 3D relative motion orbit for approximately three days. With the mean element initial conditions, the relative motion trajectory is nearly periodic. The beginning of the effect of the absolute perigee rotation is evident with the thickness of the line. Obviously, neglecting J_2 in determining the initial conditions creates a problem: The relative motion trajectory deteriorates quickly. The drift that occurs in the along-track direction, resulting from neglecting J_2 in determining the initial conditions, is also evident in Fig. 8.8. With the mean element initial conditions, there is essentially no drift. The minor drift results from the neglected higher-order terms, which are $\mathcal{O}(J_2^2)$. When assuming that the initial mean elements are osculating elements, the along-track drift occurs at a rate of about 200 meters per day; this drift would have to be negated with control. Furthermore, if the J_2 effect were neglected, the control system would return the system to the same desired differential elements and the drift would continue.

TWO CONSTRAINTS

The initial values of the chief orbital elements for the two-constraints simulations are given in Table 8.2. The maximum cross-track motion is to be approximately 1 km and created by an inclination difference, which gives $\delta i = 1/a_0$. We want to negate the cross-track drift, so the second constraint, Eq. (8.14), gives $\delta e = 0.001366$. The corresponding differential semimajor axis required

Table 8.1 Chief satellite orbital elements: three-constraints case

Desired mean orbit elements	Value	Units
a	7100	km
e	0.05	
i	50	deg
$\hbar$	0.0	deg
g	30.0	deg
l	0.0	deg

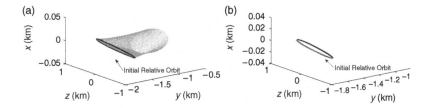

FIGURE 8.7 (a) Initial relative orbit setup in osculating elements. (b) Initial relative orbit setup in mean elements.

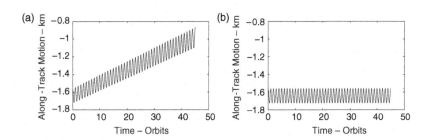

FIGURE 8.8 (a) Along-track relative motion for osculating element setup. (b) Along-track relative motion for mean element setup.

to negate the along-track drift is $\delta a = -2.5089$ m. The other differential elements are $\delta g = \delta \hbar = 0, \delta l = 2\delta e$. This choice of δl centers the formation on the chief. The along-track dimension of the formation is approximately 20.5 km, 10 times the cross-track.

Figure 8.9 shows the relative motion orbit for three days with and without the along-track constraint. The value of the constraint in maintaining the formation is evident. Figure 8.10 shows the along-track motion without and with the along-track constraint. The along-track drift, d_{drift}, per orbit is $d_{\text{drift}} = 3\pi \delta a$. Thus, not using the along-track constraint of $\delta a = -2.5089$ m should result in

Table 8.2 Chief satellite orbital elements; two-constraints case

Desired mean orbit elements	Value	Units
a	7500	km
e	0.1	
i	50	deg
$\hbar$	0.0	deg
g	0.0	deg
l	0.0	deg

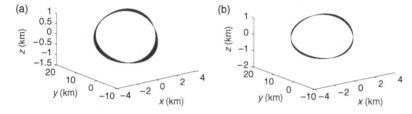

FIGURE 8.9 (a) Relative motion orbit without along-track negation. (b) Relative motion orbit with along-track negation.

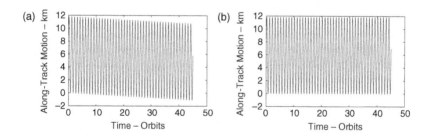

FIGURE 8.10 (a) Along-track time history without along-track constraint. (b) Along-track time history with along-track constraint.

a drift of approximately 316 m/day. The along-track drift in Fig. 8.10 is about 1 km over three days, about the expected value. The right ascension time history is shown in Fig. 8.11 and illustrates the effectiveness of the J_2 invariant orbit in suppressing the cross-track growth.

The next simulation uses the same chief elements as the previous one, but the relative motion orbit is selected to be similar to the projected circular orbit (see Eq. (5.36)) of radius 1 km with the cross-track motion created by an ascending node difference. The differential elements are $\delta e = 1/2a_0$, $\delta\hbar = 1/(a_0 \sin i_0)$, and $\delta l = \delta g = 0$. The constraints for negating the differential nodal precession and the along-track drift give $\delta a = -0.425$ m and $\delta i = 0.00013°$. Note that

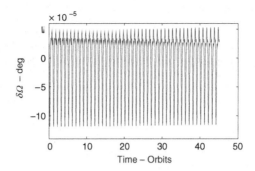

FIGURE 8.11 Differential right ascension time history.

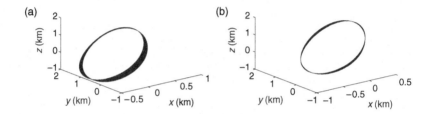

FIGURE 8.12 (a) Relative motion orbit without along-track negation. (b) Relative motion orbit with along-track negation.

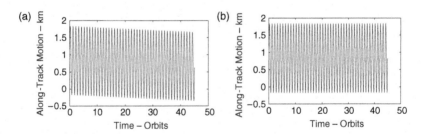

FIGURE 8.13 (a) Along-track time history without along-track constraint. (b) Along-track time history with along-track constraint.

these values are much less than the previous case. This demonstrates that δi has a much larger effect on the differential element constraints than δe. Figure 8.12 shows the 3D relative motion orbit without and with the along-track constraint. The drift is evident when the along-track constraint of $\delta a = -0.425$ m is not used. Figure 8.13 shows the along-track motion without and with the along-track constraint. The drift is much less in this case because the cross-track motion is caused by a differential right ascension, not a differential inclination. Figure 8.14 shows again the effectiveness in suppressing any cross-track growth.

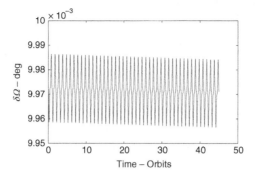

FIGURE 8.14 Differential right ascension time history.

ONE CONSTRAINT

The one-constraint case uses only the constraint on δa to negate the along-track drift. The effectiveness of using this constraint was demonstrated previously, so no separate results are presented here. Figure 8.14 shows that when the cross-track motion is created by a node difference, not an inclination difference, then the cross-track constraint is not nearly as important.

8.2 A NONLINEAR THEORY BASED ON ORBITAL ELEMENTS

The CW equations assume that the Earth is spherically symmetric, the reference or chief orbit is circular, and that the ratio of the formation size is small compared to the chief radius, so that the differential gravity terms can be linearized in the local coordinates. Each of these assumptions causes errors in the solution. In Section 8.3 the effects of these modeling errors in the various theories, including the one developed in this section, are evaluated. This evaluation shows that the effect of not modeling J_2 effects and assuming a circular chief orbit generally dominate the errors for formation sizes smaller than 20–40 km. Therefore, except for very large formations, for a nonlinear theory to be useful for formation design and analysis it needs to also include the J_2 and chief orbit eccentricity effects. This results in a formidable problem.

The first attempt at incorporating nonlinear effects was performed by London [145]. He assumed a circular reference orbit and a spherical Earth, and expanded the differential gravity terms into a power series, retaining terms through second order. He then performed a perturbation analysis with the first-order solution being the CW equations to obtain an approximate solution through second order. The problem with this approach is that in the along-track solution the first-order terms have t terms and the second-order terms contain t^2 terms. For the solution to be uniformly valid,

$$\lim_{t \to \infty} \left(\frac{x_{n+1}}{x_n} \right) = \mathcal{O}(1) \tag{8.32}$$

where x_n is the nth-order solution. For London's solution

$$\lim_{t \to \infty} \left(\frac{x_{n+1}}{x_n} \right) = \epsilon t \qquad (8.33)$$

where ϵ is a small parameter indicative of the ratio of the relative separation to the orbit radius. Thus, the solution is not uniformly valid; it is only valid for a short period of time. The perturbation approach does, however, improve the solution. Mitchell and Richardson [146] took the same approach to obtain the initial conditions for periodic solutions. Since the solution is periodic, it does not suffer from the uniform validity problem. However, it is only valid for circular reference orbits and does not include any J_2 effects. Vaddi *et al.* [147] expand the work of Mitchell and Richardson to include first-order eccentricity effects. Since they are also obtaining periodic solutions, their solution does not suffer from the uniform validity problem. The problem with the along-track secular terms is a result of the physics, not the perturbation approach. Bond [104] showed that the term leading to the secular term is actually $\sin(\Delta nt)/\Delta n$ where Δn is the difference in the mean motions of the chief and deputy. Expansion of this term gives the secular term at first-order and the higher-order terms in t at higher orders. Karlgaard and Lutze [148] take a different approach using spherical coordinates and multiple time scales to obtain a perturbation solution to second-order. Their solution is uniformly valid through second order. However, it is expected that the third-order solution would have higher-order terms in t, but the analysis has not been obtained to prove this. The reasoning is that the use of spherical coordinates results in the higher-order t terms not appearing before the third perturbation solution. The unit sphere approach of Section 7.1 is not a perturbation solution about the non-spherical Earth and does not suffer from these problems.

With the need to include gravitational effects and eccentric chief orbits, and the demonstrated evidence that differential orbital elements [149] result in the most accurate linear relative motion solution, it is advantageous to develop a perturbation solution in mean elements. In addition to the accuracy improvement with mean elements, the angles do not appear in the Hamiltonian, so there should be no $t^n, n > 1$ terms in the solution. Hence, the solution will be uniformly valid. The second-order state transition approach for propagating perturbed relative motion developed by Sengupta *et al.* [2], introduced in Section 7.7, uses an orbital elements approach but includes only the J_2 terms. The theory developed in this section includes $\mathcal{O}(J_2^2)$ terms.

We start with the averaged Hamiltonian, expanded to the second order, given by Eqs. (3.19), (3.20), and (3.23). The rates for the mean argument of latitude, argument of perigee and right ascension are given below. Since the formation is defined by differences in (a, e, i) or (L, e, i) it is preferable to have the angle rates as a function of (L, e, i) instead of (L, G, H). The angle rates as a function of (L, e, i) are:[1]

$$\dot{\lambda} = \frac{\partial \bar{\mathcal{H}}}{\partial L} + \frac{\partial \bar{\mathcal{H}}}{\partial G}$$

$$\dot{\lambda} = \frac{1}{L^3} + \epsilon \left(\frac{3}{4L^3 \eta^4} \right) \left[(1 + \eta) - (5 + 3\eta) \cos^2 i \right) + 0.5\epsilon^2 \left(\frac{3}{64L^{11} \eta^8} \right)$$
$$\times \left[\left(-35 + 9\eta + 41\eta^2 + 25\eta^3 \right) + \left(90 - 162\eta - 222\eta^2 - 90\eta^3 \right) \cos^2 i \right.$$
$$\left. + \left(385 + 465\eta + 189\eta^2 + 25\eta^3 \right) \cos^4 i \right] \tag{8.34}$$

$$\dot{g} = \dot{\omega} = \frac{\partial \bar{\mathcal{H}}}{\partial G}$$

$$\dot{\omega} = \epsilon \left(\frac{3}{4L^7 \eta^4} \right) \left(1 - 5 \cos^2 i \right) - 0.5\epsilon^2 \left(\frac{15}{320L^{11} \eta^8} \right)$$
$$\times \left[\left(35 - 24\eta - 25\eta^2 \right) + \left(-90 + 192\eta + 126\eta^2 \right) \cos^2 i \right.$$
$$\left. - \left(385 + 360\eta + 45\eta^2 \right) \cos^4 i \right] \tag{8.35}$$

$$\dot{h} = \dot{\Omega} = \frac{\partial \bar{\mathcal{H}}}{\partial H}$$

$$\dot{\Omega} = \epsilon \left(\frac{3}{2L^7 \eta^4} \right) \cos i - 0.5\epsilon^2 \left(\frac{3}{16L^{11} \eta^8} \right)$$
$$\left[\left(5 - 12\eta - 9\eta^2 \right) \cos i + \left(35 + 36\eta + 5\eta^2 \right) \cos^2 i \right] \tag{8.36}$$

The differential angle rates between two satellites in a formation are given below where the subscript "0" is used to represent the elements of the chief satellite. Since this is a nonlinear theory the Taylor series expansion is carried to second order, in contrast to the first-order expansion in the linear theories.

$$\delta \dot{\lambda} = \frac{\partial \dot{\lambda}}{\partial L} \delta L + \frac{\partial \dot{\lambda}}{\partial \eta} \delta \eta + \frac{\partial \dot{\lambda}}{\partial i} \delta i + \frac{1}{2} \frac{\partial^2 \dot{\lambda}}{\partial L^2} \delta L^2 + \frac{1}{2} \frac{\partial^2 \dot{\lambda}}{\partial \eta^2} \delta \eta^2$$
$$+ \frac{1}{2} \frac{\partial^2 \dot{\lambda}}{\partial i^2} \delta i^2 + \frac{\partial^2 \dot{\lambda}}{\partial L \partial \eta} \delta L \delta i + \frac{\partial^2 \dot{\lambda}}{\partial L \partial i} \delta L \delta i + \frac{\partial^2 \dot{\lambda}}{\partial L \partial \eta} \delta \eta \delta i \tag{8.37}$$

$$\delta \dot{\omega} = \frac{\partial \dot{\omega}}{\partial L} \delta L + \frac{\partial \dot{\omega}}{\partial \eta} \delta \eta + \frac{\partial \dot{\omega}}{\partial i} \delta i + \frac{1}{2} \frac{\partial^2 \dot{\omega}}{\partial L^2} \delta L^2 + \frac{1}{2} \frac{\partial^2 \dot{\omega}}{\partial \eta^2} \delta \eta^2$$
$$+ \frac{1}{2} \frac{\partial^2 \dot{\omega}}{\partial i^2} \delta i^2 + \frac{\partial^2 \dot{\omega}}{\partial L \partial \eta} \delta L \delta i + \frac{\partial^2 \dot{\omega}}{\partial L \partial i} \delta L \delta i + \frac{\partial^2 \dot{\omega}}{\partial L \partial \eta} \delta \eta \delta i \tag{8.38}$$

$$\delta \dot{\Omega} = \frac{\partial \dot{\Omega}}{\partial L} \delta L + \frac{\partial \dot{\Omega}}{\partial \eta} \delta \eta + \frac{\partial \dot{\Omega}}{\partial i} \delta i + \frac{1}{2} \frac{\partial^2 \dot{\Omega}}{\partial L^2} \delta L^2 + \frac{1}{2} \frac{\partial^2 \dot{\Omega}}{\partial \eta^2} \delta \eta^2$$
$$+ \frac{1}{2} \frac{\partial^2 \dot{\Omega}}{\partial i^2} \delta i^2 + \frac{\partial^2 \dot{\Omega}}{\partial L \partial \eta} \delta L \delta i + \frac{\partial^2 \dot{\Omega}}{\partial L \partial i} \delta L \delta i + \frac{\partial^2 \dot{\Omega}}{\partial L \partial \eta} \delta \eta \delta i \tag{8.39}$$

---

[1] We actually use η instead of e, where $\eta^2 = (1 - e^2)$.

Now obtain the partial derivatives of the angle rates given in Eqs. (8.34)–(8.36) and let

$$\frac{\partial \dot{\lambda}}{\partial L} = a_{10} + \epsilon a_{11} + 0.5\epsilon^2 a_{12}, \quad \frac{\partial \dot{\lambda}}{\partial \eta} = \epsilon a_{21} + 0.5\epsilon^2 a_{22}$$

$$\frac{\partial \dot{\lambda}}{\partial i} = \epsilon a_{31} + 0.5\epsilon^2 a_{32}, \quad \frac{\partial^2 \dot{\lambda}}{\partial L^2} = a_{40} + \epsilon a_{41} + 0.5\epsilon^2 a_{42}$$

$$\frac{\partial^2 \dot{\lambda}}{\partial \eta^2} = \epsilon a_{51} + 0.5\epsilon^2 a_{52}, \quad \frac{\partial^2 \dot{\lambda}}{\partial i^2} = \epsilon a_{61} + 0.5\epsilon^2 a_{62} \qquad (8.40)$$

$$\frac{\partial^2 \dot{\lambda}}{\partial L \partial \eta} = \epsilon a_{71} + 0.5\epsilon^2 a_{72}, \quad \frac{\partial^2 \dot{\lambda}}{\partial L \partial i} = \epsilon a_{81} + 0.5\epsilon^2 a_{82}$$

$$\frac{\partial^2 \dot{\lambda}}{\partial \eta \partial i} = \epsilon a_{91} + 0.5\epsilon^2 a_{92}$$

$$\frac{\partial \dot{\omega}}{\partial L} = \epsilon b_{11} + 0.5\epsilon^2 b_{12}, \quad \frac{\partial \dot{\omega}}{\partial \eta} = \epsilon b_{21} + 0.5\epsilon^2 b_{22}$$

$$\frac{\partial \dot{\omega}}{\partial i} = \epsilon b_{31} + 0.5\epsilon^2 b_{32}, \quad \frac{\partial^2 \dot{\omega}}{\partial L^2} = \epsilon b_{41} + 0.5\epsilon^2 b_{42}$$

$$\frac{\partial^2 \dot{\omega}}{\partial \eta^2} = \epsilon b_{51} + 0.5\epsilon^2 b_{52}, \quad \frac{\partial^2 \dot{\omega}}{\partial i^2} = \epsilon b_{61} + 0.5\epsilon^2 b_{62} \qquad (8.41)$$

$$\frac{\partial^2 \dot{\omega}}{\partial L \partial \eta} = \epsilon b_{71} + 0.5\epsilon^2 b_{72}, \quad \frac{\partial^2 \omega}{\partial L \partial i} = \epsilon b_{81} + 0.5\epsilon^2 b_{82}$$

$$\frac{\partial^2 \dot{\omega}}{\partial \eta \partial i} = \epsilon b_{91} + 0.5\epsilon^2 b_{92}$$

$$\frac{\partial \dot{\Omega}}{\partial L} = \epsilon c_{11} + 0.5\epsilon^2 c_{12}, \quad \frac{\partial \dot{\Omega}}{\partial \eta} = \epsilon c_{21} + 0.5\epsilon^2 c_{22}$$

$$\frac{\partial \dot{\Omega}}{\partial i} = \epsilon c_{31} + 0.5\epsilon^2 c_{32}, \quad \frac{\partial^2 \dot{\Omega}}{\partial L^2} = \epsilon c_{41} + 0.5\epsilon^2 c_{42}$$

$$\frac{\partial^2 \dot{\Omega}}{\partial \eta^2} = \epsilon c_{51} + 0.5\epsilon^2 c_{52}, \quad \frac{\partial^2 \dot{\Omega}}{\partial i^2} = \epsilon c_{61} + 0.5\epsilon^2 c_{62} \qquad (8.42)$$

$$\frac{\partial^2 \dot{\Omega}}{\partial L \partial \eta} = \epsilon c_{71} + 0.5\epsilon^2 c_{72}, \quad \frac{\partial^2 \Omega}{\partial L \partial i} = \epsilon c_{81} + 0.5\epsilon^2 a_{82}$$

$$\frac{\partial^2 \dot{\Omega}}{\partial \eta \partial i} = \epsilon c_{91} + 0.5\epsilon^2 c_{92}$$

The a_{ij}, b_{ij}, c_{ij} coefficients are given in Appendix H.

Now consider the size of the relative motion orbit. If $\delta i = \mathcal{O}(\epsilon)$ and/or $\delta e = \mathcal{O}(\epsilon)$, then the size of the relative motion orbit can be several hundred kilometers. Therefore, assume $\delta i = \mathcal{O}(\epsilon)$ and $\delta e = \mathcal{O}(\epsilon)$. If they are smaller, that is no problem. Since $\eta\delta\eta = -e\delta e$, the order of magnitude of $\delta\eta$ depends on the value of e and δe. We will assume that $\delta\eta = \mathcal{O}(\epsilon)$. Again, if it is smaller there is no problem. With these assumptions, and based on the J_2 invariant orbits research and Eq. (8.37), we see that $\delta L = \mathcal{O}(\epsilon^2)$. Therefore, set

$$\delta L = \epsilon^2 \delta L_2 + \epsilon^3 \delta L_3$$
$$\delta\eta = \epsilon\delta\eta_1 + \epsilon^2\delta\eta_2 \qquad (8.43)$$
$$\delta i = \epsilon\delta i_1 + \epsilon^2\delta i_2$$

Now substitute Eqs. (8.42)–(8.43) into Eqs. (8.34)–(8.36) and equate terms of like powers of ϵ.

$$\delta\dot{\lambda} = (a_{10}\delta L_2 + a_{21}\delta\eta_1 + a_{31}\delta i_1)\,\epsilon^2 + (a_{11}\delta L_2 + a_{10}\delta L_3 + a_{21}\delta\eta_2$$
$$+ a_{31}\delta i_2 + 0.5a_{22}\delta\eta_1 + 0.5a_{32}\delta i_1 + 0.5a_{51}\delta\eta_1^2$$
$$+ 0.5a_{61}\delta i_1^2 + a_{91}\delta\eta_1\delta i_1\Big) \qquad (8.44)$$

$$\delta\dot{\omega} = (b_{21}\delta\eta_1 + b_{31}\delta i_1)\,\epsilon^2 + (b_{11}\delta L_2 + b_{21}\delta\eta_2 + b_{31}\delta i_2$$
$$+ 0.5b_{22}\delta\eta_1 + 0.5b_{32}\delta i_1 + 0.5b_{51}\delta\eta_1^2 + 0.5b_{61}\delta i_1^2 + b_{91}\delta\eta_1\delta i_1\Big) \quad (8.45)$$

$$\delta\dot{\Omega} = (c_{21}\delta\eta_1 + c_{31}\delta i_1)\,\epsilon^2 + (c_{11}\delta L_2 + c_{21}\delta\eta_2 + c_{31}\delta i_1$$
$$+ 0.5c_{22}\delta\eta_1 + 0.5c_{32}\delta i_1 + 0.5c_{51}\delta\eta_1^2 + 0.5c_{61}\delta i_1^2 + c_{91}\delta\eta_1\delta i_1\Big) \quad (8.46)$$

Setting any of these three equations equal to zero provides a constraint on the relative motion. The impact of each of the $\mathcal{O}(\epsilon^2)$, linear theory, constraints on the relative motion and design of satellite formations was investigated in Section 8.1. The effect of one, two and three constraints were investigated with the $\delta\dot{\lambda} = 0$ constraint being replaced by $\delta\dot{\lambda} + \delta\dot{\Omega}\cos i = 0$, which is the no along-track drift condition. Just as with the linear theory, the only solution with all three constraints is the trivial solution, $\delta L = \delta\eta = \delta i = 0$. However, the no along-track drift constraint is a linear theory constraint, not a nonlinear theory constraint. The linear no along-track drift constraint from Eqs. (8.44)–(8.46) is

$$\delta L_2 = -\epsilon\frac{(3\eta_0 + 4)}{4L_0^3\eta_0^5}\left[\left(1 - 3\cos^2 i_0\right)\delta\eta - (\eta_0\sin 2i_0)\,\delta i\right] \qquad (8.47)$$

or

$$\delta a = 0.5J_2 a_0 \left(\frac{R_e}{a_0}\right)^2 \left(\frac{3\eta_0 + 4}{2\eta_0^5}\right)$$
$$\times \left[\left(1 - 3\cos^2 i_0\right)\delta\eta - (\eta_0\sin 2i_0)\,\delta i\right] \qquad (8.48)$$

The second-order correction for the semimajor axis for the linear no along-track drift constraint is

$$
\begin{aligned}
a_{10}\delta L_3 = {} & -(a_{11} + c_{11}\cos i)\,\delta L_2 \\
& -(a_{21} + c_{21}\cos i)\,\delta\eta_2 - (a_{31} + c_{31}\cos i)\,\delta i_2 \\
& -0.5\,(a_{22} + c_{22}\cos i)\,\delta\eta_1 - 0.5\,(a_{32} + c_{32}\cos i)\,\delta i_1\delta\eta_1^2 \\
& -0.5\,(a_{31} + c_{31}\cos i) - 0.5\,(a_{61} + c_{61}\cos i)\,\delta i_1^2 \\
& -(a_{91} + c_{91}\cos i)\,\delta\eta_1\delta i_1
\end{aligned}
\tag{8.49}
$$

Again, it is important to point out that this is not a nonlinear constraint, it is the linear constraint that includes the higher-order terms in the expansion. The nonlinear no along-track drift condition is an open question at this time.

The constraint for no out-of-plane drift is $\delta\dot{\Omega} = 0$. The first-order constraint from Eq. (8.46) is

$$
\delta\eta_1 = -(0.25\eta_0\tan i_0)\delta i_1
\tag{8.50}
$$

which is identical to Eq. (8.14) in Section 8.1. As shown in Section 8.1, if the out-of-plane motion is created by a differential inclination and the chief orbit is near-circular or near-polar, then the required eccentricity will result in a large formation, and is likely not feasible. Substituting Eq. (8.50) into Eq. (8.46) and setting it equal to zero gives the solution for $\delta\eta_2$ and δi_2:

$$
\begin{aligned}
c_{21}\delta\eta_2 + c_{31}\delta i_2 = {} & -c_{11}\delta L_2 - 0.5c_{22}\delta\eta_1 - 0.5c_{32}\delta i_1 - 0.5c_{51}\delta\eta_1^2 \\
& -0.5c_{61}\delta i_1^2 - c_{91}\delta\eta_1\delta i_1
\end{aligned}
\tag{8.51}
$$

All of the potential cases are beyond the scope of discussion, but for the use of this theory some discussion is in order. Consider first the case of one constraint, the no along-track drift constraint, and that δi and $\delta\eta$ are defined by the desired formation. Also assume that the no along-track drift constraint is the linear theory constraint $\delta\dot{\lambda} + \delta\dot{\Omega}\cos i = 0$. Then set $\delta\eta_1 = \delta\eta$, $\delta i_1 = \delta i$ and $\delta\eta_2 = \delta i_2 = 0$ and compute δL_2 and δL_3 using Eqs. (8.47) and (8.49). These provide the mean momenta and then with the mean differential angles $(\delta l, \delta\omega, \delta\Omega)$ defined by the formation obtain the mean deputy elements and transform to osculating elements to obtain the initial conditions. Now consider the two constraint case and assume the second constraint is negation of the out-of-plane drift, $\delta\dot{\Omega} = 0$. Also assume that the formation defines δe. Then compute $\delta\eta$ from

$$
\delta\eta = \sqrt{1 - (e_0 + \delta e)^2} - \sqrt{1 - e_0^2}
\tag{8.52}
$$

Set $\delta\eta_1 = \delta\eta$ and $\delta\eta_2 = 0$. Then compute δi_1 using Eq. (8.50), δL_2 using Eq. (8.47), δi_2 using Eq. (8.51), and δL_3 using Eq. (8.49). Then,

$$
\delta L = \epsilon^2\delta L_2 + \epsilon^3\delta L_3, \quad \delta i = \epsilon\delta i_1 + \epsilon^2\delta i_2
\tag{8.53}
$$

With the angles defined by the formation, the mean elements are now defined. Transform to osculating elements to obtain the osculating elements and corresponding position and velocity.

To investigate the effect of adding the nonlinear effects to the model, consider the following example.

Example 8.1. *Choose a projected circular orbit (PCO) with the out-of-plane motion created by a differential inclination. The initial mean elements of the chief are:*

$$a_0 = 8000 \text{ km}, \quad e_0 = 0.01, \quad i_0 = 50° \tag{8.54}$$

$$M_0 = \omega_0 = \Omega_0 = 0$$

Using Ref. [123], the initial differential mean elements of the deputy are:

$$\delta q_1 = 0, \quad \delta q_2 = -\frac{\rho}{2a_0} \tag{8.55}$$

$$\delta i = \frac{\rho}{a_0}, \quad \delta\Omega = 0, \quad \delta\lambda = 0$$

To evaluate the accuracy of the no along-track drift computation as a function of the formation size, the PCO radius is varied from 800 m to 160 km. The differential semimajor axis to negate the along-track drift is computed in four ways: (i) using the first-order no along-track drift condition, Eq. (8.48); (ii) using the second-order correction, Eqs. (8.47) and (8.49) without the J_2^2 term in the Hamiltonian; (iii) using the second-order correction, Eqs. (8.47) and (8.49) with the J_2^2 term in the Hamiltonian; and (iv) a numerical method. The numerical method is performed by numerically integrating the equations of motion of the chief and deputy for 10 orbits, differencing them to obtain the along-track motion, averaging it and then iterating to find the differential semimajor axis that negates the along-track secular drift in the averaged along-track motion. More details are given in Ref. [150].

Table 8.3 shows the results. The along-track drift per orbit is given by $d = 3\pi \delta a$. With the numerical δa as the value that results in no drift, the drift that occurs with each of the approximations can be obtained by using for δa in the drift equation the difference between the numerical δa and the δa for the approximation. For an 8 km PCO radius the error in δa using the first-order or linear theory is only 3.6 cm. Including the second-order effects, but without including the J_2^2 terms in the Hamiltonian only reduces the error to 2 cm, whereas including the terms in the Hamiltonian reduces the error to less than 0.1 mm. For the 160 km PCO, the error in δa from the nonlinear theory when including the J_2^2 terms in the Hamiltonian is only 1.5 cm, but the linear along-track drift condition results in an error of 2.6 m, which leads to a drift of 25 m/orbit. It should be noted that using the CW equations results in a δa error equal to that in the last column because $\delta a = 0$ for no drift when using the CW equations. Keeping in mind that these results were computed for the more stressing condition where the out-of-plane motion is caused completely

Table 8.3 δa (m) For the along-track drift

ρ (km)	First-order	Second-order without J_2^2	Second-order with J_2^2	Numerical
0.80	−0.1879	−1.8968	−1.8948	−1.8948
1.60	−3.7958	−3.7934	−3.7895	−3.7895
4.00	− 9.4895	−9.4828	−9.4730	−9.4730
8.00	− 18.9790	−18.9632	−18.9435	−18.9435
16.00	− 37.9580	−37.9167	−37.8775	−37.8774
40.00	− 94.8950	−94.7189	−94.6218	−94.6199
80.00	− 189.7901	−189.1948	−189.0039	−188.9892
160.00	− 379.5801	−377.4179	−377.0487	−376.9339

by a differential inclination, one can conclude that (a) the linear no along-track drift condition probably provides satisfactory results for formations up to at least 8 km in size; (b) the addition of the nonlinear terms without including the J_2^2 in the Hamiltonian provides some, but not substantial improvement; and (c) including the J_2^2 in the Hamiltonian provides much better results than not including them. It must be kept in mind that when using the nonlinear theory it may be necessary to include other zonal harmonics in the theory since they are of the same order of magnitude as the J_2^2 terms. This has not been done here, but is straightforward given the averaged Hamiltonian and using the approach presented herein. In addition, it should be noted that the long-periodic terms appear as secular terms of $\mathcal{O}(J_2^2)$ over a period of time much shorter than the perigee rotation period. Since the numerical procedure does not average over a long period of time, the computed δa would vary depending on which portion of the period was being calculated. The long-periodic terms are proportional to the eccentricity, and in this particular example $e = 0.01$; consequently, the effect would be very small.

8.3 DYNAMIC MODEL ERROR EFFECT COMPARISON

When designing control and navigation subsystems for a satellite formation, one of the key – and challenging – questions that must be answered is which relative motion theory should be used. The answer to this question is dependent on the reference, or chief, orbit, the size and type of the formation and the required accuracy of the state knowledge and control.

Numerous linear and nonlinear theories have been presented in the literature. The theories compared in this section are the CW and TH equations, discussed in Chapter 5; the GA theory [75], presented in Section 7.3; a small eccentricity version of the GA theory; and two nonlinear theories, the unit-sphere approach, presented in Section 7.1, and the Yan–Alfriend theory [123], discussed in the previous section. The equations for the small-eccentricity theory are given in Appendix G. To help the designer, this section presents a comparison of the errors that occur with various theories. However, the results should not be considered as a theory comparison, but as a comparison of the errors associated

Table 8.4 Theory comparison

Theory	Eccentricity	J_2	Nonlinearity
CW	$e = 0$	No	No
Lawden	$0 \leq e < 1$	No	No
Gim–Alfriend	$0 \leq e < 1$	Yes	No
Small e	$e \ll 1$	Yes	No
Unit sphere	$0 \leq e < 1$	Yes	Yes
Yan–Alfriend	$0 \leq e < 1$	Yes	Yes

with not modeling the effects of the chief eccentricity, J_2 and nonlinearities. Table 8.4 summarizes the effects modeled with these theories.

For comparing the effect of not modeling the various effects a modeling error index is introduced. This index does not require linearization of the equations of motion so nonlinear theories can also be evaluated. The index can be thought of as proportional to the percentage error, consequently the smaller the index the more accurate the theory. The two key parameters in the evaluation are the eccentricity of the reference orbit and the size measure of the relative motion orbit. This nonlinear index is a variation of one introduced by Junkins [151] for the comparison of linear theories.

Consider the nonlinear differential equations of motion

$$\dot{\mathbf{x}} = \mathbf{f}(\mathbf{x}, t), \quad \mathbf{x}(t_0) = \mathbf{x}_0 \tag{8.56}$$

Junkins [151] used as the nonlinearity index

$$\nu = \sup_{i=1,\dots,N} \frac{\| \Phi_i (t, t_0) - \bar{\Phi}(t, t_0) \|}{\| \bar{\Phi}(t, t_0) \|} \tag{8.57}$$

where $\bar{\Phi} (t, t_0)$ is a state transition matrix that is obtained from Eq. (8.56) with the expected initial condition $\bar{x}(t, t_0)$; $\Phi (t, t_0)$ is a state transition matrix that is obtained from Eq. (8.56) with a worst-case distribution of initial conditions neighboring the expected initial conditions. For example, if the problem is a breakup or explosion, the worst-case initial conditions would be the points on the surface of the 3D ellipsoid of the relative velocity created by the breakup. This approach was used in Ref. [151] to compare the accuracy of three linear theories for a circular reference orbit: The CW equations, the linear theory using polar coordinates and one obtained from differential orbital elements. The objective was to determine which theory best captured the nonlinear effects. The comparison showed that differential orbital elements were the most accurate. The objective here is different; it is to evaluate the effect of the modeling error even when the theory is nonlinear, and to compare the accuracy of the theories for specific types of relative motion orbits. Thus, a modification of the Junkins nonlinearity index is needed. Let $\bar{\mathbf{x}}_i(t)$ be the solution for the initial condition $\bar{\mathbf{x}}_i(t_0)$ at the corresponding points and let $\mathbf{x}_i(t)$ be the solution for the model being considered. These need not be linearized solutions. It is important that the states be in dimensionless variables or a weighting matrix used. Let $\mathbf{W}$ be a weighting matrix that non-dimensionalizes $\mathbf{x}$, that is

$$\mathbf{y} = \mathbf{W}\mathbf{x} \tag{8.58}$$

Now define

$$\nu = \max_{i=1,\ldots,N} \nu_i \tag{8.59}$$

$$\nu_i = \left(\frac{\bar{\mathbf{y}}_i^T \bar{\mathbf{y}}_i}{\mathbf{y}_i^T \mathbf{y}_i} - 1 \right) \tag{8.60}$$

Note that if there is only one state and we let $y_i = \bar{y}(1 + \gamma)$ then $\nu = 2|\gamma|$. Therefore, the index is proportional to the percentage error.

The index provides a method for comparing the errors that occur for the various theories for a specific formation. As shown earlier, in the one-dimensional case the index is representative of twice the percentage error. In the n-dimensional case, the acceptable value for the index can only be determined by the acceptable errors for the mission. Many factors, such as time between thruster firings, allowable error growth and allowable error impact on the decision of which specific model to use. The method presented here only provides a guideline for determining the dynamic effects that need to be included in the model for a specific problem. Also, keep in mind that the index represents the maximum error over all the initial conditions for the PCO. For a specific formation, the index should be evaluated over the range of possible initial conditions.

There are formations for which the modeling errors are minimal. For example, because the differential J_2 effects are caused primarily by the inclination difference for small eccentric orbits, the effect of not modeling J_2 is small if there is no out-of-plane motion, or if the out-of-plane motion is caused by a right ascension difference. The differential eccentricity effect is caused by δe^2. Therefore, its effect for near-circular orbits is small, but for highly-eccentric orbits, such as MMS (cf. Section 1.7), it can have as much impact as the inclination difference.

The effect of the modeling error is a function of the relative motion orbit. For example, a differential inclination causes out-of-plane drift, so any relative motion orbit that does not have a differential inclination will experience different types of errors. For the theory comparison, we have chosen the PCO. This is one of the more stressing cases since it involves out-of-plane motion. Actually, the PCO does not exist if the chief eccentricity is not zero. However, by correctly choosing the initial conditions, a periodic orbit which is a deformed PCO can be found for a spherical Earth. Finding the initial conditions is not simple in Cartesian coordinates, but finding them is straightforward using differential orbital elements.

For the non-spherical Earth model, it is best to use mean elements because the angular rates are constant, which means that the constraints to minimize

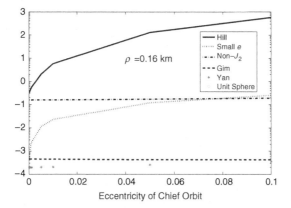

FIGURE 8.15 Index comparison for $\rho = 0.16$ km.

or prevent drift between the satellites are a function of only the momenta or (a, e, i). When including J_2, the differential elements given in Eq. (6.69) should be mean elements and as mentioned in Section 7.1, the $\delta a = 0$ is replaced by

$$
\delta \bar{a} = -0.5 J_2 \bar{a}_0 \left(\frac{R_e}{\bar{a}_0} \right)^2 \left(\frac{3\bar{\eta}_0 + 4}{\bar{\eta}_0} \right)
$$
$$
\times \left[\left(1 - 3 \cos^2 \bar{i}_0 \right) \frac{\bar{q}_{10} \delta \bar{q}_1 + \bar{q}_{20} \delta \bar{q}_2}{\bar{\eta}_0^2} + \delta \bar{i} \sin 2\bar{i}_0 \right] \tag{8.61}
$$

This constraint serves the same function with J_2 as $\delta a = 0$ does with no J_2; it negates the along-track drift. The derivation of Eq. (8.61) is discussed in more detail in Section 8.1.

The orbital elements of the chief used for the error effect evaluation are:

$$
\bar{a}_0 = 8000 \text{ km}, \quad \bar{i}_0 = 50°, \quad \bar{\Omega}_0 = \bar{\omega}_0 = \bar{M}_0 = 0 \tag{8.62}
$$

First, consider the effect of the eccentricity of the reference orbit. Figures 8.15–8.17 show the effect of the modeling errors for the various models as a function of the chief eccentricity for PCO sizes of 0.16, 12 and 40 km, respectively. From these figures, the following conclusions are evident: (i) the CW equations are inaccurate, even for a small chief eccentricity. The index increases quickly and at $\bar{e}_0 = 0.1$ is almost two orders of magnitude greater than the index for the non-J_2 theory (TH equations); (ii) not including J_2 has a significant effect. The index for the TH equations is more than two orders of magnitude larger than that for the GA theory. In fact, the small e theory that includes first-order eccentricity effects in the non-J_2 terms and assumes $\bar{e}_0 = 0$ in the J_2 terms is more accurate than TH equations until $\bar{e}_0 \approx 0.06$; (iii) the unit sphere and the Yan–Alfriend nonlinear theories provide equivalent accuracy, although the two theories are different. The Yan-Alfriend theory includes J_2^2 terms, and by virtue of using orbital elements and the expansion in powers of J_2, includes nonlinear

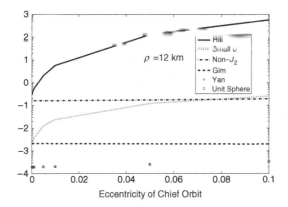

FIGURE 8.16 Index comparison for $\rho = 12$ km.

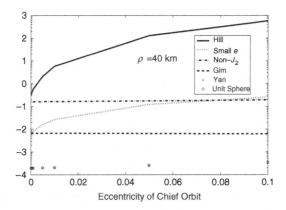

FIGURE 8.17 Index comparison for $\rho = 40$ km.

effects. By mapping the relative motion onto the unit sphere, the unit-sphere method better models the nonlinear effects, but does not include J_2^2 effects. These differences appear to balance each other, and result in the equivalent accuracy seen for the investigated examples.

Figures 8.18–8.20 show the effect of the modeling errors for the various models as a function of formation size for chief eccentricities of 0.001, 0.01 and 0.1, respectively. From these figures the following conclusions are evident. (i) The errors caused by not including the chief eccentricity and J_2 in the CW equations dominate any error caused by the linearization. The index is flat for formations up to 160 km in size. (ii) The errors caused by not including J_2 in the TH equations dominate any error caused by the linearization. The index is flat for formations up to 160 km in size, but smaller than the index for the CW equations because the chief eccentricity is included in the model. (iii) The unit sphere and Yan–Alfriend nonlinear theories provide equivalent accuracy. (iv) Even though $\nu < 0.01$ for the GA theory, for formation sizes less than 40 km it is still an order of magnitude larger than the nonlinear theories.

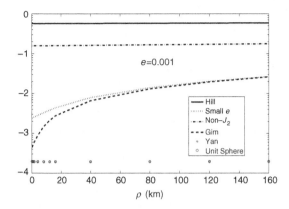

FIGURE 8.18 Index comparison for $e = 0.001$.

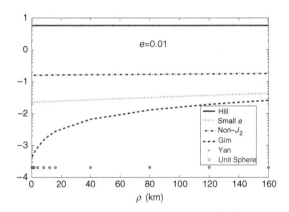

FIGURE 8.19 Index comparison for $e = 0.01$.

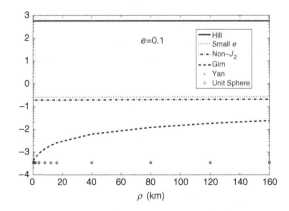

FIGURE 8.20 Index comparison for $e = 0.1$.

The properties of the perturbed relative orbit, under the no along-track drift constraint, are developed further in the following sections. The initial conditions on the differential orbital elements for the PCO relative orbit are related to its shape and size parameters. Furthermore, a basis for designing an intelligent control law for formation maintenance and *fuel balancing* is presented by exploiting the properties of the natural relative motion. We will use the classical orbital element representation in the rest of this chapter.

8.4 PERTURBED FUNDAMENTAL FREQUENCIES FOR FORMATIONS IN NEAR-CIRCULAR ORBITS

The *fundamental frequency* of the unperturbed periodic relative motion in the two-body problem is the mean motion of the chief (or the deputy). In general, the in-plane and cross-track fundamental frequencies are not the same for perturbed relative motion, due to perigee rotation and nodal precession. Knowledge of the perturbed frequencies due to J_2 aids in the design and tuning of the controllers for mitigating its adverse effects efficiently. The perturbations of the in-plane and cross-track frequencies associated with formations in near-circular orbits is treated in this section [139].

The linearized relative motion equations expressed in terms of the differential orbital elements were derived in Section 7.1. Inspection of Eq. (7.18) reveals that in the absence of differential nodal precession, the fundamental cross-track frequency is $\dot{\lambda}_1 = \dot{M}_1 + \dot{\omega}_1$, the rate of change of the mean argument of latitude of the deputy. This is no longer true if $\delta\Omega \neq 0$. Equations (7.16)–(7.17) can be simplified further by expanding r and θ to first-order in eccentricity and neglecting terms involving frequencies $2M$ and higher:

$$r \approx a(1 - e\cos M) \tag{8.63}$$

$$\theta \approx \omega + M + 2e\sin M \tag{8.64}$$

$$\sin\theta \approx \sin\lambda - e\sin\omega \tag{8.65}$$

$$\cos\theta \approx \cos\lambda - e\cos\omega \tag{8.66}$$

For orbits with small eccentricities and even for small formations, δM and $\delta\omega$ can be large, but $\delta\lambda$ remains small. Substitution of Eqs. (8.63)–(8.66) in Eqs. (7.16)–(7.18) results in

$$x \approx \delta a + a_0[(e_1\sin\delta M)\sin M_0 + (e_0 - e_1\cos\delta M)\cos M_0] \tag{8.67}$$

$$\begin{aligned} y \approx{}& a_0[\delta\lambda + \delta\Omega\cos i_0 - e_0(e_1\sin\delta M)] \\ &+ 2a_0[-(e_0 - e_1\cos\delta M)\sin M_0 + (e_1\sin\delta M)\cos M_0] \end{aligned} \tag{8.68}$$

$$z \approx a_0\left[\delta i\sin\lambda_0 - \sin i_0\delta\Omega\cos\lambda_0 - \frac{3}{2}e_0(\delta i\sin\omega_0 - \sin i_0\delta\Omega\cos\omega_0)\right] \tag{8.69}$$

It has been shown previously in Eq. (8.15) that δa, appearing in Eq. (8.67), is required to be $\mathcal{O}(J_2)$ for preventing along-track secular drift [143]. Equations (8.67)–(8.69) reveal, upon further analysis, the frequencies of motion and a means to set up the initial conditions for formations in near-circular orbits.

8.4.1 In-plane and cross-track frequencies

The analysis to follow is based on the mean orbital elements. At this point, the secular drift rates $\delta\dot\lambda$ and $\delta\dot\Omega$ can be substituted into Eqs. (8.67)–(8.68) to obtain

$$x \approx \delta a + \rho_x \sin(M_0 + \alpha_x) \tag{8.70}$$

$$\begin{aligned} y \approx a_0[\delta\lambda(0) + \delta\Omega(0)\cos i_0 - e_0(e_1 \sin \delta M(0))] \\ + a_0[\delta\dot\lambda + \delta\dot\Omega \cos i_0]t + 2\rho_x \cos(M_0 + \alpha_x) \end{aligned} \tag{8.71}$$

where

$$\rho_x = a_0\sqrt{(e_1 \sin \delta M)^2 + (e_0 - e_1 \cos \delta M)^2} \tag{8.72}$$

is the amplitude of the radial periodic motion and

$$\alpha_x = -\frac{\pi}{2} + \tan^{-1}\left[-\frac{e_1 \sin \delta M}{e_0 - e_1 \cos \delta M}\right] \tag{8.73}$$

is the in-plane phase angle. Equations (8.72)–(8.73) show that for circular reference orbits, $\rho_x = a_0 e_1$ is a constant and $\alpha_x = -\frac{\pi}{2} + \delta M$. Hence, since $\delta M = \delta M(0) + \delta\dot M t$, the in-plane frequency is $n_{xy} = \dot M_0 + \delta\dot M = \dot M_1$, the mean motion of the deputy. For small formations, the differential inclination itself is $\mathcal{O}(J_2)$. Hence, the approximation

$$n_{xy} \approx \dot M_0 \tag{8.74}$$

is valid.

Equation (8.69) can also be represented as follows:

$$z \approx \rho_z \sin(\lambda_0 + \alpha_z) - \frac{3}{2}\rho_z e_0 \sin(\omega_0 + \alpha_z) \tag{8.75}$$

where

$$\rho_z \approx a_0\sqrt{\delta i^2 + \sin^2 i_0(\delta\Omega(0) + \delta\dot\Omega t)^2} \tag{8.76}$$

and

$$\alpha_z = \tan^{-1}\left[-\frac{\sin i_0(\delta\Omega(0) + \delta\dot\Omega t)}{\delta i}\right] \tag{8.77}$$

It can be shown by differentiating Eq. (8.77) that

$$\dot\alpha_z = \frac{-\sin i_0 \delta\dot\Omega \delta i}{\delta i^2 + (\sin i_0 \delta\Omega)^2} \tag{8.78}$$

As can be seen from the above equation, $\dot\alpha_z$ is not a constant due to the action of differential nodal precession. However, focusing attention on a relatively small time-scale and ignoring the effect of $\delta\dot\Omega$ in the denominator of the above equation, an estimate of the rate of change of the phase angle is

$$\dot\alpha_z \approx \frac{-\sin i_0 \delta\dot\Omega \delta i}{\delta i^2 + (\sin i_0 \delta\Omega(0))^2} \tag{8.79}$$

The above analysis shows that the cross-track natural frequency, n_z, can be approximated as

$$n_z \approx \dot M_0 + \dot\omega_0 - \frac{\sin i_0 \delta\dot\Omega \delta i}{\delta i^2 + (\sin i_0 \delta\Omega(0))^2} \tag{8.80}$$

The result presented above is accurate to $\mathcal{O}(J_2)$ and valid for near-circular orbits, over a finite period of time ($\approx$ 1 day).

8.5 SELECTION OF THE PCO INITIAL CONDITIONS FOR NEAR-CIRCULAR ORBITS

The procedures for determining the initial conditions for perturbed PCO and GCO have been presented in Refs. [152,153] and also in Eqs. (6.69). The development herein utilizes the approximate solutions to the relative motion variables given by Eqs. (8.67)–(8.69), which are valid for near-circular orbits. The along-track and cross-track initial conditions for a PCO can be parameterized as

$$y(0) = \rho(0)\cos(\lambda_0(0) + \alpha(0)) \tag{8.81}$$

$$z(0) = \rho(0)\left[\sin(\lambda_0(0) + \alpha(0)) - \frac{3}{2}e_0\sin(\omega_0 + \alpha(0))\right] \tag{8.82}$$

where $\alpha(0)$ is the desired initial phase angle and $\rho(0)$ is the initial radius of the PCO, in the yz plane. Note that Eqs. (8.81) and (8.82) are valid at the initial time and do not describe the ensuing motion. Hence, the same initial phase angle is used for $y(0)$ and $z(0)$. Upon comparing Eq. (8.71) with Eq. (8.81) and similarly, Eq. (8.75) with Eq. (8.82), it is observed that $\rho_x = 0.5\rho(0)$ and $\rho_z = \rho(0)$.

The initial differential orbital elements satisfying the conditions of Eqs. (8.81)–(8.82) can be obtained from Eqs. (8.68)–(8.69) as

$$\delta\lambda(0) + \delta\Omega(0)\cos i_0 = \frac{\rho(0)e_0}{2a_0}\cos(\omega_0(0) + \alpha(0)) \tag{8.83}$$

$$e_1\sin\delta M(0) = \frac{\rho(0)}{2a_0}\cos(\omega_0(0) + \alpha(0)) \tag{8.84}$$

$$e_1\cos\delta M(0) = e_0 - \frac{\rho(0)}{2a_0}\sin(\omega_0(0) + \alpha(0)) \tag{8.85}$$

$$\delta\Omega(0) = -\frac{\rho(0)}{a_0\sin i_0}\sin\alpha(0) \tag{8.86}$$

$$\delta i = \frac{\rho(0)}{a_0}\cos\alpha(0) \tag{8.87}$$

Equations (8.83)–(8.87) do not contain small divisors, except for the case of equatorial orbits. An eccentricity-induced bias term is present in the RHS of Eq. (8.83). Equations (8.84) and (8.85) provide a means to calculate e_1 and $\delta M(0)$, thus enabling the computation of $\delta e = e_1 - e_0$. For circular orbits, $\delta e(0) = \rho(0)/(2a_0)$ and $\tan(\delta M(0)) = -\cot(\omega_0(0) + \alpha(0))$. Finally, $\delta\lambda(0)$ can be determined from Eqs. (8.86) and (8.83). The required change in the semimajor axis is obtained from the no along-track constraint, Eq. (8.15).

Example 8.2. *Consider the example of setting up of the PCOs with the following mean orbital elements of the chief:*

$$a_0 = 7092 \text{ km}, \quad e_0 = 0, \quad i_0 = 70° \\ \Omega_0 = 45°, \quad \omega_0 = 0°, \quad M_0 = 0° \tag{8.88}$$

The initial conditions of the deputy for the example mean elements of the chief are obtained from Eqs. (8.83)–(8.87) and Eq. (8.15). Evolutions of four PCOs, with $\rho = 1$ km and $\alpha(0)$ values ranging from $0°$ to $90°$, are shown in Fig. 8.21 for 50 orbits of the chief. The relative orbits do remain bounded along the y-axis, but also present in each figure, to varying extents, is secular growth along the z-axis and precession, due to non-commensurate in-plane and out-of-plane frequencies. Note also the variations of the precession patterns for different values of $\alpha(0)$. The evolution of the error in the radius of the PCO for a deputy with $\alpha(0) = 0$ is shown over ten orbital periods of the chief in Fig. 8.22. The secular error over this period reaches approximately 45 m, i.e., the rate of growth is 4.5 m/orbit.

For circular reference orbits, the first-order differential drift rates are functions of δa and δi:

$$\delta\dot{\lambda} = \frac{\partial\dot{\lambda}}{\partial a}\delta a + \frac{\partial\dot{\lambda}}{\partial i}\delta i \tag{8.89a}$$

$$\delta\dot{\Omega} = \frac{\partial\dot{\Omega}}{\partial i}\delta i \tag{8.89b}$$

and they can be evaluated by using the following partial derivatives:

$$\frac{\partial\dot{\lambda}}{\partial i} = -6Jn_0 \sin 2\bar{i}_0 \tag{8.90a}$$

$$\frac{\partial\dot{\lambda}}{\partial a} = -\frac{3n_0}{2\bar{a}_0} \tag{8.90b}$$

$$\frac{\partial\dot{\Omega}}{\partial i} = \frac{3}{2}Jn_0 \sin \bar{i}_0 \tag{8.90c}$$

where J is as defined in Section 7.5. Even though the mean elements of the chief are used in the partial derivatives shown in Eq. (8.90), these equations are valid for estimating the differential drift rates of the deputies as well, since a first-order approximation is being adopted.

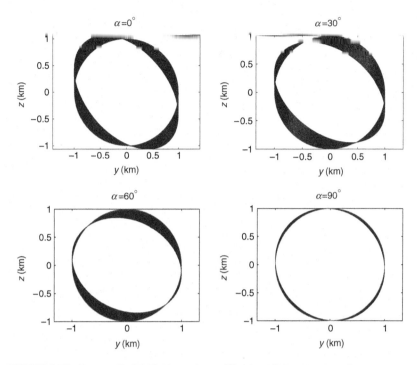

FIGURE 8.21 Uncontrolled PCOs for various $\alpha(0)$ values (50 orbits shown).

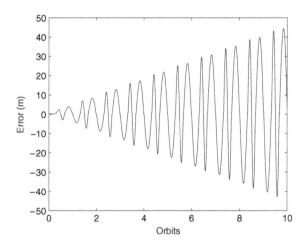

FIGURE 8.22 Evolution of the error in the radius of the PCO for an uncontrolled satellite with $\alpha(0) = 0$.

8.6 MATCHING THE IN-PLANE AND CROSS-TRACK FUNDAMENTAL FREQUENCIES

A mismatch between the in-plane and cross-track frequencies causes precession of the relative orbit, which may not be desirable for some applications. The condition under which the two frequencies are equal is derived in this section. Equations (8.74) and (8.80) show that the difference between the in-plane and cross-track frequencies, over a short time interval (compared to the period of the differential nodal precession rate) is

$$n_z - n_{xy} = \dot{\omega}_0 - \frac{\delta\dot{\Omega}\delta i \sin i_0}{\delta i^2 + (\delta\Omega(0)\sin i_0)^2} \tag{8.91}$$

The rates of perigee rotation and differential nodal precession, valid for circular orbits, can respectively, be written as

$$\dot{\omega}_0 = -k\left(2 - \frac{5}{2}\sin^2 i_0\right) \tag{8.92}$$

and

$$\delta\dot{\Omega} = -k\sin i_0\delta i \tag{8.93}$$

where

$$k = -1.5J_2\left(\frac{R_e}{a_0}\right)^2 n_0 \tag{8.94}$$

Thus, the frequency mismatch is estimated to be

$$n_z - n_{xy} = k\sin^2 i_0\left(\frac{5}{2} + \frac{\delta i^2}{\delta i^2 + (\delta\Omega(0)\sin i_0)^2}\right) - 2k \tag{8.95}$$

For the special case of the PCO, substitution of Eqs. (8.86)–(8.87) into Eq. (8.95) leads to the following result:

$$i_0^* = \sin^{-1}\left(\sqrt{\frac{2}{2.5 + \cos^2\alpha(0)}}\right) \tag{8.96}$$

Thus, the frequency-matching condition is satisfied by two possible values of the chief's orbit inclination for any $\alpha(0)$. The inclinations for $\alpha(0) = 0$ are $i_0^* = 49.11°$ and $130.89°$ and those for $\alpha(0) = 90°$ are the critical inclination values: $i_0^* = 63.43°$ and $116.57°$. The numerical results presented in Refs. [154, 155] agree very closely with the special inclination values determined above.

Figure 8.23 shows 15 relative motion orbits for a 7100 km circular reference orbit with $\alpha(0) = 0$ and $i_0 = 49.11°$. The effect of the frequency-matching condition is apparent from a comparison of Fig. 8.23 and Fig. 8.21. The mismatch in the in-plane and cross-track frequencies results in the precession of the PCO shown in Fig. 8.21. If allowed to continue in this manner indefinitely, without control, the PCO of Fig. 8.23 will also distort significantly due to the effect of differential nodal precession.

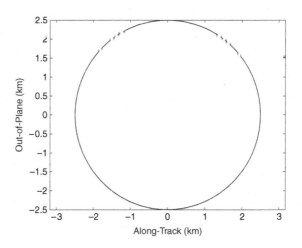

FIGURE 8.23 Frequency matched relative orbits for $i_0 = 49.11°$ (15 orbits).

It can be shown by substituting Eq. (8.93) into Eq. (8.79) and from Eqs. (8.86)–(8.87) that for a PCO formation, the natural rate of change of the cross-track phase angle as a function of its initial phase angle is given by

$$\dot{\alpha}_z = k \sin^2 i_0 \cos^2 \alpha(0) \qquad (8.97)$$

Considering an infinite number of satellites distributed in a PCO ($\alpha(0) \in [0, 2\pi]$), the average rate of change of the phase angle can be obtained as

$$(\dot{\alpha}_z)_{ave} = \frac{1}{2} k \sin^2 i_0 \qquad (8.98)$$

Note that $\dot{\alpha}_z$ and $(\dot{\alpha}_z)_{ave}$ do not depend on the size of the relative orbit.

8.6.1 Amplitude considerations

This section treats the effects of the parameters, i_0 and $\alpha(0)$ on the cross-track amplitude growth-rate. The in-plane amplitude variation is negligible, as long as Eq. (8.15) is satisfied.

Equation (8.76) can be expressed directly in terms of i_0 and $\alpha(0)$ via Eqs. (8.86), (8.87), and (8.93). The result obtained is

$$\rho_z \approx \rho \sqrt{1 + k^2 t^2 \sin^4 i_0 \cos^2 \alpha(0) + kt \sin^2 i_0 \sin 2\alpha(0)} \qquad (8.99)$$

Equation (8.99) shows that for non-equatorial orbits, cross-track amplitude growth rate is zero for the special case: $\alpha(0) = 90°, 270°$ or equivalently, for $\delta i = 0$. Another special case of zero linear-growth is obtained, corresponding to $\alpha(0) = 0, 180°$, by neglecting the t^2 term in Eq. (8.99). This assumption is reasonable for a time period of one day. Equations (8.96)–(8.99), together,

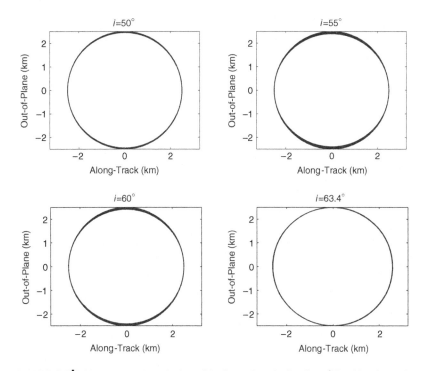

FIGURE 8.24 Non-precessing relative orbits for various inclinations (15 orbits shown for each inclination).

explain the existence of the so-called "magic inclinations" for obtaining non-precessing or "frozen" relative orbits [154,155].

Relative orbits for other values of i_0, set up with the corresponding values of $\alpha(0)$ given by Eq. (8.96), are shown in Fig. 8.24. The orbits in these figures do not precess, but show localized, cross-track growth for inclinations corresponding to $\alpha(0) \neq 0, 90°$.

8.7 PCO FORMATION MAINTENANCE BASED ON THE MODIFIED CW EQUATIONS

In this section, expressions for the components of the J_2-induced perturbing acceleration vector are derived for circular orbits ($e_0 = 0$) by retrofitting a modified set of CW equations to the analytical solutions given by Eqs. (8.67)–(8.69). This objective is achieved by differentiating each of Eqs. (8.67)–(8.69) twice, while holding δa, δe, and δi constant. Furthermore, since δa is $\mathcal{O}(J_2)$, the term $\dot{M}_0^2 \delta a$ is approximated by $n_0^2 \delta a$ where n_0 is the unperturbed two-body mean motion of the chief. The results are

$$\ddot{x} - 2\dot{M}_0\dot{y} - 3\dot{M}_0^2 x = -3n_0^2\delta a - 2n_0 a_0 \left(\delta\dot{\lambda} + \delta\dot{\Omega}\cos i_0\right) \quad (8.100a)$$

$$\ddot{y} + 2\dot{M}_0\dot{x} = 0 \quad (8.100b)$$

$$\ddot{z} + \dot{\lambda}_0^2 z = 2a_0\dot{\lambda}_0\delta\dot{\Omega}\sin i_0 \sin\lambda_0 \quad (8.100c)$$

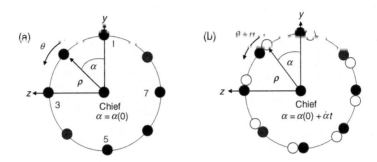

FIGURE 8.25 (a) PCO formation with $\alpha = \alpha(0)$ when the chief is at the ascending node. (b) PCO formation with $\alpha = \alpha(0) + \dot\alpha t$ when the chief is at the ascending node.

The terms involving $\delta\ddot\Omega$ and $\delta\ddot\lambda$ have been neglected in Eqs. (8.100). The secular variation of $\delta\dot\Omega$ can also be incorporated into these equations. The control magnitude required for PCO formation maintenance is $\rho n_0^2 \mathcal{O}(J_2)$. Hence, neglecting terms of $\mathcal{O}(J_2^2)$, Eqs. (8.100) can be modified as

$$\ddot{x} - 2\dot{M}_0 \dot{y} - 3\dot{M}_0^2 x = -3n_0^2 \delta a + u_{x_c} \tag{8.101a}$$

$$\ddot{y} + 2\dot{M}_0 \dot{x} = u_{y_c} \tag{8.101b}$$

$$\ddot{z} + \dot{M}_0^2 z = -2n_0\dot\omega_0 z - 2\rho(0)kn_0 \sin i_0^2 \sin \lambda_0 \cos \alpha(0) + u_{z_c} \tag{8.101c}$$

where u_{x_c}, u_{y_c}, and u_{z_c} are the control accelerations along their respective axes.

An impractical alternative for formation maintenance in the presence of the J_2-perturbation is to cancel its effect by the application of thrust. A better approach is to use a control-induced formation phase rotation rate $\dot\alpha$ of $\mathcal{O}(J_2)$, instead of enforcing the CW-result: $\alpha = \alpha(0)$. The additional benefit of inter-satellite fuel balancing due to such a scheme has been demonstrated in Refs. [139,153,156]. In essence, this scheme modifies the disturbance term of Eq. (8.101c) by requiring the PCO differential orbital elements of Eqs. (8.83)–(8.87) to depend on a time-varying phase angle, given by $\alpha = \alpha(0) + \dot\alpha t$. The result of this modification is shown pictorially in Fig. 8.25. Figure 8.25(a) shows the snapshot of the ideal PCO formation when the chief is at the ascending node. Deputies 1 and 5 have the maximum δi and $\delta\Omega = 0$, whereas deputies 3 and 7 have the maximum $\delta\Omega$ and $\delta i = 0$. Hence, deputies 1 and 5 will require the maximum fuel to mitigate the effects of differential nodal precession and the pair 3 and 7 will require the least fuel for orbit maintenance. Figure 8.25(b) shows that the effect of the formation phase rotation is to perturb the orbit rates of all the deputies ever so slightly so that they appear as shown by the small white circles at a subsequent ascending nodal crossing of the chief. As this process progresses over time, gradually, each deputy will spend equal time in "good" and "bad" locations, thereby balancing the fuel required for formation maintenance among the deputies. Another alternative for fuel balancing using the virtual center approach is discussed in Refs. [5,7].

Therefore, the following reference trajectories are selected to estimate the control requirements for formation maintenance:

$$x_r(t) = 0.5\rho(0)\sin(\lambda_0(0) + \alpha(0) + (\dot{M}_0 + \dot{\alpha})t) + \delta a \qquad (8.102)$$

$$y_r(t) = \rho(0)\cos(\lambda_0(0) + \alpha(0) + (\dot{M}_0 + \dot{\alpha})t) \qquad (8.103)$$

$$z_r(t) = \rho(0)\sin(\lambda_0(0) + \alpha(0) + (\dot{M}_0 + \dot{\alpha})t) \qquad (8.104)$$

The choice of the frequency, $\dot{M}_0$, in Eqs. (8.102)–(8.104) is not unique. Perigee rotation of the chief can also be accounted for in the reference orbit. However, this ambiguity in the choice of the reference frequency can also be resolved by absorbing $\dot{\omega}$ into $\dot{\alpha}$. The natural motion variables, as given by Eqs. (8.70), (8.71), and (8.75), differ from their respective reference trajectories given by Eqs. (8.102)–(8.104). Hence, control inputs are required for trajectory following.

The control accelerations obtained by substituting Eqs. (8.102)–(8.104) into Eqs. (8.101) are

$$u_{x_c} \approx \rho(0)n_0\dot{\alpha}\sin(\lambda_0(0) + \alpha(0) + (\dot{M}_0 + \dot{\alpha})t) \qquad (8.105)$$

$$u_{y_c} \approx -\rho(0)n_0\dot{\alpha}\cos(\lambda_0(0) + \alpha(0) + (\dot{M}_0 + \dot{\alpha})t) \qquad (8.106)$$

$$u_{z_c} \approx 2n_0(\dot{\omega}_0 - \dot{\alpha})\rho(0)\sin(\lambda_0(0) + \alpha(0) + (\dot{M}_0 + \dot{\alpha})t)$$
$$+ 2\rho(0)kn_0\sin^2 i_0\cos(\alpha(0))\sin\lambda_0 \qquad (8.107)$$

Eqs. (8.105)–(8.106) indicate that the magnitudes of the in-plane thrust acceleration components for formation maintenance depend linearly on $\dot{\alpha}$. Further insight on the dependence of the cross-track control acceleration magnitude on $\dot{\alpha}$ is obtained by squaring Eq. (8.107) and averaging the result over an orbit of the chief, thereby eliminating the short-periodic variations. The mean square cross-track control acceleration is given below:

$$u_{z_c}^2 = 2\rho^2(0)n_0^2\left[(\dot{\omega}_0 - \dot{\alpha} + k\sin^2 i_0)^2\cos^2(\alpha(0)) + (\dot{\omega}_0 - \dot{\alpha})^2\sin^2\alpha(0)\right] \qquad (8.108)$$

The minimum value of $u_{z_c}^2$ is obtained for

$$\dot{\alpha} = \dot{\omega}_0 + k\sin^2 i_0\cos^2\alpha(0) = \dot{\omega}_0 + \dot{\alpha}_z \qquad (8.109)$$

This is not a surprising result, considering the expression for the natural cross-track phase angle rotation rate provided by Eq. (8.97). For an orbit with inclination satisfying Eq. (8.96), the optimal control-induced phase rotation rate is zero, since $\dot{\omega}_0 = -\dot{\alpha}_z$. Therefore, orbit control is not required for such a satellite, at least over a period of one day.

8.7.1 Formation maintenance without radial thrust

An important special case of control without the use of radial thrust is examined next. Successive differentiations of the in-plane equations and imposition of the zero-radial thrust condition result in the following:

$$\ddddot{y} + M_0^4 \ddot{y} = \ddot{u}_{y_c} - 3M_0^2 u_{y_c} \qquad (8.110)$$

Substitution of Eq. (8.103) into Eq. (8.110) results in the following:

$$u_{x_c} = 0 \qquad (8.111)$$

$$u_{y_c} = -0.5 n_0 \rho(0) \dot{\alpha} \cos(\lambda_0(0) + \alpha(0) + (\dot{M}_0 + \dot{\alpha})t) \qquad (8.112)$$

$$x(t) = 0.5 \left(1 + \frac{0.5\dot{\alpha}}{n_0}\right) \rho(0) \sin(\lambda_0(0) + \alpha(0) + (\dot{M}_0 + \dot{\alpha})t) + x_{bias} \quad (8.113)$$

where, as will be shown in Chapter 10, x_{bias} is a constant, related to δa. In the absence of radial thrust, $x(t)$ does not follow the reference solution of Eq. (8.102) exactly. A comparison of Eqs. (8.111)–(8.112) with Eqs. (8.105)–(8.106) shows a 50% reduction in the net in-plane thrust acceleration achieved by not using radial thrust and relaxing the tracking requirement for the radial component of motion. The difference between the two expressions in Eq. (8.113) and Eq. (8.102) is $\mathcal{O}(J_2)$.

8.8 FUEL MINIMIZATION AND BALANCING

Balancing the rate of fuel-consumption among identical satellites in a formation results in a common ballistic coefficient (cf. Section 1.3.4) for all the satellites. In the absence of radial thrust, the mean square control acceleration, averaged over an orbit, can be evaluated as shown below:

$$\mathcal{J} = \frac{n_0}{2\pi} \int_0^{2\pi/n_0} (u_{y_c}^2 + u_{z_c}^2) dt \qquad (8.114)$$

The performance index defined above simplifies the analysis, but it does not accurately represent the fuel requirement. However, $\mathcal{J}$ is directly proportional to the fuel consumption for power-limited, low-thrust propulsion. Substituting Eqs. (8.107) and (8.112) into Eq. (8.114) and evaluating the integral gives

$$\mathcal{J} = (\rho n_0)^2 \left[\frac{1}{8} \dot{\alpha}^2 + 2(\dot{\alpha} - \dot{\omega}_0)^2 + 2k^2 \sin^4 i_0 \cos^2 \alpha(0) \right.$$

$$\left. - 4(\dot{\alpha} - \dot{\omega}_0)k \sin^2 i_0 \cos^2 \alpha(0) \right] \qquad (8.115)$$

The averaged cost per satellite, considering an infinite number of satellites over an orbit of the chief, can be represented as

$$\mathcal{J}_{Formation} = \frac{n_0}{4\pi^2} \int_0^{2\pi/n_0} \int_0^{2\pi} (u_{y_c}^2 + u_{z_c}^2) d\alpha(0) dt \qquad (8.116)$$

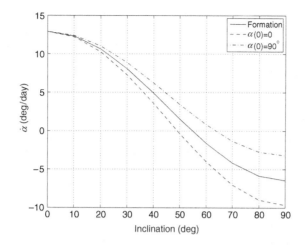

FIGURE 8.26 Optimal phase rotation rate as a function of inclination.

Evaluation of the integral Eq. (8.116) results in

$$J_{Formation} = (\rho n_0)^2 \left[\frac{1}{8}\dot{\alpha}^2 + 2(\dot{\alpha} - \dot{\omega}_0)^2 + k^2 \sin^4 i_0 - 2(\dot{\alpha} - \dot{\omega}_0)k \sin^2 i_0 \right]$$

(8.117)

Minimization of the above expression with respect to $\dot{\alpha}$ yields

$$\dot{\alpha}_{opt_{Formation}} = \frac{16}{17}\left(\dot{\omega}_0 + \frac{1}{2}k \sin^2 i_0\right) = \frac{16}{17}(\dot{\omega}_0 + \dot{\alpha}_{z_{ave}})$$

(8.118)

Note that the results of Eqs. (8.94) and (8.98) have been used in the derivation of Eq. (8.118). For each deputy in the formation, the optimal phase rotation rate as a function of its initial phase angle is given by

$$\dot{\alpha}_{opt_{Satellite}} = \frac{16}{17}(\dot{\omega}_0 + k \sin^2 i_0 \cos^2 \alpha(0)) = \frac{16}{17}(\dot{\omega}_0 + \dot{\alpha}_z)$$

(8.119)

Equations (8.119) and (8.109) are closely related. Whereas the result of Eq. (8.109) is obtained by minimizing the cross-track thrust acceleration only, the result of Eq. (8.119) minimizes the total cost. Hence, the optimal value of $\dot{\alpha}$ is dictated predominantly by the cross-track dynamics. The optimal rate of phase shift for a formation is independent of the radius of the relative orbit and it is $\mathcal{O}(J_2)$.

Figure 8.26 shows the variation of $\dot{\alpha}$ for a formation as well as two individual satellites, one with $\alpha(0) = 0$ and the other with $\alpha(0) = 90°$. There exist multiple i_0 and $\alpha(0)$ pairs satisfying Eq. (8.96) for which $\dot{\alpha} = 0$ is optimal. For a formation $\dot{\alpha} = 0$ if $i_0 = 54.73°$ or its supplement, the values for which the perturbed mean motion is equal to the two-body mean motion. For such a formation, $\dot{\omega}_0$ is equal and opposite $\dot{\alpha}_{z_{ave}}$.

The costs for individual satellites, as well as the average formation maintenance cost, can also be minimized with respect to inclination. Simultaneous minimization of $\mathcal{J}$, given by Eq. (8.115), with respect to l_0 and $\alpha(0)$, results in the following special relationship for the inclination:

$$i_0^{**} = \sin^{-1}\left(\sqrt{\frac{20 + 8\cos^2\alpha(0)}{88\cos^2\alpha(0) - 64\cos^4\alpha(0) + 25}}\right) \qquad (8.120)$$

Equation (8.120) shows that for a satellite with $\alpha(0) = 0$, $i_0^{**} = i_0^* = 49.11°$. Similarly, the critical inclination value is optimal for a satellite with $\alpha(0) = 90°$. The value of i_0^{**} for a formation can be obtained from Eqs. (8.117)–(8.118), or directly, by substituting in Eq. (8.120), $\cos^2\alpha(0) = 0.5$. The result is $i_0^{**} = 42.3°$.

Control acceleration for formation-keeping with respect to the reference trajectory chosen is required to mitigate two effects: frequency mismatch and cross-track amplitude variation. The i_0-$\alpha(0)$ constraint given by Eq. (8.96) automatically accounts for the first effect, but not the second. Equations (8.96) and (8.120) are simultaneously satisfied by $i_0 = 49.11°$ and the critical inclination.

SUMMARY

In mean-element space, the Hamiltonian is a function of only the momenta, i.e., the angles are ignorable coordinates. This means that the angle rates are functions of only the momenta, and are therefore constant. Thus, one can analytically investigate the effect of the momenta differences and initial angle values on the design and maintenance of formations. This fact led to three important developments presented in this section.

Firstly, the design of relative motion orbits is more straightforward in mean-element space. It is analogous to the design of formations in orbital element space for a spherical Earth. The differential inclination, differential eccentricity and differential initial right ascension define the size of the formation, and the differential mean anomaly and argument of perigee define the phasing and center location of the formation. One then transforms the mean elements to osculating elements to obtain the initial position and velocity.

Secondly, constraints for minimizing the drift in the angles resulting from the J_2 perturbation were derived and demonstrated by simulation. These are constraints on the mean differential momenta, or $(\delta a, \delta a, \delta e)$. To minimize the along-track drift in the presence of the J_2 perturbation, an analytical expression for a semimajor axis difference was derived. This constraint should always be used as it does not impact on the formation design just as the $\delta a = 0$ constraint does not impact on the formation design for a spherical Earth. For a formation of approximately 1 km in size, this semimajor axis difference is typically only a few meters.

In addition, analytical expressions for negating the differential nodal precession or differential perigee rotation were derived and demonstrated by simulation. If the orbit design allows the freedom to choose either the inclination difference or eccentricity difference, then either the differential nodal precession or the differential argument of perigee rotation can be negated. Using these constraints will reduce the amount of fuel required to maintain the formation. The types of formations that can be obtained based on using one, two or three constraints were identified.

Thirdly, the effect of an energy difference on formation design was addressed. It was shown that in contrast to the spherical-Earth problem, equal energy does not prevent drift, except for the case $\delta a = \delta e = \delta i = 0$.

Finally, a method for evaluating modeling error effects for formations has been presented. Results using the method have been presented for a specific PCO or its equivalent for eccentric chief orbits, for a range of eccentricities and formation sizes. For formations that involve out-of-plane motion caused by an inclination difference, one can conclude that not including the chief eccentricity and first-order J_2 effects has a significant effect on predicting relative motion and will waste fuel. The nonlinear effects do not become significant until the formation size is 30–40 km if curvilinear coordinates are used, as is done with the small e theory, the GA theory and the unit sphere nonlinear theory. The Yan–Alfriend theory uses orbital elements.

Analysis of the solutions to the motion variables for near-circular reference orbits showed that unlike for the two-body problem, the fundamental in-plane and cross-track frequencies of perturbed relative motion are in general not the same. A special inclination-phase angle relationship was obtained for which the in-plane and out-of-plane fundamental frequencies are equal, over an extended period of time. Such relative orbits do not precess and, under further restrictions, do not distort in the presence of differential nodal precession. Special values of the orbit inclination were identified for minimizing control requirements for individual satellites as well as a formation. An intelligent control methodology for formation maintenance and inter-satellite fuel-balancing was also presented and the advantage gained by not utilizing radial thrust for formation maintenance near circular orbits was highlighted. This control concept will be validated further in Chapter 10, via detailed numerical simulations using the nonlinear models of relative motion. Mitigating the effects of differential perturbations without the use of thrust involves pitting one disturbance against the other.

Even though the discussion presented in this chapter centered on accommodating J_2 by suitable choices of the initial conditions, other approaches to perturbation mitigation are also possible. For example, Williams and Wang [157] show that secular effects of oblateness can be countered by differential solar radiation pressure for dissimilar satellites in a formation.

Chapter | nine

Rotation-Translation Coupling

Earliest morning, switching all the tracks
that cross the sky from cinder star to star,
coupling the ends of streets
to trains of light.

Elizabeth Bishop (1911–1979)

Thus far, we have discussed point-mass models for relative spacecraft translational motion, including nonlinear modeling in the Cartesian LVLH frame, discussed in Chapter 4, linear models, described in Chapter 5, and orbital-elements-based modeling, elaborated upon in Chapters 6–8. In other words, we focused on three degrees of freedom (DOF) only. Obviously, designing a space mission that consists of several space vehicles requires modeling the relative rotational motion in addition to the relative translation, i.e., six-degrees-of-freedom models.

It has been well-known that there exists an analogy between orbital dynamics and rigid-body dynamics [129]. We have pointed out this analogy in Sections 2.4, 7.1 and 7.2, where tools from attitude parameterizations, namely Euler parameters, were used for modeling relative translational dynamics. However, models for the relative motion of six-DOF spacecraft have gained attention in the literature only in recent years. Among the first to model the spacecraft relative angular velocity in the framework of spacecraft formation flying were Pan and Kapila [158], who addressed the coupled translational and rotational dynamics of two spacecraft. By defining two body-fixed reference frames, one attached to the chief and the other attached to the deputy, Pan and Kapila proposed using a two-part relative motion model: One that accounts for the relative translational dynamics of the body-fixed coordinate frame origins, and another that captures the relative attitude dynamics of the two body-fixed frames. A similar modeling approach was used for relative motion estimation [159]. In addition, tensorial equations of motion for a formation consisting of N spacecraft, each modeled as a rigid body, were derived [160], and **227**

Spacecraft Formation Flying; ISBN: 9780750685337

coordination architectures relying on six-DOF models were developed for spacecraft interferometry applications [161]

In general, rigid-body dynamics can be represented as translation of the center-of-mass (CM) and rotation about the CM [162]. Thus, spacecraft relative motion must be composed by combining the relative translational and rotational dynamics of arbitrary points on the spacecraft. Whenever one of these points, termed *feature points*, does not coincide with the spacecraft's CM, a kinematic coupling between the rotational and translational dynamics of these points occurs.

The purpose of this chapter is to model the kinematic coupling effect and to show that this effect is key for high-precision modeling of tight formation flying, rendezvous, and docking. This effect, originally pointed out by Segal and Gurfil [163], is also important in vision-based relative attitude and position control, where arbitrary feature points on a target vehicle are tracked. Given two rigid-body spacecraft, the model presented herein is formulated in a general manner that describes the motion between any two arbitrary points on the spacecraft. Hence, the relative translational motion between feature points on different satellites is a function of their orbital and attitude motions.

We will also provide a CW-like approximation of the relative motion that includes the kinematic coupling effect. This approximation is aimed at alleviating an apparent contradiction in linearized relative motion theories: To obtain linear equations of motion, the spacecraft are assumed to operate in close proximity. However, if the spacecraft are close to each other (e.g. rendezvous), then they can no longer be treated as point masses, since the spacecraft shape and size affects the relative translation between off-CM points. This effect is accentuated as the distances between spacecraft decrease.

9.1 RELATIVE DYNAMICS

As in previous chapters, we consider two rigid-body spacecraft orbiting the Earth: One of them is a chief and the other is a deputy. We will use the standard coordinate systems introduced in Chapter 2, namely $\mathscr{I}$, the ECI frame, $\mathscr{L}_0$, the chief-fixed LVLH frame, and $\mathscr{L}_1$, the deputy-fixed LVLH frame. In addition, we will assume that the orbital frames $\mathscr{L}_0$ and $\mathscr{L}_1$ coincide with the body-fixed frames of the chief and deputy spacecraft, respectively.[1] In Chapter 4, we developed the equations of relative orbital motion (4.14)–(4.16). In the absence of orbital perturbations, $\dot{\theta}_0 = \dot{f}_0$, so Eqs. (4.14)–(4.16) become

$$\ddot{x} - 2\dot{f}_0\dot{y} - \ddot{f}_0 y - \dot{f}_0^2 x = -\frac{\mu(r_0 + x)}{[(r_0 + x)^2 + y^2 + z^2]^{\frac{3}{2}}} + \frac{\mu}{r_0^2} \qquad (9.1a)$$

$$\ddot{y} + 2\dot{f}_0\dot{x} + \ddot{f}_0 x - \dot{f}_0^2 y = -\frac{\mu y}{[(r_0 + x)^2 + y^2 + z^2]^{\frac{3}{2}}} \qquad (9.1b)$$

---

[1] In reality, gravity gradient effects can cause attitude oscillations. For elliptic orbits, there is an inherent orbit and attitude coupling. For spin-stabilized satellites with inertial pointing, the body axes and the LVLH axes are quite different. We are neglecting these effects here, assuming that the attitude controller is maintaining LVLH pointing.

$$\ddot{z} = -\frac{\mu z}{[(r_0 + x)^2 + y^2 + z^2]^{\frac{3}{2}}} \tag{9.1c}$$

We define the rotational angular velocity of the deputy relative to the chief, $\delta\boldsymbol{\omega} \in \mathbb{R}^3$ as

$$\delta\boldsymbol{\omega} \triangleq \boldsymbol{\omega}_1 - \boldsymbol{\omega}_0 \tag{9.2}$$

where $\boldsymbol{\omega}_0$ and $\boldsymbol{\omega}_1$ are the respective angular velocities of the deputy and the chief, respectively, in some given reference frame.

The relative attitude can be parameterized using a rotation matrix $D \in SO(3)$, which transforms a vector from the body-fixed frame $\mathscr{L}_1$ to the body-fixed frame $\mathscr{L}_0$. The relative attitude will be parameterized using the relative Euler parameters, which we defined by Eqs. (2.52)–(2.54), and used in Section 7.2. Recall that [164]

$$\delta\beta_1 \triangleq e_1 \sin\frac{\wp}{2}, \delta\beta_2 \triangleq e_2 \sin\frac{\wp}{2}, \delta\beta_3 \triangleq e_3 \sin\frac{\wp}{2}, \delta\beta_0 \triangleq \cos\frac{\wp}{2} \tag{9.3}$$

subject to the constraint

$$\delta\beta_1^2 + \delta\beta_2^2 + \delta\beta_3^2 + \delta\beta_0^2 = 1 \tag{9.4}$$

In Eq. (9.3), $[e_1, e_2, e_3]^T$ constitutes a vector along the axis of rotation, and $\wp$ is the angle of the rotation from $\mathscr{L}_0$ to $\mathscr{L}_1$. As we mentioned in Sections 7.1 and 7.2, the Euler parameters form a quaternion, $\delta\boldsymbol{\beta} = [\delta\beta_0, \delta\beta_1, \delta\beta_2, \delta\beta_3]^T$, where $[\delta\beta_1, \delta\beta_2, \delta\beta_3]^T$ is the vector part and $\delta\beta_0$ is the scalar part. In the subsequent discussion, we will drop the δ symbol before the relative quantities for the ease of notation; we thus write $\delta\beta_i \equiv \beta_i$ and $\delta\boldsymbol{\omega} \equiv \boldsymbol{\omega}$.

The rotation matrix, D, can be expressed in terms of the quaternion as

$$D(\boldsymbol{\beta}) = \begin{bmatrix} \beta_1^2 - \beta_2^2 - \beta_3^2 + \beta_0^2 & 2(\beta_1\beta_2 - \beta_3\beta_0) & 2(\beta_1\beta_3 + \beta_2\beta_0) \\ 2(\beta_1\beta_2 + \beta_3\beta_0) & -\beta_1^2 + \beta_2^2 - \beta_3^2 + \beta_0^2 & 2(\beta_2\beta_3 - \beta_1\beta_0) \\ 2(\beta_1\beta_3 - \beta_2\beta_0) & 2(\beta_2\beta_3 + \beta_1\beta_0) & -\beta_1^2 - \beta_2^2 + \beta_3^2 + \beta_0^2 \end{bmatrix} \tag{9.5}$$

Using $D(\boldsymbol{\beta})$, the relative angular velocity vector, $\boldsymbol{\omega}$, can be calculated in the body-fixed frames $\mathscr{L}_0$ and $\mathscr{L}_1$, respectively, as follows:

$$[\boldsymbol{\omega}]_{\mathscr{L}_0} = D(\boldsymbol{\beta})[\boldsymbol{\omega}_1]_{\mathscr{L}_1} - [\boldsymbol{\omega}_0]_{\mathscr{L}_0} \tag{9.6}$$

Utilizing $\boldsymbol{\omega}$ and $\boldsymbol{\beta}$, the attitude kinematics of the deputy relative to the chief can be described using the quaternion kinematic equations of motion

$$\dot{\boldsymbol{\beta}} = \frac{1}{2}Q(\boldsymbol{\beta})[\boldsymbol{\omega}]_{\mathscr{L}_1} \tag{9.7}$$

where

$$Q(\beta) = \begin{bmatrix} -\beta_1 & -\beta_2 & -\beta_3 \\ \beta_0 & -\beta_3 & \beta_2 \\ \beta_3 & \beta_0 & -\beta_1 \\ -\beta_2 & \beta_1 & \beta_0 \end{bmatrix} \tag{9.8}$$

The attitude dynamics of the deputy relative to the chief, expressed in $\mathscr{L}_0$, will now be derived according to the guidelines of Ref. [158]. First, a differentiation of Eq. (9.2) with respect to the inertial frame leads to

$$\frac{\mathrm{d}^{\mathscr{I}} \boldsymbol{\omega}}{\mathrm{d}t} = \frac{\mathrm{d}^{\mathscr{I}} \boldsymbol{\omega}_1}{\mathrm{d}t} - \frac{\mathrm{d}^{\mathscr{I}} \boldsymbol{\omega}_0}{\mathrm{d}t} \tag{9.9}$$

Expressing Eq. (9.9) in the body-fixed frame $\mathscr{L}_0$ yields

$$\left(\frac{\mathrm{d}^{\mathscr{I}} \boldsymbol{\omega}}{\mathrm{d}t}\right)_{\mathscr{L}_0} = D(\boldsymbol{\beta}) \left(\frac{\mathrm{d}^{\mathscr{I}} \boldsymbol{\omega}_1}{\mathrm{d}t}\right)_{\mathscr{L}_1} - \left(\frac{\mathrm{d}^{\mathscr{I}} \boldsymbol{\omega}_0}{\mathrm{d}t}\right)_{\mathscr{L}_0} \tag{9.10}$$

On the other hand,

$$\frac{\mathrm{d}^{\mathscr{I}} \boldsymbol{\omega}}{\mathrm{d}t} = \left(\frac{\mathrm{d}\boldsymbol{\omega}}{\mathrm{d}t}\right)_{\mathscr{L}_0} + \boldsymbol{\omega}_0 \times \boldsymbol{\omega} \tag{9.11}$$

Expressing Eq. (9.11) in $\mathscr{L}_0$ provides us with the expression

$$\left(\frac{\mathrm{d}^{\mathscr{I}} \boldsymbol{\omega}}{\mathrm{d}t}\right)_{\mathscr{L}_0} = \left(\frac{\mathrm{d}\boldsymbol{\omega}}{\mathrm{d}t}\right)_{\mathscr{L}_0} + [\boldsymbol{\omega}_0]_{\mathscr{L}_0} \times [\boldsymbol{\omega}]_{\mathscr{L}_0} \tag{9.12}$$

Comparing Eq. (9.10) to Eq. (9.12) yields

$$\left(\frac{\mathrm{d}\boldsymbol{\omega}}{\mathrm{d}t}\right)_{\mathscr{L}_0} = D(\boldsymbol{\beta}) \left(\frac{\mathrm{d}^{\mathscr{I}} \boldsymbol{\omega}_1}{\mathrm{d}t}\right)_{\mathscr{L}_1} - \left(\frac{\mathrm{d}^{\mathscr{I}} \boldsymbol{\omega}_0}{\mathrm{d}t}\right)_{\mathscr{L}_0} - [\boldsymbol{\omega}_0]_{\mathscr{L}_0} \times [\boldsymbol{\omega}]_{\mathscr{L}_0} \tag{9.13}$$

The differentiation of the angular velocity in a fixed frame or in a body frame gives the same result. Using this fact and multiplying Eq. (9.13) by the inertia tensor of the chief, $\mathbb{I}_0 \in \mathbb{R}^{3 \times 3}$, gives

$$\mathbb{I}_0 \left(\frac{\mathrm{d}^{\mathscr{L}_0} \boldsymbol{\omega}}{\mathrm{d}t}\right)_{\mathscr{L}_0} = \mathbb{I}_0 D(\boldsymbol{\beta}) \left(\frac{\mathrm{d}^{\mathscr{L}_1} \boldsymbol{\omega}_1}{\mathrm{d}t}\right)_{\mathscr{L}_1} - \mathbb{I}_0 \left(\frac{\mathrm{d}^{\mathscr{L}_0} \boldsymbol{\omega}_0}{\mathrm{d}t}\right)_{\mathscr{L}_0}$$
$$- \mathbb{I}_0 [\boldsymbol{\omega}_0]_{\mathscr{L}_0} \times [\boldsymbol{\omega}]_{\mathscr{L}_0} \tag{9.14}$$

Eq. (9.14) relates the derivative of the relative angular velocity to the angular velocities rates of the deputy and the chief, thus yielding the relative rotational dynamic equations. The relative dynamics in terms of $\mathscr{L}_0$-quantities can be obtained by writing

$$\frac{d\mathbf{H}}{dt} = \mathbf{N} \tag{9.15}$$

where $\mathbf{H}$ is the total angular momentum and $\mathbf{N}$ is an external torque. Expressing (9.15) in body axes yields

$$\frac{d^{\mathscr{I}}\mathbf{H}_1}{dt} = \frac{d^{\mathscr{L}_1}\mathbf{H}_1}{dt} + \boldsymbol{\omega}_1 \times \mathbf{H}_1 = \mathbf{N}_1 \tag{9.16}$$

$$\frac{d^{\mathscr{I}}\mathbf{H}_0}{dt} = \frac{d^{\mathscr{L}_0}\mathbf{H}_0}{dt} + \boldsymbol{\omega}_0 \times \mathbf{H}_0 = \mathbf{N}_0 \tag{9.17}$$

Since $\mathbf{H}_0 = \mathbb{I}_0\boldsymbol{\omega}_0$ and $\mathbf{H}_1 = \mathbb{I}_1\boldsymbol{\omega}_1$, with $\mathbb{I}_1 \in \mathbb{R}^{3\times3}$ being the inertia tensor of the deputy,

$$\mathbb{I}_1\frac{d^{\mathscr{I}}\boldsymbol{\omega}_1}{dt} = \mathbb{I}_1\frac{d^{\mathscr{L}_1}\boldsymbol{\omega}_1}{dt} + \boldsymbol{\omega}_1 \times \mathbb{I}_1\boldsymbol{\omega}_1 = \mathbf{N}_1 \tag{9.18}$$

and

$$\mathbb{I}_0\frac{d^{\mathscr{I}}\boldsymbol{\omega}_0}{dt} = \mathbb{I}_0\frac{d^{\mathscr{L}_0}\boldsymbol{\omega}_0}{dt} + \boldsymbol{\omega}_0 \times \mathbb{I}_0\boldsymbol{\omega}_0 = \mathbf{N}_0 \tag{9.19}$$

where $\mathbf{N}_0$ and $\mathbf{N}_1$ are external torques acting on the chief and deputy, respectively. Now, substituting Eqs. (9.18) and (9.19) into Eq. (9.14) gives

$$\mathbb{I}_0\left(\frac{d^{\mathscr{L}_0}\boldsymbol{\omega}}{dt}\right)_{\mathscr{L}_0} = \mathbb{I}_0 D(\boldsymbol{\beta})\mathbb{I}_1^{-1}\{\mathbf{N}_1 - [\boldsymbol{\omega}_1]_{\mathscr{L}_1} \times \mathbb{I}_1[\boldsymbol{\omega}_1]_{\mathscr{L}_1}\} - \mathbb{I}_0[\boldsymbol{\omega}_0]_{\mathscr{L}_0} \times [\boldsymbol{\omega}]_{\mathscr{L}_0}$$
$$- \{\mathbf{N}_0 - [\boldsymbol{\omega}_0]_{\mathscr{L}_0} \times \mathbb{I}_0[\boldsymbol{\omega}_0]_{\mathscr{L}_0}\} \tag{9.20}$$

Finally, using Eq. (9.6) yields the equation for the relative attitude dynamics expressed using relative and $\mathscr{L}_0$-related angular velocities, i.e., without the deputy's angular velocities:

$$\mathbb{I}_0\left(\frac{d^{\mathscr{L}_0}\boldsymbol{\omega}}{dt}\right)_{\mathscr{L}_0} = \mathbb{I}_0 D(\boldsymbol{\beta})\mathbb{I}_1^{-1}\left\{\mathbf{N}_1 - D(\boldsymbol{\beta})^T([\boldsymbol{\omega}]_{\mathscr{L}_0} + [\boldsymbol{\omega}_0]_{\mathscr{L}_0}) \times \mathbb{I}_1 D(\boldsymbol{\beta})^T\right.$$
$$\left. \cdot ([\boldsymbol{\omega}]_{\mathscr{L}_0} + [\boldsymbol{\omega}_0]_{\mathscr{L}_0})\right\} - \mathbb{I}_0[\boldsymbol{\omega}_0]_{\mathscr{L}_0} \times [\boldsymbol{\omega}]_{\mathscr{L}_0}$$
$$- \left\{\mathbf{N}_0 - [\boldsymbol{\omega}_0]_{\mathscr{L}_0} \times \mathbb{I}_0[\boldsymbol{\omega}_0]_{\mathscr{L}_0}\right\} \tag{9.21}$$

Omitting the subscripts $\mathscr{L}_0$ and $\mathscr{L}_1$ and using the conventional notation $\dot{\boldsymbol{\omega}}$ to imply that $\boldsymbol{\omega}$ is both differentiated and expressed in $\mathscr{L}_0$ gives

$$\mathbb{I}_0\dot{\boldsymbol{\omega}} = \mathbb{I}_0 D\mathbb{I}_1^{-1}\left[\mathbf{N}_1 - D^T([\boldsymbol{\omega}] + [\boldsymbol{\omega}_0]) \times \mathbb{I}_1 D^T([\boldsymbol{\omega}] + [\boldsymbol{\omega}_0])\right]$$
$$- \mathbb{I}_0\boldsymbol{\omega}_0 \times \boldsymbol{\omega} - [\mathbf{N}_0 - \boldsymbol{\omega}_0 \times \mathbb{I}_0\boldsymbol{\omega}_0] \tag{9.22}$$

Thus, the relative rotational kinematics and dynamics are described by Eqs. (9.7) and (9.22) for a seven-dimensional state vector $[\omega^T, \mu^T]^T$.

The complete relative motion is described by both the relative rotational and relative translational dynamics, which yields a system comprising a 15-element relative state vector $[\rho^T, \dot{\rho}^T, f, \dot{f}, \omega^T, \beta^T]^T$. The coupling between the rotational and translation dynamics arises because of either *external torques* (the most obvious example is the gravity gradient torque, which depends on altitude) or *internal coupling*, which is external-perturbations independent. The internal coupling stems from the fact that relative motion equations can be written for any point on the spacecraft, not necessarily the CM. Thus, an apparent translational motion of points on the deputy spacecraft (that do not coincide with the CM) will result from rotation of the deputy about its CM.

9.2 KINEMATICALLY-COUPLED RELATIVE SPACECRAFT MOTION MODEL

Consider two arbitrary feature points located on the chief and deputy spacecraft, as shown in Fig. 9.1. The spacecraft in this discussion are assumed to be rigid. Let P_0^j be a point on the chief. Then, $\mathbf{P}_0^j = [P_{x_0}^j, P_{y_0}^j, P_{z_0}^j]^T$ is a vector directed from the origin of the coordinate system $\mathscr{L}_0$ to the point P_0^j. In the special case where $\mathbf{P}_0^j$ coincides with the chief's CM, $\mathbf{P}_0^0 = [0, 0, 0]^T$. Similarly, P_1^i is an arbitrary point on the deputy. Thus, $\mathbf{P}_1^i = [P_{x_1}^i, P_{y_1}^i, P_{z_1}^i]^T$ is a vector directed from the origin of the coordinate system $\mathscr{L}_1$ to the point P_1^i, and $\mathbf{P}_1^0 = [0, 0, 0]^T$ is the vector description of the deputy's CM. Let ρ_{ij} denote the relative position vector between point j on the chief and point i on the deputy and ρ be, as before, the relative position vector between the CMs of the spacecraft. By observing Fig. 9.1, one can note that the following relationship holds:

$$\mathbf{P}_0^j + \rho_{ij} = \rho + \mathbf{P}_1^i \tag{9.23}$$

Thus,

$$\rho_{ij} = \rho + \mathbf{P}_1^i - \mathbf{P}_0^j \tag{9.24}$$

The first- and second-order time derivatives of ρ_{ij} are

$$\dot{\rho}_{ij} = \dot{\rho} + \dot{\mathbf{P}}_1^i - \dot{\mathbf{P}}_0^j \tag{9.25}$$

$$\ddot{\rho}_{ij} = \ddot{\rho} + \ddot{\mathbf{P}}_1^i - \ddot{\mathbf{P}}_0^j \tag{9.26}$$

where in frame $\mathscr{L}_0$,

$$\dot{\mathbf{P}}_0^j = \ddot{\mathbf{P}}_0^j = 0 \tag{9.27}$$

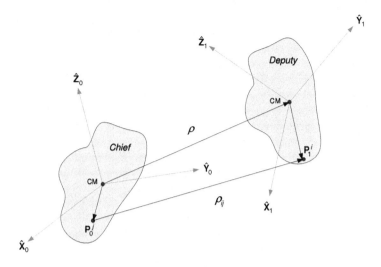

FIGURE 9.1 Two rigid-body spacecraft with body-fixed reference frames.

since the chief is assumed a rigid body. Calculating $\dot{\mathbf{P}}_1^i$ and $\ddot{\mathbf{P}}_1^i$ with respect to the rotating frame $\mathscr{L}_0$ results in

$$\left[\dot{\mathbf{P}}_1^i\right]_{\mathscr{L}_0} = \left[\dot{\mathbf{P}}_1^i\right]_{\mathscr{L}_1} + \boldsymbol{\omega} \times \mathbf{P}_1^i \tag{9.28}$$

$$\left[\ddot{\mathbf{P}}_1^i\right]_{\mathscr{L}_0} = \left[\ddot{\mathbf{P}}_1^i\right]_{\mathscr{L}_1} + 2\left\{\left[\boldsymbol{\omega} \times \dot{\mathbf{P}}_1^i\right]_{\mathscr{L}_1}\right\} + \dot{\boldsymbol{\omega}} \times \mathbf{P}_1^i + \boldsymbol{\omega} \times (\boldsymbol{\omega} \times \mathbf{P}_1^i) \tag{9.29}$$

Since the deputy is assumed a rigid body,

$$\left[\dot{\mathbf{P}}_1^i\right]_{\mathscr{L}_1} = \left[\ddot{\mathbf{P}}_1^i\right]_{\mathscr{L}_1} = 0$$

This leads us to the following equations:

$$\dot{\boldsymbol{\rho}}_{ij} = \dot{\boldsymbol{\rho}} + \boldsymbol{\omega} \times \mathbf{P}_1^i \tag{9.30}$$

$$\ddot{\boldsymbol{\rho}}_{ij} = \ddot{\boldsymbol{\rho}} + \dot{\boldsymbol{\omega}} \times \mathbf{P}_1^i + \boldsymbol{\omega} \times (\boldsymbol{\omega} \times \mathbf{P}_1^i) \tag{9.31}$$

Expressing the vectors using their components along frame $\mathscr{L}_0$, i.e. $\boldsymbol{\rho} = [x, y, z]^T$, $\boldsymbol{\rho}_{ij} = [x_{ij}, y_{ij}, z_{ij}]^T$, $\mathbf{P}_1^i = [P_{x_1}^i, P_{y_1}^i, P_{z_1}^i]^T$, $\mathbf{P}_0^j = [P_{x_0}^j, P_{y_0}^j, P_{z_0}^j]^T$, $\boldsymbol{\omega} = [\omega_x, \omega_y, \omega_z]^T$ and substituting into Eqs. (9.24), (9.30), (9.31) gives the following relationships for the relative position components:

$$x_{ij} = x + P_{x_1}^i - P_{x_0}^j \tag{9.32a}$$

$$y_{ij} = y + P_{y_1}^i - P_{y_0}^j \tag{9.32b}$$

$$z_{ij} = z + P_{z_1}^i - P_{z_0}^j \tag{9.32c}$$

the relative velocity components:

$$\dot{x}_{ij} = \dot{x} + \omega_y P_{z_1}^i - \omega_z P_{y_1}^i \tag{9.33a}$$

$$\dot{y}_{ij} = \dot{y} + \omega_z P_{x_1}^i - \omega_x P_{z_1}^i \tag{9.33b}$$

$$\dot{z}_{ij} = \dot{z} + \omega_x P_{y_1}^i - \omega_y P_{x_1}^i \tag{9.33c}$$

and the relative acceleration components:

$$\ddot{x}_{ij} = \ddot{x} + \omega_y(\omega_x P_{y_1}^i - \omega_y P_{x_1}^i) - \omega_z(\omega_z P_{x_1}^i - \omega_x P_{z_1}^i) \\ + \dot{\omega}_y P_{z_1}^i - \dot{\omega}_z P_{y_1}^i \tag{9.34a}$$

$$\ddot{y}_{ij} = \ddot{y} + \omega_z(\omega_y P_{z_1}^i - \omega_z P_{y_1}^i) - \omega_x(\omega_x P_{y_1}^i - \omega_y P_{x_1}^i) \\ + \dot{\omega}_z P_{x_1}^i - \dot{\omega}_x P_{z_1}^i \tag{9.34b}$$

$$\ddot{z}_{ij} = \ddot{z} + \omega_x(\omega_z P_{z_1}^i - \omega_x P_{x_1}^i) - \omega_y(\omega_y P_{z_1}^i - \omega_z P_{y_1}^i) \\ + \dot{\omega}_x P_{y_1}^i - \dot{\omega}_y P_{x_1}^i \tag{9.34c}$$

The translational motion model describing the relative motion between the chief and deputy spacecraft CMs, i.e. a model for the case where $\mathbf{P}_1^i = \mathbf{P}_0^j = \mathbf{0}$, is usually written in the form

$$\ddot{\boldsymbol{\rho}} = \mathbf{g}_0(\boldsymbol{\rho}, r_0, \dot{f}_0, \ddot{f}_0)$$

as was shown in Chapter 4 and repeated in Eqs. (9.1). We have now obtained a more general model, in which

$$\ddot{\boldsymbol{\rho}}_{ij} = \mathbf{g}(\boldsymbol{\rho}_{ij}, r_0, \dot{f}_0, \ddot{f}_0, \boldsymbol{\omega}, \dot{\boldsymbol{\omega}}, \mathbf{P}_1^i, \mathbf{P}_0^j)$$

This model is obtained by substituting Eqs. (9.32)–(9.34) into Eqs. (9.1), to yield the following general description of the translational motion between any arbitrary points $\mathbf{P}^i$ and $\mathbf{P}^j$ in the absence of perturbing forces:

$$\ddot{x}_{ij} - \left[\omega_y(\omega_x P_{y_1}^i - \omega_y P_{x_1}^i) + \omega_z(\omega_z P_{x_1}^i - \omega_x P_{z_1}^i) \right] - \dot{\omega}_y P_{z_1}^i + \dot{\omega}_z P_{y_1}^i \\ - 2\dot{f}_0 \left[\dot{y}_{ij} - (\omega_z P_{x_1}^i + \omega_x P_{z_1}^i) \right] - \ddot{f}_0(y_{ij} - P_{y_1}^i + P_{y_0}^j) - \dot{f}_0^2(x_{ij} - P_{x_1}^i + P_{x_0}^j) \\ = \frac{-\mu(r_0 + x_{ij} - P_{x_1}^i + P_{x_0}^j)}{[(r_0 + x_{ij} - P_{x_1}^i + P_{x_0}^j)^2 + (y_{ij} - P_{y_1}^i + P_{y_0}^j)^2 + (z_{ij} - P_{z_1}^i + P_{z_0}^j)^2]^{\frac{3}{2}}} \\ + \frac{\mu}{r_0^2} \tag{9.35}$$

$$\ddot{y}_{ij} - \left[\omega_z(\omega_y P_{z_1}^i - \omega_z P_{y_1}^i) + \omega_x(\omega_x P_{y_1}^i - \omega_y P_{x_1}^i) \right] - \dot{\omega}_z P_{x_1}^i + \dot{\omega}_x P_{z_1}^i \\ + 2\dot{f}_0 \left[\dot{x}_{ij} - (\omega_y P_{z_1}^i + \omega_z P_{y_1}^i) \right] + \ddot{f}_0(x_{ij} - P_{x_1}^i - P_{x_0}^j) - \dot{f}_0^2(y_{ij} - P_{y_1}^i + P_{y_0}^j) \\ = \frac{-\mu(y_{ij} - P_{y_1}^i + P_{y_0}^j)}{[(r_0 + x_{ij} - P_{x_1}^i + P_{x_0}^j)^2 + (y_{ij} - P_{y_1}^i + P_{y_0}^j)^2 + (z_{ij} - P_{z_1}^i + P_{z_0}^j)^2]^{\frac{3}{2}}} \tag{9.36}$$

$$\ddot{z}_{ij} - \left[\omega_x(\omega_z P^i_{x_1} - \omega_x P^i_{z_1}) + \omega_y(\omega_y P^i_{z_1} - \omega_z P^i_{y_1})\right] - \dot{\omega}_x P^i_{y_1} + \dot{\omega}_y P^i_{x_1}$$

$$= \frac{-\mu(z_{ij} - P^i_{z_1} + P^j_{z_0})}{[(r_0 + x_{ij} - P^i_{x_1} + P^j_{x_0})^2 + (y_{ij} - P^i_{y_1} + P^j_{y_0})^2 + (z_{ij} - P^i_{z_1} + P^j_{z_0})^2]^{\frac{3}{2}}} \tag{9.37}$$

These equations are coupled to the rotational motion equations (9.7) and (9.22) through the components of the relative angular velocity vector, $\boldsymbol{\omega}$. Therefore, the six-DOF description of the rigid-body relative spacecraft motion is given by the following set of nonlinear coupled differential equations: Eqs. (9.35), (9.36), (9.37), Eqs. (9.7) and (9.22), together with the expressions for the true anomaly rate, given in Eq. (2.29).

An approximated set of translational equations of motion can be obtained using the same assumptions leading to the CW equations, i.e., a circular reference orbit and a small relative distance compared to the orbital radius. It should be noted that these assumptions lead to differential equations that will be linear with respect to the relative position, and nonlinear with respect to the components of the relative angular velocity vector. Thus, by applying the CW rationale on Eqs. (9.35), (9.36), and (9.37), the following approximate equations are obtained:

$$\dot{\mathbf{x}}_{tr} = A\mathbf{x}_{tr} + \mathbf{p} \tag{9.38}$$

where $\mathbf{x}_{tr} = [\boldsymbol{\rho}^T, \dot{\boldsymbol{\rho}}^T]^T$, $\dot{f}_0 = n = \text{const.}$,

$$A = \begin{bmatrix} 0 & 0 & 0 & 1 & 0 & 0 \\ 0 & 0 & 0 & 0 & 1 & 0 \\ 0 & 0 & 0 & 0 & 0 & 1 \\ 3n^2 & 0 & 0 & 0 & 2n & 0 \\ 0 & 0 & 0 & -2n & 0 & 0 \\ 0 & 0 & -n^2 & 0 & 0 & 0 \end{bmatrix} \tag{9.39}$$

and $\mathbf{p} = [0, 0, 0, p_1, p_2, p_3]^T$ is defined by

$$p_1 \triangleq 3n^2(P^j_{x_0} - P^i_{x_1}) - 2n(\omega_z P^i_{x_1} - \omega_x P^i_{z_1})$$
$$+ \left[\omega_y(\omega_x P^i_{y_1} - \omega_y P^i_{x_1}) + \omega_z(\omega_z P^i_{x_1} - \omega_x P^i_{z_1})\right] + \dot{\omega}_y P^i_{z_1} - \dot{\omega}_z P^i_{y_1}$$

$$p_2 \triangleq 2n(\omega_y P^i_{z_1} + \omega_z P^i_{y_1})$$
$$+ \left[\omega_z(\omega_y P^i_{z_1} - \omega_z P^i_{y_1}) + \omega_x(\omega_x P^i_{y_1} - \omega_y P^i_{x_1})\right] + \dot{\omega}_z P^i_{x_1} - \dot{\omega}_x P^i_{z_1}$$

$$p_3 \triangleq n^2(P^i_{z_1} - P^j_{x_0}) + \left[\omega_x(\omega_z P^i_{x_1} - \omega_x P^i_{z_1}) + \omega_y(\omega_y P^i_{z_1} - \omega_z P^i_{y_1})\right]$$
$$+ \dot{\omega}_x P^i_{y_1} - \dot{\omega}_y P^i_{x_1} \tag{9.40}$$

These terms result from the coupling to the rotational dynamics, and can be treated as *kinematic perturbations*. However, these perturbations will always be present, regardless of the orbital altitude and external perturbations.

The kinematic coupling effect will be accentuated when the relative distances become small.

Example 9.1. *A chief is orbiting the Earth in an elliptic orbit with eccentricity* $e_0 = 0.05$, *semimajor axis* $a_0 = 7170$ *km and inclination* $i_0 = 15°$. *The argument of perigee of the chief satisfies* $\omega_0 = 340°$, *the right ascension of the ascending node is* $\Omega_0 = 0°$ *and the initial true anomaly is* $f_0(0) = 20°$. *The nonlinear coupled-motion model is used for simulating the relative distances between three feature points on the deputy,* $\mathbf{P}_1^i$, $i = 0, 1, 2$, *and the chief's CM,* $\mathbf{P}_0^0$. *These points are* $\mathbf{P}_0^0 = [0, 0, 0]^T$, $\mathbf{P}_1^0 = [0, 0, 0]^T$, $\mathbf{P}_1^1 = [1.5, 1.5, 0]^T$ *m and* $\mathbf{P}_1^2 = [-1.5, -1.5, 0]^T$ *m. The initial conditions are as follows:*

$$\rho(0) = [25, 25, 50]^T \text{ m}, \quad \dot{\rho}(0) = [0, -0.0555, 0]^T \text{ m/s}$$
$$\omega(0) = [0.1\dot{f}_0(0), 0.1\dot{f}_0(0), 2\dot{f}_0(0)]^T, \quad \beta(0) = [0, 0, 0, 1]^T \quad (9.41)$$

where $\dot{f}_0(0) = 0.0656$ *deg/s. In addition,*

$$\mathbb{I}_0 = \mathbb{I}_1 = \mathbb{I} = \text{diag}\{500, 550, 600\} \text{ kgm}^2$$

It can be readily verified that the given initial conditions were chosen to satisfy the energy matching condition (4.36):

$$\frac{1}{2}\left\{[\dot{x}(0) - \dot{f}_0(0)y(0) + \dot{r}_0(0)]^2 + [\dot{y}(0) + \dot{f}_0(0)(x(0) + r_0(0))]^2 + \dot{z}^2(0)\right\}$$
$$-\frac{\mu}{\sqrt{[x(0) + r_0(0)]^2 + y^2(0) + z^2(0)}} = -\frac{\mu}{2a_0}$$

guaranteeing bounded relative motion between the spacecraft in the formation.

The initial conditions for the feature points on the deputy that do not coincide with the CM are calculated by applying Eqs. (9.24) and (9.30), which results in

$$\rho_{10}(0) = [26.5, 26.5, 50]^T \text{ m}, \quad \rho_{20}(0) = [23.5, 23.5, 50]^T \text{ m} \quad (9.42)$$

The results of the formation flying simulation for a single orbital period of the chief are depicted by Figs. 9.2 and 9.3. Figure 9.2 shows the position components of the selected feature point on the deputy's body relative to the chief's CM, i.e. $\rho_{10} = \rho + \mathbf{P}_1^1 - \mathbf{P}_0^0$, $\rho_{20} = \rho + \mathbf{P}_1^2 - \mathbf{P}_0^0$ and ρ (the last vector represents the position of the deputy's CM relative to the chief's CM). Figure 9.3 shows the deviation in the relative position of the feature points due to the coupling effect. These deviations are defined as $\Delta\rho_1 = \rho_{10} - \rho$ and $\Delta\rho_2 = \rho_{20} - \rho$.

Figure 9.2 clearly demonstrates that the relative position between the chief's CM and the chosen feature points on the deputy depends upon the chosen points' location in the deputy's body frame. Moreover, Fig. 9.3 shows that the relative motion between the chief's CM and feature points on the deputy spacecraft that do not coincide with the deputy's CM exhibit harmonic oscillations whose frequency is determined by the relative angular velocity. The magnitude

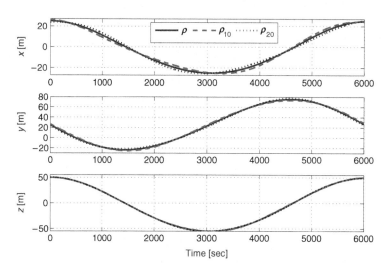

FIGURE 9.2 Time histories of the position components of several feature points on the deputy spacecraft relative to the chief's spacecraft center-of-mass.

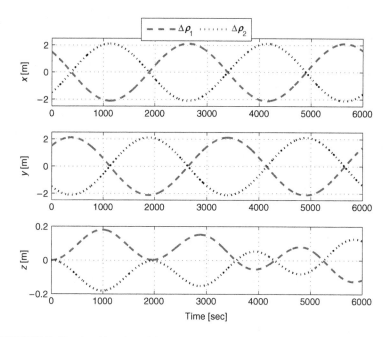

FIGURE 9.3 The time histories of the relative displacement components of points on the deputy spacecraft relative to the line joining the spacecraft centers of masses, showing that the kinematic coupling induces relative translational motion due to rotation.

of these oscillations depends on the location of the point in the deputy's body frame. Evidently, this result is different from standard six-DOF models, which

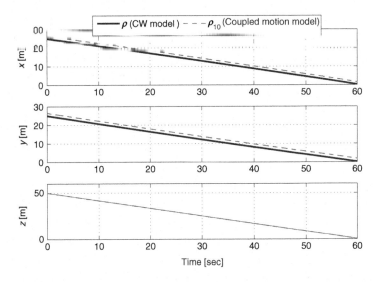

FIGURE 9.4 A comparison between the relative position components as modeled by the CW equations and the approximate model (9.38) for a rendezvous mission.

do not take the kinematic coupling into consideration, i.e., neglect the effect of the relative rotation on the relative translation.

Example 9.2. *A chief spacecraft is orbiting the Earth in a circular orbit with radius* $a_0 = 7170$ *km. The approximate model (9.38) is used to simulate the relative distance between two feature points,* $\mathbf{P}_1^i, i = 0, 1$ *on the deputy spacecraft and the chief spacecraft's CM,* $\mathbf{P}_0^0 = [0, 0, 0]^T$. *These points are selected as follows: The deputy's CM,* $\mathbf{P}_1^0 = [0, 0, 0]^T$, *and* $\mathbf{P}_1^1 = [1.5, 1.5, 0]^T$ *m. The initial conditions of the relative position and relative velocity were chosen so as to nullify the relative position* ρ *at the specified time* $t_1 = 0.01T = 60.42$ *sec, where* $T = 2\pi\sqrt{\frac{a_0^3}{\mu}} = 100.7$ *min. The resulting initial conditions are*

$$\rho(0) = [25, 25, 50]^T \text{ m}, \quad \dot{\rho}(0) = [-0.3889, -0.4392, -0.8264]^T \text{ m/s} \quad (9.43)$$

$$\rho_{10}(0) = [26.5, 26.5, 50]^T \text{ m},$$

$$\dot{\rho}_{10}(0) = [-0.392, -0.4361, -0.8264]^T \text{ m/s} \quad (9.44)$$

and

$$\omega(0) = [0.1n, 0.1n, 2n]^T \quad (9.45)$$

where $n = \sqrt{\frac{a_0^3}{\mu}} = 0.0010402$ *rad/s.*

The results of this rendezvous example are shown in Figs. 9.4 and 9.5. Figure 9.4 depicts the relative positions ρ and ρ_{10}. Although the relative distance between the spacecrafts' CMs is decreasing to zero, the distance from

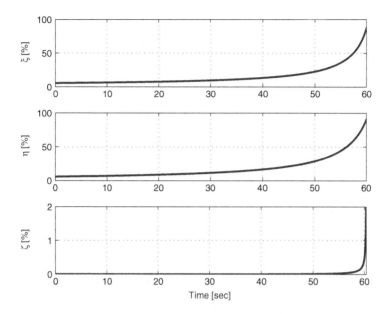

FIGURE 9.5 Although the CW model is widely used for the final approach phase of rendezvous missions, this figure shows that the effect of rotational-translation kinematic coupling impairs the validity of the CW model.

the chief's CM to the point $\mathbf{P}_1^1$ is not. Moreover, the normalized differences between model (9.38) and the CW model (Eqs. (5.4)–(5.6)), defined as

$$\xi = \frac{x_{10} - x}{x_{10}}, \eta = \frac{y_{10} - y}{y_{10}}, \zeta = \frac{z_{10} - z}{z_{10}} \tag{9.46}$$

reach 100% as $t \to t_f$, as can be seen in Fig. 9.5. Thus, this example clearly demonstrates that at the final approach phase of a rendezvous, the CW model will be invalid due to the kinematic coupling of rotation and translation.

SUMMARY

In this chapter, a kinematically coupled relative rotational and translational motion model, describing the six-degrees-of-freedom relative dynamics between two rigid-body spacecraft, was developed. This model generalizes the nonlinear relative translation models discussed in Chapter 4 to include trajectories of points that are not located on the spacecraft centers-of-mass (CMs). The relative motion of the non-CM points was illustrated by using two examples: A formation flying scenario and a rendezvous mission. The main observation is that neglecting the relative translation induced by the relative rotation can lead to considerable errors.

It was demonstrated that the points that are not located at the deputy's CM oscillate harmonically with respect to the chief's CM. This motion cannot be detected using traditional relative motion models. It was also shown that if the kinematic coupling is neglected the CW equation can lead to considerable errors when applied to the modeling of rendezvous and docking.

Orbital perturbations discussed in Chapters 4 and 7 may induce differential dynamics that are more significant than the kinematic coupling effect. However, the latter effect is always present, even in short operation times (typical, e.g., to orbital rendezvous), where orbital perturbations are less dominant. In fact, the kinematic coupling effect does not depend on environmental perturbations and is an inherent part of the nominal relative motion equations. As such, it should be taken into account in the linearized relative-translation equations.

Chapter ten

Formation Control

Methinks if I should kiss thee, no control
Within the thrilling brain could keep afloat
The subtle spirit.

Lord Alfred Tennyson (1809–1892)

In this chapter, we present a brief discussion of methods for continuous and impulsive formation control: Formation establishment, maintenance, and reconfiguration. Formation maintenance is a relatively long-duration activity compared to establishment or reconfiguration. Hence, it is imperative that the reference relative orbit for formation maintenance be chosen as close as possible to that supported by the physics of the perturbed relative motion problem. Otherwise, the fuel required for control will be prohibitive, since perturbations due to J_2, drag, and solar radiation pressure are persistent. In this regard, as much of the non-periodic effects of the perturbations as practically possible must be accommodated into the reference trajectory, a task made easy by establishing either the proper mean or averaged differential orbital elements. It may be recalled that the averaged elements are obtained from the mean elements by including the orbit averaged effects of the short-periodic variations. As shown previously in Chapter 8, selection of the proper initial mean differential elements is also important to achieve naturally bounded formations, thus minimizing the need for control.

10.1 CONTINUOUS CONTROL

The commonly-used continuous feedback control methods are based on Control Lyapunov Functions (CLF) and the Linear Quadratic Regulator (LQR); both are discussed in Chapter 3. While the CLF method is applicable to linear as well as nonlinear models, as the name implies, the LQR is valid only for linear models. However, LQR-based designs may be applicable to nonlinear systems but the margin of stability is limited. A discussion of robust stabilization methods is beyond the scope of this book.

10.1.1 The CLF approach

In this section, the CLF approach – presented in Section 3.6 – is used in the context of formation maintenance or tracking a reference orbit. Let the   **241** equations of motion be represented in terms of the CW equations, perturbed by unmodeled nonlinear effects denoted by $\mathbf{N}$:

$$\ddot{\mathbf{x}} = F_1\mathbf{x} + F_2\dot{\mathbf{x}} + \mathbf{N} + \mathbf{u} \tag{10.1}$$

where

$$F_1 = \begin{bmatrix} 3n_0{}^2 & 0 & 0 \\ 0 & 0 & 0 \\ 0 & 0 & -n_0{}^2 \end{bmatrix} \tag{10.2}$$

and

$$F_2 = \begin{bmatrix} 0 & 2n_0 & 0 \\ -2n_0 & 0 & 0 \\ 0 & 0 & 0 \end{bmatrix} \tag{10.3}$$

Note that F_2 is a skew-symmetric matrix. We will use a natural solution to the CW equations as $\mathbf{x}_r$, our reference solution to be tracked. The control effort required to track the CW solution will be nonzero in the steady state due to the fact that $\mathbf{x}_r$ does not satisfy Eq. (10.1) exactly. Nevertheless, $\mathbf{x}_r$ does satisfy the following equation:

$$\ddot{\mathbf{x}}_r = F_1\mathbf{x}_r + F_2\dot{\mathbf{x}}_r \tag{10.4}$$

and the tracking error, given by $\mathbf{e} = \mathbf{x} - \mathbf{x}_r$, satisfies

$$\ddot{\mathbf{e}} = F_1\mathbf{e} + F_2\dot{\mathbf{e}} + \mathbf{N} + \mathbf{u} \tag{10.5}$$

A simple quadratic test Lyapunov function of the form

$$V = \frac{n_0{}^2}{2}\mathbf{e}^T\mathbf{e} + \frac{1}{2}\dot{\mathbf{e}}^T\dot{\mathbf{e}} \tag{10.6}$$

is selected and its derivative is obtained as shown below:

$$\dot{V} = n_0{}^2\dot{\mathbf{e}}^T\mathbf{e} + \dot{\mathbf{e}}^T\ddot{\mathbf{e}} \tag{10.7}$$

Equation (10.5) can be substituted in Eq. (10.7) to eliminate $\ddot{\mathbf{e}}$. A choice for the control law which cancels the nonlinear effects and induces asymptotic closed-loop stability of the error dynamics is

$$\mathbf{u} = -(n_0{}^2 + F_1)\mathbf{e} - \mathbf{N} - kn_0\dot{\mathbf{e}} \tag{10.8}$$

where k is a positive constant (it can also be a positive definite matrix), which can be selected to achieve a desired speed of response. Furthermore, the closed-loop system satisfies

$$\ddot{\mathbf{e}} = -n_0{}^2\mathbf{e} - kn_0\dot{\mathbf{e}} \tag{10.9}$$

Hence, the tracking error approaches zero, asymptotically, but the steady-state control input will be equal to $\mathbf{N}(\mathbf{x}_r)$. Note that F_2 does not appear in Eq. (10.8) due to its skew-symmetric property. A discussion of the merits of including F_2 in the control law for redundancy and robustness is beyond the scope of this book.

A better reference trajectory can be obtained by applying the nonlinearity correction to the PCO and GCO solutions [106] and including the effects of J_2. However, the simplest and most effective approach to formation maintenance is via the no along-track drift condition. For the two-body problem this is just a condition on matching the energies of the satellites or their semimajor axes. For the J_2 problem, the required $\delta \bar{a}$ is given by Eq. (8.15). The CLF can be written as

$$V = \frac{1}{2}(\delta\bar{a} - \delta\bar{a}_d)^2 \tag{10.10}$$

where $\delta\bar{a}_d$ is the desired mean differential semimajor axis. The first derivative of the CLF is

$$\dot{V} = (\delta\bar{a} - \delta\bar{a}_d)(\delta\dot{\bar{a}} - \delta\dot{\bar{a}}_d) = (\delta\bar{a} - \delta\bar{a}_d)\dot{\bar{a}}_1 \tag{10.11}$$

where the subscripts 0 and 1 stand for the chief and the deputy, respectively. Furthermore, $\delta\bar{a}_d$ has been assumed to be constant and the chief is assumed to be non-maneuvering and acted upon only by the J_2 perturbation. An approximation is also introduced by the direct substitution of the derivative of the osculating semimajor axis of the deputy from Gauss' variational equations (2.107) in Eq. (10.11):

$$\dot{a} = \frac{2a^2 e \sin f}{h}u_x + \frac{2a^2 p}{hr}u_y \tag{10.12}$$

where u_x and u_y are, respectively, the radial and along-track control accelerations. This approximation is consistent with the $\mathcal{O}(J_2)$ approximation of the mean-to-osculating element transformation adopted in this book. Eq. (10.12) suggests that u_y is more effective in controlling a than does u_x, especially for orbits with very small eccentricities. Assuming the use of tangential acceleration only and neglecting the error due to the mean-to-osculating element conversion, a control law for achieving the desired objective of no along-track drift is obtained from Eq. (10.11) as

$$\frac{2a_1^2 p_1}{h_1 r_1}u_{y_1} = -k(\delta a - \delta a_d) \tag{10.13}$$

where k is a positive constant. The along-track control law of Eq. (10.13) is expressed in the $\mathcal{L}_1$ frame of reference of the deputy. In a nonlinear $\mathcal{I}$ frame simulation, this control has to be expressed in the $\mathcal{I}$ frame by using the $\mathcal{L}_1$-$\mathcal{I}$ coordinate transformation matrix, similar to Eq. (4.99), resulting in

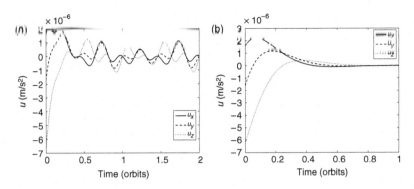

FIGURE 10.1 (a) ECI control components, osculating no along-track drift controller. (b) ECI control components, mean no along-track drift controller.

a control vector with three components. Furthermore, for practical application, the required instantaneous quantities such as a_0, a_1, p_1, and h_1 can be calculated from the instantaneous position and velocity vectors of the two satellites. Such an implementation will be termed the *osculating element control law*. A simple modification of the above procedure by the use of $\bar{a}_0$ and $\bar{a}_1$ in the computation of $\delta\bar{a}$ results in the *mean element control law*. The advantage of using mean elements in the implementation of Eq. (10.13) is demonstrated on a simple example problem.

Example 10.1. *Consider the following mean nonsingular orbital elements of the chief:*

$$\bar{a}_0 = 7100 \text{ km}, \quad \bar{\theta}_0 = 0 \text{ rad}, \quad \bar{i}_0 = 70°$$
$$\bar{q}_{10} = 0.01, \quad \bar{q}_{20} = 0, \quad \bar{\Omega}_0 = 45° \tag{10.14}$$

Assume that five of the six initial mean differential elements are in accordance with the PCO conditions, but the semimajor axis is arbitrarily chosen as $\delta a = 100$ m, violating the no along-track drift condition.

The gain k is set equal to n_0 in Eq. (10.13). The $\mathscr{I}$ frame components of the control histories are shown for the osculating and mean element controllers in Fig. 10.1(a) and Fig. 10.1(b), respectively. The initial transients in the controls seen in these figures are due to the error in $\delta\bar{a}$. The osculating element-based controls do not reach their steady states but show persistent activity; however, the mean element-based controls asymptotically approach zero within one orbit of the chief.

A discussion of the CLF approach to formation flying using Cartesian coordinates as well as orbital elements is provided in Ref. [29]. Sengupta and Vadali [31] use the CLF approach in conjunction with an Euler-parameter-based formation control law. De Queiroz *et al.* [165] present a CLF-based adaptive control technique for formation maintenance in the presence of uncertainties in the spacecraft masses and perturbations.

10.1.2 LQR control based on averaged orbital elements

Applications of the continuous-time LQR (cf. Section 3.7) to the CW model can be found in Refs. [154,166]. A direct application of the CW model-based control is inefficient for the J_2 problem since it does not account for the perturbed mean motion or quasi-periodic nature of the relative orbits [153, 156]. It has also been recognized that radial thrust is inefficient for formation maintenance near a circular orbit [152,167]. Fortunately, the CW model is controllable with only the along-track and cross-track control components.

In Section 8.7, we derived the modified CW equation, appropriate for modeling the formation maintenance problem. These equations are repeated again for ready reference:

$$\ddot{x} - 2\dot{\bar{M}}_0\dot{y} - 3\dot{\bar{M}}_0^2 x = -3n_0^2 x_{bias} + u_x \tag{10.15a}$$

$$\ddot{y} + 2\dot{\bar{M}}_0\dot{x} = u_y \tag{10.15b}$$

$$\ddot{z} + (\dot{\bar{M}}_0^2 + 2n_0\dot{\bar{\omega}}_0)z = -2\rho(0)kn_0 \sin \bar{i}_0^2 \sin \bar{\lambda}_0 \cos \alpha(0) + u_z \tag{10.15c}$$

The reference solutions selected in Section 8.7 are

$$x_r(t) = 0.5 \left(1 + \frac{0.5\dot{\alpha}}{n_0}\right)\rho(0)\sin(\bar{\lambda}_0(0) + \alpha(0)$$
$$+ (\dot{\bar{M}}_0 + \dot{\alpha})t) + x_{bias} \tag{10.16a}$$

$$y_r(t) = \rho(0)\cos(\bar{\lambda}_0(0) + \alpha(0) + (\dot{\bar{M}}_0 + \dot{\alpha})t) \tag{10.16b}$$

$$z_r(t) = \rho(0)\sin(\bar{\lambda}_0(0) + \alpha(0) + (\dot{\bar{M}}_0 + \dot{\alpha})t) \tag{10.16c}$$

where x_{bias} is determined from Eq. (7.137) after substituting, respectively, for $\delta\bar{a}$ and $\delta\bar{i}$ from Eqs. (8.15) and (8.87) as

$$x_{bias} \approx -\frac{5}{4}J_2\rho(0)\left(\frac{R_e}{\bar{a}_0}\right)^2 \sin(2\bar{i}_0)\cos\alpha(0) \tag{10.17}$$

The reference controls are

$$u_{x_r} = 0 \tag{10.18a}$$

$$u_{y_r} \approx -\rho(0)n_0\dot{\alpha}\cos(\bar{\lambda}_0(0) + \alpha(0) + (\dot{\bar{M}}_0 + \dot{\alpha})t) \tag{10.18b}$$

$$u_{z_r} \approx 2n_0(\dot{\bar{\omega}}_0 - \dot{\alpha})\rho(0)\sin(\bar{\lambda}_0(0) + \alpha(0) + (\dot{\bar{M}}_0 + \dot{\alpha})t)$$
$$+ 2\rho(0)kn_0 \sin^2 \bar{i}_0 \cos(\alpha(0))\sin\bar{\lambda}_0 \tag{10.18c}$$

The reference states and controls, respectively given by Eqs. (10.16a)–(10.16c) and Eqs. (10.18a)–(10.18c), also satisfy Eqs. (10.15). Hence the six-dimensional error vector, $\mathbf{e} = \mathbf{x} - \mathbf{x}_r$, satisfies

$$\dot{\mathbf{e}} = A\mathbf{e} + B \begin{vmatrix} u_x \\ u_y - u_{y_r} \\ u_z - u_{z_r} \end{vmatrix} \tag{10.19}$$

where

$$A = \begin{bmatrix} 0 & 0 & 0 & 1 & 0 & 0 \\ 0 & 0 & 0 & 0 & 1 & 0 \\ 0 & 0 & 0 & 0 & 0 & 1 \\ 3\dot{M}_0^2 & 0 & 0 & 0 & 2\dot{M}_0 & 0 \\ 0 & 0 & 0 & -2\dot{M}_0 & 0 & 0 \\ 0 & 0 & -\dot{M}_0^2 - 2n_0\dot{\omega}_0 & 0 & 0 & 0 \end{bmatrix} \tag{10.20}$$

and

$$B = \begin{bmatrix} 0_{3\times3} \\ I_{3\times3} \end{bmatrix} \tag{10.21}$$

Thus, with the appropriate choices of Q and R, the LQR tracking control law for formation maintenance can be determined in terms of the error coordinates as

$$\mathbf{u} = -K\mathbf{e} \tag{10.22}$$

where K is the LQR gain matrix. There is no appreciable difference between the solutions for K with or without the term $2n_0\dot{\omega}_0 z$ included in Eq. (10.20). Similarly, the solution of the ARE is not significantly affected by replacing $\dot{M}_0$ in Eq. (10.20) by n_0. Hence, K is based on the solution of the ARE for the CW model. The linear J_2 model presented in Section 7.5 has periodic coefficients and its use will require the gain matrix K to be periodic. Moreover, its state variables are influenced by the short-periodic variations in the differential orbital elements. In the implementation being discussed in this section the state variables are filtered by the averaging process of Section 7.4 and then used in the control law equation (10.22).

NUMERICAL RESULTS

Results of the application of the LQR for formation maintenance and fuel balancing without the use of radial thrust are presented in this section. We consider a formation of seven satellites in a 1 km, PCO configuration with initial phase angles ranging from $0°$ to $90°$. The chief is assumed to be in a circular orbit with $\bar{a}_0 = 7100$ km. The control weight matrix is diagonal with the entries: $[1, 1]/n_0^4$; the state weight matrix is also diagonal with unit weights for the position errors and the rate error weights of n_0^{-2} for each axis. The simulation time period is 1500 orbits of the chief (equivalent to approximately 100 days). The $\dot{\alpha}$ value for each case considered is optimal for the selected inclination of the reference orbit.

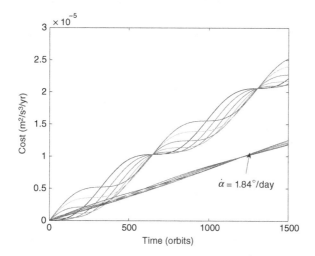

FIGURE 10.2 Cost vs. time with and without $\dot{\alpha}$, $i_0 = 49.11°$, 1500 orbits.

Figure 10.2 shows, the individual satellite cost functions for $i_0 = 49.11°$, obtained by evaluating Eq. (8.114) for each deputy and extrapolating the value over a period of one year. The variations of the costs for the satellites are shown for two cases, with and without fuel balancing. Each curve in this figure represents the cost for a single satellite. The fuel-balanced cost curves are close to each other and have a smaller average slope compared to the unbalanced cost curves. Hence, the average rate of fuel consumption with $\dot{\alpha} = 0$ is higher than that with $\dot{\alpha}_{opt} = 1.84°/\text{day}$, the corresponding optimal rate. As can be seen from the figure, the use of the optimal rotation rate reduces the formation maintenance cost and simultaneously balances the inter-satellite fuel requirements to a great extent. Balancing the rate of fuel consumption results in the ballistic coefficients (cf. Section 1.3.4) of identical satellites remaining constant over the period of operation of the formation. This is important, since differential drag need not be considered as an additional perturbation in such a situation. The total formation maintenance cost (for seven satellites) with fuel balancing is 8.49×10^{-5} m^2/s^3/year compared to 1.59×10^{-4} m^2/s^3/year, obtained by holding α constant for each satellite.

Figure 10.3 shows that, as discussed in Chapter 8, natural fuel balancing takes place for a formation with $i_0 = 54.73°$ and $\dot{\alpha} = 0$, i.e., without any additional control induced formation rotation. The total formation maintenance cost for this case is 1.08×10^{-4} m^2/s^3/year. This value of the inclination is ideal for fuel balancing, but not for the best fuel economy, since the out-of-plane formation maintenance cost increases with inclination. Figure 10.4 shows a cost comparison with and without fuel-balancing for $i_0 = 70°$. The fuel balanced cost curves are very close to each other and can be easily distinguished from their counterparts. The optimal rotation-rate for this case is $\dot{\alpha}_{opt} = -4.1827°/\text{day}$. The total cost with fuel balancing is 2.12×10^{-4} m^2/s^3/year compared to 5.56×10^{-4} m^2/s^3/year, without.

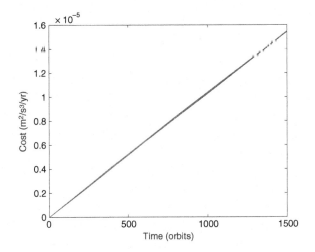

FIGURE 10.3 Cost vs. time with $\dot{\alpha} = 0$, $i_0 = 54.73°$, 1500 orbits.

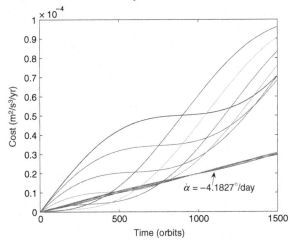

FIGURE 10.4 Cost vs. time with and without $\dot{\alpha}$, $i_0 = 70°$, 1500 orbits.

These results show that longer-term extrapolations of fuel requirements for individual satellites, based on short-term simulations can be erroneous, especially when the individual costs are not well-balanced. However, it is reasonable to extrapolate the total fuel consumption for a PCO formation involving satellites with uniformly distributed initial phase angles over extended periods of time, since the process of cost-averaging virtually eliminates the periodic effects. Figures 10.2–10.4 show that the cost curves for the individual satellites have periodic variations, superimposed over their mean variations, for $\dot{\alpha} = 0$. The periodic components of the cost curves are relatively small compared to their respective linear growth terms when the fuel consumption is well-balanced.

Figures 10.5 and 10.6 show the plots of the along-track and cross-track components of the total and feedforward controls for a deputy with $\alpha(0) = 0$ and for

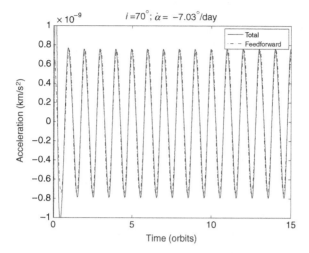

FIGURE 10.5 Comparison of the total and feedforward along-track controls.

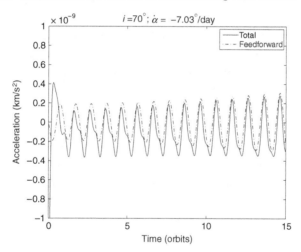

FIGURE 10.6 Comparison of the total and feedforward cross-track controls.

$i_0 = 70°$ for a period of 15 orbits. The fundamental frequencies of the controls for each axis are the same. However, there is also a twice-per-orbit component in the total control which is not present in the feedforward, as can be ascertained from Fig. 10.7, which shows the differences between the two acceleration components for each axis. A small bias component is also visible in the cross-track acceleration error. These effects are caused by the short-periodic effects.

10.2 DISCRETE-TIME LQR CONTROL

The discrete-time LQR (DLQR) formulation, discussed in Section 3.7, is also applicable to the formation maintenance problem. Sherwood and Vassar [168] used the DLQR formulation with impulsive controls for the in-line formation

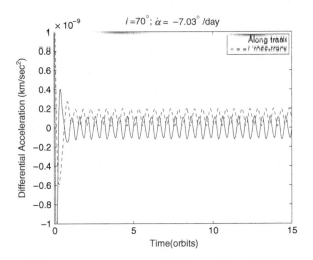

FIGURE 10.7 Differences between the total and feedforward accelerations.

maintenance of dissimilar satellites in a geostationary orbit, under the action of differential solar radiation pressure. The impulse application intervals for in-plane and out-of-plane corrections considered in this study are one and four hours, respectively. Ulybyshev [169] also presents such an application of the DLQR method for long-term formation keeping of a satellite constellation in LEO, using purely tangential maneuvers. We will treat the DLQR approach (cf. Section 3.7) with impulsive control in this section.

For the CW model, under the impulsive control assumption, the discrete-time state and control matrices are

$$A_d = e^{A\Delta T} \tag{10.23}$$

and

$$B_d = A_d B \tag{10.24}$$

where A and B are the state and control matrices for the CW equations and ΔT is the sample time (time between impulses). The reference solutions of Eqs. (10.16a)–(10.16c) do not satisfy the discrete-time model of the CW equations. There is a rigorous means of deriving the feedback gains for the tracking problem when the reference trajectories are known a priori. However, the implementation of a such a control law requires a time-varying gain matrix. The results for the example problems treated in this section are based on the approximate control law of the form

$$\mathbf{u}(i) = -K(\mathbf{x}(i) - \mathbf{x}_r(i)) \tag{10.25}$$

with $\mathbf{x}_r(i)$, the reference trajectory, obtained by sampling Eqs. (10.16a)–(10.16c) at each sample time.

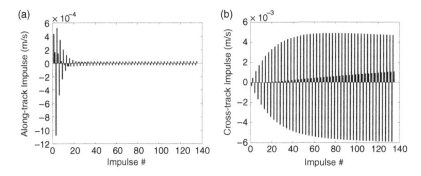

FIGURE 10.8 (a) Along-track impulses for $\dot\alpha = 0$. (b) Cross-track impulses for $\dot\alpha = 0$.

10.2.1 Numerical results

Results presented in this section are for the following mean nonsingular elements of the chief:

$$\bar{a}_0 = 7100 \text{ km}, \quad \bar\theta_0 = 0 \text{ rad}, \quad \bar{i}_0 = 70°$$
$$\bar{q}_{10} = 0, \qquad \bar{q}_{20} = 0, \quad \bar\Omega_0 = 0 \tag{10.26}$$

The initial conditions of the deputy are selected for a PCO with $\rho = 1$ km and $\alpha(0) = 0$. Two cases are considered: $\dot\alpha = 0$ and $\dot\alpha = -7.03°/\text{day}$. The reference solution chosen for tracking is the same as that used for the continuous-time case. The state weight matrix Q is a diagonal matrix with entries 0.01 for the position errors and 0 for the velocity errors. The control weights for the two controls are n_0^{-2}. The simulations are carried out by integrating the nonlinear model. The integration process is stopped at each impulse application time and carried forward after the ECI velocity components of the deputy are updated by the impulses. The ECI components of the impulses are obtained from their LVLH counterparts given by Eq. (10.25). Figure 10.8 shows the along-track and cross-track impulse magnitudes for $\dot\alpha = 0$ for the case of three impulses per orbit, i.e., $\Delta T = \frac{2\pi}{3n_0}$. It is interesting to note that the along-track impulses rapidly decay after reaching a peak of approximately 6×10^{-4} m/s. The maximum cross-track impulse magnitude is 6×10^{-3} m/s. These figures show data for a duration of 3 days, too short for drawing any general conclusions regarding impulse magnitudes required for formation maintenance. Results for the second example, with $\dot\alpha = -7.03°/\text{day}$, are shown in Fig. 10.9. Figure 10.9(a) shows a higher level of along-track impulse compared to that for the previous case and a reverse trend for the cross-track impulse due to the nonzero $\dot\alpha$. Simulations for a one-year period show that the cost for formation maintenance with $\dot\alpha = 0$ is 42.3 m/s/year and that for $\dot\alpha = -7.03°/\text{day}$ is 36.5 m/s/year. No significant differences in the states or the impulse requirements were found with or without the use of the averaging filter.

These examples show the applications of the DLQR approach. However, the DLQR approach has a deficiency in that a large number of the impulses

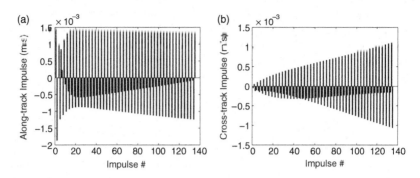

FIGURE 10.9 (a) Along-track impulses for $\dot{\alpha} = -7.03°/\text{day}$. (b) Cross-track impulses for $\dot{\alpha} = -7.03°/\text{day}$.

may be too small in magnitude for a practical implementation and the impulse application times are chosen a priori and not based on any optimality criterion. An alternative approach for formation maintenance with impulsive thrust is presented in the next section.

10.3 IMPULSIVE CONTROL BASED ON GAUSS' VARIATIONAL EQUATIONS

Gauss' variational equations (GVE), incorporating mean elements and the secular J_2 effects, offer a convenient model for computing solutions to multiple-impulse orbit transfer problems. This is so, because mean $a, e,$ and i remain constant and the mean $\Omega, \omega,$ and M vary linearly during the coasting phases. Approximate GVE for the mean classical orbital elements, including the effects of secular variations due to J_2, are (cf. Eqs. (2.107) and Ref. [29])

$$\dot{\boldsymbol{\alpha}} = \mathbf{A}(\boldsymbol{\alpha}) + B(\boldsymbol{\alpha})\mathbf{u} \qquad (10.27)$$

where $\boldsymbol{\alpha} = [a\ e\ i\ \Omega\ \omega\ M]^T$, $\mathbf{u} = [u_x\ u_y\ u_z]^T$,

$$\mathbf{A}(\boldsymbol{\alpha}) = \begin{bmatrix} 0 \\ 0 \\ 0 \\ -1.5 J_2 \left(\dfrac{R_e}{p}\right)^2 n \cos i \\ 0.75 J_2 \left(\dfrac{R_e}{p}\right)^2 n \left(5 \cos^2 i - 1\right) \\ n + 0.75 J_2 \sqrt{1 - e^2} \left(\dfrac{R_e}{p}\right)^2 n \left(3 \cos^2 i - 1\right) \end{bmatrix} \qquad (10.28)$$

and

$$
B(\boldsymbol{\alpha}) =
\begin{bmatrix}
\dfrac{2a^2 e \sin f}{h} & \dfrac{2a^2 p}{hr} & 0 \\[3mm]
\dfrac{p \sin f}{h} & \dfrac{(p+r)\cos f + re}{h} & 0 \\[3mm]
0 & 0 & \dfrac{r \cos \theta}{h} \\[3mm]
0 & 0 & \dfrac{r \sin \theta}{h \sin i} \\[3mm]
\dfrac{-p \cos f}{he} & \dfrac{(p+r)\sin f}{he} & \dfrac{-r \sin \theta \cos i}{h \sin i} \\[3mm]
\dfrac{\eta(p \cos f - 2re)}{he} & \dfrac{-\eta(p+r)\sin f}{he} & 0
\end{bmatrix}
\tag{10.29}
$$

The true anomaly, f, is related to the mean anomaly through the following equations:

$$
M = E - e \sin E \tag{10.30}
$$

$$
\tan \frac{f}{2} = \sqrt{\frac{1+e}{1-e}} \tan \frac{E}{2} \tag{10.31}
$$

10.3.1 Formation establishment

The differential mean elements, δa, δe, and δi for each deputy, must be modified according to any one of the constraint criteria discussed in Chapter 8 to establish a formation. In addition, the mean $\delta\Omega$, $\delta\omega$, and δM must also satisfy certain shape and orientation requirements, such as those for PCO and GCO formations. The orbital elements undergo jump discontinuities at the impulse application times. During the coasting phases, they can be integrated analytically by using their mean secular rates. The true anomaly and the other time-varying quantities in the $B(\mathbf{e})$ matrix can be calculated by using the equations presented in Chapter 2. The impulse application times and the $\Delta\mathbf{v}$ magnitude/directions become the free parameters for optimization. Although the maximum number of impulses required for an optimal transfer is unknown a priori, it is not more than six, the number of orbital elements. Typically, 2–4 impulses are adequate for most orbit establishment maneuvers. Computation of a two-impulse maneuver requires the determination of six impulse components (three per impulse), the two impulse application times, and a final coast interval, leading to a total of nine parameters.

Two different performance measures can be constructed for the fuel-optimal transfer problem. If each impulse is assumed to be produced by a single thruster, with the capability for its pointing provided either by attitude control of the satellite or by a gimbal mechanism, then the performance index is

$$
\mathcal{J}_1 = \sum_{j=1}^{N} \|\Delta\mathbf{v}_j\| = \sum_{j=1}^{N} \sqrt{\Delta v_{x_j}^2 + \Delta v_{y_j}^2 + \Delta v_{z_j}^2} \tag{10.32}
$$

Special care must be taken for formulating the optimization problem for impulsive thrust maneuvers for the three-thruster case, since the fuel cost function is related to the absolute value of each impulse component. Following the approach of Ref. [170], which uses a procedure developed in the Linear Programming literature for the minimization of absolute value functions, each impulse magnitude is represented by two variables as follows:

$$\Delta v = \Delta v^+ - \Delta v^- \tag{10.33}$$

with Δv^+ and Δv^- bounded by

$$0 \leq \Delta v^+ \leq \Delta v_{max} \tag{10.34a}$$
$$0 \leq \Delta v^- \leq \Delta v_{max} \tag{10.34b}$$

where Δv_{max} is the maximum impulse magnitude. With the above representation, the performance index for an N-impulse maneuver can be written as

$$\mathcal{J}_2 = \sum_{j=1}^{N} (\Delta v_{x_j}^+ + \Delta v_{x_j}^- + \Delta v_{y_j}^+ + \Delta v_{y_j}^- + \Delta v_{z_j}^+ + \Delta v_{z_j}^-) \tag{10.35}$$

Note that either Δv^+ or Δv^- is always zero, resulting in $\mathcal{J}_2$ being the sum of the impulse magnitudes. Each component of an impulse can either be positive or negative, by virtue of Eqs. (10.33) and (10.34).

Example 10.2. *Consider an example with the classical mean elements of the chief selected as*

$$\text{œ}_0 = [\ 7100 \text{ km} \ \ .05 \ \ 48° \ \ 0 \ \ 0 \ \ 0 \ ]^T \tag{10.36}$$

The initial mean differential elements are all zero, except for $\delta \bar{M} = \rho / \bar{a}_0$, with $\rho_i = 1$ km; indicating an in-line chief-deputy configuration with an initial separation of 1 km. The desired orbit element differences at the completion of the maneuver correspond to a PCO formation of $\rho_f = 2$ km and $\alpha = 0$, as given by Eqs. (8.83)–(8.87) and Eq. (8.15). The jth impulse application time is denoted by t_j. The time between the application of any two impulses is restricted to a maximum of one orbit. The optimal solutions obtained from the software package SNOPT [171] are presented in Table 10.1. The 2-impulse maneuver is carried out primarily by radial and cross-track thrust impulses; the along-track impulses required are much smaller. The impulses are separated by half the orbital period. Note that for the 2-impulse maneuver, $\mathcal{J}_1 < \mathcal{J}_2$ due to the triangle inequality. Estimates for circular orbits are $|\Delta v_{xy}| = \frac{1}{2}\rho_f n_0 = 1.05$ m/s for the in-plane impulse and $|\Delta v_z| = \rho_f n_0 = 2.1$ m/s. These numbers match well with the data in Table 10.1.

It has been shown for continuous control, with three independent thrusters, that radial thrust is inefficient for formation maintenance. This assertion can be

Table 10.1 Impulse requirements for reconfiguration

	Δv_x m/s	Δv_y m/s	Δv_z m/s	Total m/s	t_j orbits
2-Imp	$\begin{bmatrix} 0.797 \\ -0.257 \end{bmatrix}$	$\begin{bmatrix} -0.002 \\ -0.002 \end{bmatrix}$	$\begin{bmatrix} 1.347 \\ -0.789 \end{bmatrix}$	$\mathcal{J}_1 = 2.395$	$\begin{bmatrix} 0 \\ 0.499 \end{bmatrix}$
2-Imp	$\begin{bmatrix} -0.257 \\ 0.796 \end{bmatrix}$	$\begin{bmatrix} -0.003 \\ 0 \end{bmatrix}$	$\begin{bmatrix} -2.008 \\ 0 \end{bmatrix}$	$\mathcal{J}_2 = 3.064$	$\begin{bmatrix} 0.5 \\ 1 \end{bmatrix}$
4-Imp	0	$\begin{bmatrix} 0 \\ -0.078 \\ 0.261 \\ -0.186 \end{bmatrix}$	$\begin{bmatrix} -2.008 \\ 0 \\ 0 \\ 0 \end{bmatrix}$	$\mathcal{J}_2 = 2.533$	$\begin{bmatrix} 0.498 \\ 1.236 \\ 1.76 \\ 2.236 \end{bmatrix}$

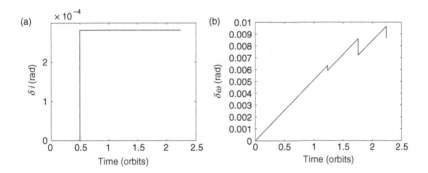

FIGURE 10.10 (a) Change in δi. (b) Change in $\delta \omega$.

verified for maneuvers available with at least three impulses. For a 4-impulse maneuver performed with independent along-track and cross-track thrusters (no radial thrust), the number of free parameters are the eight impulse components, four impulse application times, and a final coast time. The results for this case are shown in the third row of Table 10.1. It is seen that there indeed is a reduction of the in-plane impulse by nearly 50% for the 4-impulse case. The example maneuver can also be performed with three impulses, but the required maneuver time will necessarily have to be longer in order to achieve the level of performance of the 4-impulse solution. Longer maneuver durations allow for the exploitation of the along-track drift caused by a differential semimajor axis. Figure 10.10(a) shows the variation in δi during the maneuver and the jump is a result of the cross-track impulse. On the other hand, Fig. 10.10(b) shows that $\delta \omega$ has secular growth as well as jump discontinuities. Finally, it is noted that for maneuvers performed over one or two orbits, the effect of J_2 is not significant and the impulse magnitudes and maneuver times can be well estimated by the two-body model.

Table 10.2 Impulse components and locations for a two-impulse reconfiguration scheme [3]

	First impulse	Second impulse
Orbit position	$\theta(t_1) = 2\pi - \alpha_i$	$\theta(t_2) = \theta(t_1) + \pi$
Radial	$-\dfrac{\sqrt{\delta q_1^2 + \delta q_2^2}}{2\gamma}$	$\dfrac{\sqrt{\delta q_1^2 + \delta q_2^2}}{2\gamma}$
Along-track	0	0
Cross-track	$\dfrac{\sqrt{\delta i^2 + \delta \Omega^2 \sin^2 i}}{\gamma}$	0

10.4 TWO-IMPULSE FORMATION RECONFIGURATION FOR CIRCULAR ORBITS

For impulse levels of the order of $\rho n_0 \times \mathcal{O}(J_2)$ and circular reference orbit s, the first-order approximations to Gauss' variational equations for the nonsingular elements introduced in Section 2.4, with $\lambda = M + \omega$ and $\theta = f + \omega$, are [3]:

$$\delta a \approx \frac{2}{n} \Delta v_y \tag{10.37a}$$

$$\delta \lambda \approx -2\gamma \Delta v_x - \gamma \sin\theta \cot i \, \Delta v_z \tag{10.37b}$$

$$\delta i \approx \gamma \cos\theta \Delta v_z \tag{10.37c}$$

$$\delta q_1 \approx \gamma \sin\theta \Delta v_x + 2\gamma \cos\theta \Delta v_y \tag{10.37d}$$

$$\delta q_2 \approx -\gamma \cos\theta \Delta v_x + 2\gamma \sin\theta \Delta v_y \tag{10.37e}$$

$$\delta \Omega \approx \frac{\gamma \sin\theta}{\sin i} \Delta v_z \tag{10.37f}$$

where Δv_x, Δv_y, and Δv_z are the magnitudes of the impulse components in the radial, tangential, and out-of-plane directions, respectively, and $\gamma = \sqrt{a/\mu}$. These equations form algebraic constraints for the formation reconfiguration and maintenance problems. As a matter of convenience, we drop the ($\bar{\cdot}$) notation in this section while dealing with the mean elements.

Vaddi et al. [3] have developed a two-impulse scheme for formation establishment and reconfiguration based on Eqs. (10.37a)–(10.37f); their results are summarized in Table 10.2 in terms of the desired differential nonsingular elements, which can be obtained from Eqs. (10.41a)–(10.41e). In this scheme, the impulse application times for each satellite are determined by its initial phase angle α_i and they are separated in latitude angle by π. A single cross-track impulse and two equal and opposite radial impulses are applied for each satellite. Figure 10.11 shows two PCOs and the slot indicated by "1" shows the initial position of the deputy corresponding to $\alpha_i = 0$, when the chief is at the equator. Satellites from the inner PCO are to be reconfigured to the outer PCO and can be placed in any one of the slots indicated on the outer circle. The solution to the reconfiguration problem provides the fuel-optimal impulse magnitudes and the latitude angle of the chief at which they must be applied, and it performs the

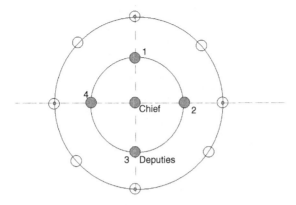

FIGURE 10.11 PCO-PCO reconfiguration (figure courtesy of S. Vaddi).

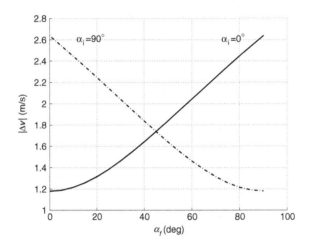

FIGURE 10.12 Impulse requirements for a PCO-PCO reconfiguration.

optimal slot selection process in the new PCO. For an example circular orbit of $a = 7100$ km and $i = 70°$ and for reconfiguration from $\rho_i = 1$ km to $\rho_f = 2$ km, the impulse magnitudes are plotted as functions of the final phase angles for two satellites, with initial phase angles of 0 and 90°, in Fig. 10.12.

The cost curves for the two satellites show that the optimal solution for slot selection in the new PCO is to maintain the initial phase angle for each satellite. This result has also been derived by analysis in Ref. [3].

10.5 TWO-IMPULSE-PER-ORBIT FORMATION MAINTENANCE

A simpler, two-impulse-per-orbit feedback control scheme is presented in this section. This scheme does not require any iterative computation. The reference solution chosen in this section is

$$x_r(t) = 0.5\rho \sin(\theta(t) + \alpha(t)) + \delta u \tag{10.38}$$
$$y_r(t) = \rho \cos(\theta(t) + \alpha(t)) \tag{10.39}$$
$$z_r(t) = \rho \sin(\theta(t) + \alpha(t)) \tag{10.40}$$

There is a difference in notation between Eqs. (10.38)–(10.40), which contain θ explicitly, and Eqs. (8.102)–(8.104), where $\dot{\omega}_0$ has effectively been absorbed into $\dot{\alpha}$. A procedure similar to that used for deriving Eqs. (8.83)–(8.87) leads to the following instantaneous differential mean elements, required to satisfy Eqs. (10.38)–(10.40):

$$\delta i(t) = \frac{\rho}{a_0} \cos \alpha(t) \tag{10.41a}$$

$$\delta q_1(t) = -\frac{\rho}{2a_0} \sin \alpha(t) \tag{10.41b}$$

$$\delta q_2(t) = -\frac{\rho}{2a_0} \cos \alpha(t) \tag{10.41c}$$

$$\delta \Omega(t) = -\frac{\rho}{a_0} \frac{\sin \alpha(t)}{\sin i_0} \tag{10.41d}$$

$$\delta \lambda(t) = -\delta \Omega(t) \cos i_0 \tag{10.41e}$$

The no along-track drift condition determines δa.

The desired mean differential elements at any instant t can be obtained by substituting in Eqs. (10.41a)–(10.41e) and Eq. (8.15), the instantaneous mean elements of the chief and

$$\alpha(t) = \alpha(0) + \dot{\alpha}t \tag{10.42}$$

The impulses required to produce the required changes in the elements of the deputies can be obtained from GVE.

10.5.1 Analytical solution for circular orbits

For a two-impulse-per-orbit control scheme, the impulse application times and the corresponding impulse vectors must be determined. If the impulse application times are known, then the problem boils down to the determination of six unknowns, which satisfy the six constraints imposed by Eqs. (10.37a)–(10.37f). Since the impulse application times are functions of α, which has inter-satellite as well as temporal variations, their direct, closed-form solutions are desirable. The approach presented for formation maintenance, although not necessarily optimal, offers a practical, non-iterative solution. It is an adaptation of the method proposed by Vaddi et al. [3] to the solution of the J_2 problem [4]. The principal assumption herein is the dominance of the out-of-plane impulse magnitude over that required for in-plane formation maintenance when

$\dot{\alpha} \neq 0$. Additionally, the mean elements are substituted for osculating elements in Eqs. (10.37a)–(10.37f), since the theory is accurate to $\mathcal{O}(J_2)$.

The details of the formation maintenance constraints for an N-impulse control scheme can be obtained by incorporating the differential drift rates into Eqs. (10.37a)–(10.37f). It is noted from Eqs. (10.37a) and (10.37c) that δa and δi are piecewise-constant functions, with jumps at the impulse application points. On the other hand, $\delta \lambda$ and $\delta \Omega$ can have jumps as well as linear variation with respect to time. Each instant at which δi jumps due to an impulse, the drift rates, $\delta \dot{\lambda}$ and $\delta \dot{\Omega}$, also undergo changes, as given by Eq. (8.89a). The effect of perigee rotation of the chief influences the evolution of δq_1 and δq_2. These details have been incorporated into the following constraint functions:

$$\delta a_d(t_f) - \delta a(t_0) = \frac{2}{n_0} \sum_{j=1}^{N} \Delta v_{y_j} \tag{10.43a}$$

$$\delta \lambda(t_f) - \delta \lambda(t_0) = \Delta t_{f0} \left(\frac{\partial \dot{\lambda}}{\partial i} \delta i(t_0) + \frac{\partial \dot{\lambda}}{\partial a} \delta a(t_0) \right)$$
$$+ \sum_{j=1}^{N} \left(-2\gamma \Delta v_{x_j} + \frac{2}{\gamma} \frac{\partial \dot{\lambda}}{\partial a} \Delta t_{fj} \Delta v_{y_j} \right.$$
$$\left. - \gamma \{ \cot i_0 \sin \theta_j - \frac{\partial \dot{\lambda}}{\partial i} \cos \theta_j \Delta t_{fj} \} \Delta v_{z_j} \right) \tag{10.43b}$$

$$\delta i_d(t_f) - \delta i(t_0) = \gamma \sum_{j=1}^{N} \cos \theta_j \Delta v_{z_j} \tag{10.43c}$$

$$\begin{Bmatrix} \delta q_{1_d}(t_f) \\ \delta q_{2_d}(t_f) \end{Bmatrix} - \begin{bmatrix} \cos(\dot{\omega}\Delta t_{f0}) & -\sin(\dot{\omega}\Delta t_{f0}) \\ \sin(\dot{\omega}\Delta t_{f0}) & \cos(\dot{\omega}\Delta t_{f0}) \end{bmatrix} \begin{Bmatrix} \delta q_1(t_0) \\ \delta q_2(t_0) \end{Bmatrix}$$
$$= \gamma \sum_{j=1}^{N} \begin{bmatrix} \sin(\theta_j + \dot{\omega}\Delta t_{fj}) & 2\cos(\theta_j + \dot{\omega}\Delta t_{fj}) \\ -\cos(\theta_j + \dot{\omega}\Delta t_{fj}) & 2\sin(\theta_j + \dot{\omega}\Delta t_{fj}) \end{bmatrix} \begin{Bmatrix} \Delta v_{x_j} \\ \Delta v_{y_j} \end{Bmatrix} \tag{10.43d}$$

$$\delta \Omega_d(t_f) - \delta \Omega(t_0) = \Delta t_{f0} \frac{\partial \dot{\Omega}}{\partial i} \delta i(t_0)$$
$$+ \gamma \sum_{j=1}^{N} \left(\frac{\partial \dot{\Omega}}{\partial i} \cos \theta_j \Delta t_{fj} + \frac{\sin \theta_j}{\sin i_0} \right) \Delta v_{z_j} \tag{10.43e}$$

where $(.)_d$ indicates the desired differential orbital element, t_f is the final time, θ_j is the latitude angle at the instant of the jth impulse application, and $\Delta t_{kl} = t_k - t_l$. The coupling between the in-plane and out-of-plane dynamics is affected by the presence of Δv_z in Eq. (10.43b). Equations (10.43a)–(10.43e) can be applied for calculating the impulse magnitudes during each orbit.

Equations (10.43) can be simplified with the knowledge that the change in $\delta \Omega$ over an orbit, due to changes in δi, is small compared to that produced directly by the impulses. This approximation in Eq. (10.43e) results in the following equation for determining the two values of the latitude angle, separated by π, at which it is ideal to perform the out-of-plane maneuvers for each satellite:

$$\tan \theta_{1,2} = \frac{\sin i_0 [\delta \dot{\lambda}_d(t_f) - \delta \Omega(i_0) - \Delta t_{f0} \frac{\partial \dot{\Omega}}{\partial i} \delta i(t_0)]}{\delta i_d(t_f) - \delta i(t_0)} \tag{10.44}$$

The impulse magnitudes can also be solved for from the simplified forms of Eq. (10.43c) and Eq. (10.43e). The total impulse magnitude is distributed equally among the two impulses:

$$\Delta v_{z_{1,2}} =$$
$$\pm \frac{1}{2\gamma} \sqrt{\sin^2 i_0 \, [\delta \Omega_d(t_f) - \delta \Omega(t_0) - \Delta t_{f0} \frac{\partial \dot{\Omega}}{\partial i} \delta i(t_0)]^2 + [\delta i_d(t_f) - \delta i(t_0)]^2} \tag{10.45}$$

The solution obtained from Eq. (10.44) should be modulated to lie between $0°$ and $180°$, if it is negative. If the solution is positive to begin with, then the positive sign applies for Δv_z obtained from Eq. (10.45), otherwise the negative sign should be chosen. Once the particulars for the first impulse are known, then the information regarding the second impulse can be obtained as $\theta_2 = \theta_1 + \pi$ and $\Delta v_{z_2} = -\Delta v_{z_1}$. This completes the process for determining the impulse application times and out-of-plane impulse magnitudes. Equations (10.44) and (10.45) are not accurate if the times between the impulses are long (more than 10–15 orbits), due to the approximation employed to derive them.

Knowing the two impulse application times, the remaining four equations (10.43a), (10.43b), and (10.43d) can be solved to obtain the four components of the two in-plane impulses. The only remaining question is how to determine an optimal value for $\dot{\alpha}$.

10.5.2 Determination of $\dot{\alpha}$ for fuel balancing

Equations (10.43) can be simplified by using the small-angle approximation, especially if the corrections are to be applied during each and every orbit. For such a case, the out-of-plane constraint equations are

$$-\frac{\rho}{a_0} \sin \alpha(0) \dot{\alpha} \Delta t_{f0} = \gamma \sum_{j=1}^{N} \cos \theta_j \Delta v_{z_j} \tag{10.46a}$$

$$-\frac{\rho}{a_0} \cos \alpha(0) \left(\dot{\alpha} + \frac{\partial \dot{\Omega}}{\partial i} \sin i_0 \right) \Delta t_{f0} = \gamma \sum_{j=1}^{N} \sin \theta_j \Delta v_{z_j} \tag{10.46b}$$

Since the two out-of-plane impulse magnitudes are equal, and as θ_1 and θ_2 are separated by $180°$, Eqs. (10.46a) show that the total out-of-plane impulse magnitude per orbit is related to $\dot{\alpha}$ as shown below:

$$\Delta v_z = (\rho n_0) \Delta t_{f0} \sqrt{\sin^2 \alpha(0) \dot{\alpha}^2 + \cos^2 \alpha(0) \left(\dot{\alpha} + \frac{\partial \dot{\Omega}}{\partial i} \sin i_0 \right)^2} \tag{10.47}$$

Estimates of the impulse requirements are provided for a PCO with $\rho = 1$ km and a reference orbit with the following mean elements:

$$a_0 = 7.092 \text{ km}, \quad \theta_0 = 0 \text{ rad}, \quad i_0 = 70°$$
$$q_{10} = 0, \quad\quad q_{20} = 0, \quad \Omega_0 = 45° \tag{10.48}$$

The out-of-plane impulse obtained from Eq. (10.47) with $\alpha(0) = 0$ and $\dot{\alpha} = 0$ is 40.8 m/s/year. By substituting for $\frac{\partial \dot{\Omega}}{\partial i}$ from Eq. (8.90) and minimizing Eq. (10.47) with respect to $\dot{\alpha}$, we obtain

$$\dot{\alpha}_{Sat} = -\frac{\partial \dot{\Omega}}{\partial i} \sin i_0 \cos^2 \alpha(0) = -\frac{3}{2} J n_0 \sin^2 i_0 \cos^2 \alpha(0) \tag{10.49}$$

Equation (10.49) shows that $\dot{\alpha}$ is a function of $\alpha(0)$, and it is zero for $\alpha(0) = 90°$. Furthermore, Eqs. (8.90c), (10.47) and (10.49) show that $\Delta v_z = 0$ for the satellite with $\alpha(0) = 0°$, an obvious result, but of limited utility for formation fuel balancing. A more useful result is obtained by averaging the above expression over $\alpha(0) \in [0, 2\pi]$:

$$\dot{\alpha}_{Formation} = -\frac{1}{2} \frac{\partial \dot{\Omega}}{\partial i} \sin i_0 = -\frac{3}{4} J n_0 \sin^2 i_0 \tag{10.50}$$

Equation (10.50) can be substituted in Eq. (10.47) to obtain

$$\Delta v_{z_{av}} = \pi \rho \frac{\partial \dot{\Omega}}{\partial i} \sin i_0 / \text{orbit/satellite} \tag{10.51}$$

An estimate of the average out-of-plane impulse requirement per-satellite-year, based on Eq. (10.51), is 20.3 m/s. A similar estimate is obtained if the out-of-plane impulse requirements are averaged for two satellites, with $\alpha(0) = 0$ and $\alpha(0) = 90°$. Accounting for the difference in the definition of α between Eqs. (8.102)–(8.104) and Eqs. (10.38)–(10.40), it can be verified that the results of Eq. (8.118) and Eq. (10.50) are quite close to each other. Hence, the equivalent result for the optimal formation rotation rate consistent with Eqs. (10.38)–(10.40) is

$$\dot{\alpha}^*_{Formation} = -\frac{16}{17} \left(\frac{3}{4} J n_0 \sin^2 i_0 \right) - \frac{1}{17} \dot{\omega} \tag{10.52}$$

Application of Eq. (10.52) for the reference orbit data (10.48) results in $\dot{\alpha} = -2.723°/\text{day}$.

10.5.3 In-plane thrust requirements

The approximations utilized for Eqs. (10.43c) and (10.43e) can also be applied to the remaining equations in (10.43) resulting in

$$\frac{7\rho}{2} J \dot{\alpha} \sin 2i_0 \sin \alpha(0) \Delta t_{f0} = \frac{2}{n_0} \sum_{j=1}^{N} \Delta v_{y_j} \tag{10.53a}$$

$$\frac{\rho}{...}\cos\alpha(0)(\text{...} \cos i_0 + \frac{3}{...}...J \sin 2i_0)\Delta t_{f0} = \sum_{j=1}^{N}\left(-2\gamma'\Delta v_{...}\right.$$

$$+\frac{2}{\gamma}\frac{\partial\dot{\lambda}}{\partial a}\Delta t_{fj}\Delta v_{y_j} - \gamma\left\{\cot i_0\sin\theta_j - \frac{\partial\dot{\lambda}}{\partial i}\cos\theta_j\Delta t_{fj}\right\}\left.\Delta v_{z_j}\right) \quad (10.53b)$$

$$\frac{\rho}{2a_0}(\dot{\alpha}+\dot{\omega})\Delta t_{f0}\left\{\begin{array}{c}-\cos\alpha(0)\\\sin\alpha(0)\end{array}\right\}$$

$$= \gamma\sum_{j=1}^{N}\left[\begin{array}{cc}\sin(\theta_j+\dot{\omega}\Delta t_{fj}) & 2\cos(\theta_j+\dot{\omega}\Delta t_{fj})\\-\cos(\theta_j+\dot{\omega}\Delta t_{fj}) & 2\sin(\theta_j+\dot{\omega}\Delta t_{fj})\end{array}\right]\left\{\begin{array}{c}\Delta v_{x_j}\\\Delta v_{y_j}\end{array}\right\} \quad (10.53c)$$

Equations (10.53) show the complex structure of the in-plane formation maintenance and fuel balancing problem. Equation (10.53a) shows that irrespective of the value of $\alpha(0)$, the sum of the tangential impulses must equal zero, for $\dot{\alpha}=0$. Neglecting the terms containing Δt_{fj} in Eq. (10.53b), it is seen that the net radial impulse, for the most part, has to counter the effect of the out-of-plane impulses. The δq_1 and δq_2 equations (Eq. (10.53c)) show that the total in-plane impulse magnitude is bounded between the two values provided by the following equations:

$$\Delta v_x = \pi\rho(\dot{\omega}+\dot{\alpha})(\text{Radial thrust only, per orbit}) \quad (10.54a)$$

$$\Delta v_y = \frac{1}{2}\pi\rho(\dot{\omega}+\dot{\alpha})(\text{Tangential thrust only, per orbit}) \quad (10.54b)$$

For the reference orbit selected and with $\dot{\alpha}=0$, the in-plane costs, considering eccentricity and perigee constraints, can vary between 2.5–5 m/s/year; the lower value applies for the case with tangential thrust only. The inefficiency of continuous radial thrust for this particular problem has been discussed in Ref. [139]. However, the two-impulse scheme does not allow for the luxury of not using radial thrust.

For the special case of $\alpha_0 = 90°$ and $\dot{\alpha}=0$, Eqs. (10.46) show that $\Delta v_z = 0$. Hence, the assumption of the dominance of Δv_z does not apply for this case. However, when $\dot{\alpha}\neq 0$, this assumption is quite reasonable.

NUMERICAL RESULTS

A formation of seven satellites in the PCO configuration is considered with $\rho = 1$ km. The initial phase angles range from 0° to 90°, in 15° increments. The mean elements of the reference orbit are given in Eq. (10.48). The total impulse requirements are computed by summing the 1-norm of the applied impulse vectors. Simulation results are shown for a period of one year and are obtained by integrating the nonlinear equations of motion of the individual satellites. The integration process is terminated just before the application of each impulse and restarted after updating the velocity of each satellite with the velocity increment. The $\mathscr{L}$-frame impulse vectors are transformed into their $\mathscr{I}$-frame components and then applied to update the velocity vectors of the satellites.

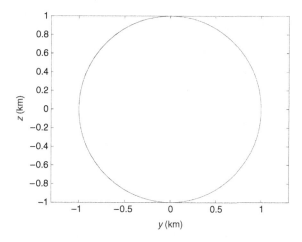

FIGURE 10.13 Resulting relative orbits for $\alpha(0) = 0$ due to the application of control, $\dot{\alpha} = 0$ (one-year simulation).

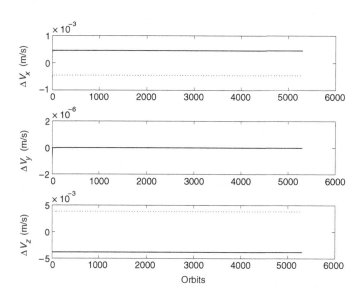

FIGURE 10.14 Impulse components required for controlling a satellite with $\alpha(0) = 0$ and $\dot{\alpha} = 0$.

The result of applying the two-impulse-per-orbit scheme on a satellite with $\alpha(0) = 0$ and $\dot{\alpha} = 0$ is shown in Fig. 10.13. This figure should be compared with the first of Fig. 8.21 to appreciate the effectiveness of the proposed control scheme. The deviation of the controlled PCO from its reference is within ± 1 m for each of the three coordinate axes. The required impulse magnitudes are shown in Fig. 10.14. The impulse magnitudes remain unchanged over time and are nearly the same for the two impulses during each orbit. The two impulse directions are opposite to each other. The out-of-plane impulse magnitudes are

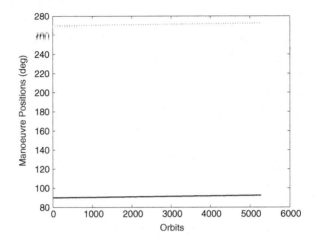

FIGURE 10.15 Impulse application points with respect to the chief's orbit for $\alpha(0) = 0$ and $\dot{\alpha} = 0$.

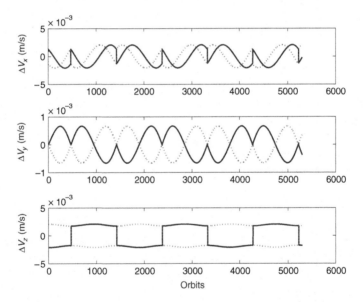

FIGURE 10.16 Impulse components for $\alpha(0) = 0$ and $\dot{\alpha} = -2.723°/$day.

of the order of 4×10^{-3} m/s. The radial impulse magnitude is approximately 0.5×10^{-3} m/s and the tangential impulse is negligible when compared to the other two components. The impulse application instants are shown with respect to the latitude angle of the chief in Fig. 10.15 and they remain very nearly constant. The first impulse is applied at $\theta_1 \approx 90°$ and the second at $\theta_2 \approx 270°$.

The next set of results are for a satellite with $\alpha(0) = 0$ and $\dot{\alpha} = -2.723°/$day, computed from Eq. (10.52). The impulse magnitudes undergo periodic changes as shown in Fig. 10.16. Note that the radial and out-of-plane impulse

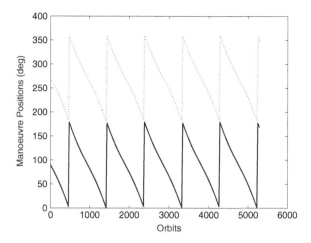

FIGURE 10.17 Impulse application points with respect to the chief's orbit for $\alpha(0) = 0$ and $\dot\alpha = -2.723°/\text{day}$.

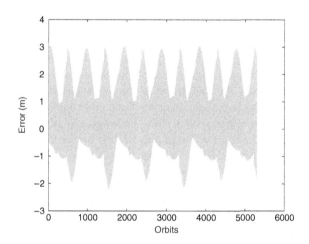

FIGURE 10.18 Evolution of the error in the radius of the PCO for $\alpha(0) = 0$ with the application of control, $\dot\alpha = -2.723°/\text{day}$ (1 year).

components have the same extrema. There is a reduction in the maximum out-of-plane impulse magnitude, coupled with an increase in that for the radial component with respect to their levels shown in Fig. 10.14. The redistribution of the impulse levels, however, leads to a degree of sub-optimality in the control, since the derivation of the control law is based on the assumption that the cross-track impulse magnitude dominates that of the in-plane impulse. The impulse application points, shown in Fig. 10.17, vary between 0–180° for the first impulse and 180–360° for the second. The error in the radius of the PCO is limited to ±3 m as indicated by Fig. 10.18. The reason for the apparent increase in the

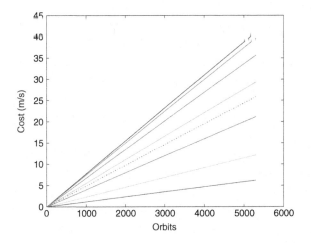

FIGURE 10.19 Formation impulse requirements vs. time with and without fuel balancing for seven satellites with α_0 uniformly distributed between $0°$ and $180°$. Dotted and solid lines indicate impulse magnitudes with and without fuel balancing, respectively.

error from that for the case with $\dot{\alpha} = 0$ is that $\alpha(0)$ is no longer a constant and hence the secular drift rates of the satellites vary with time.

Figure 10.19 shows the variations in the costs for the satellites in the formation with and without fuel balancing. Each solid line in this figure shows the cost for a single satellite, with $\dot{\alpha} = 0$. The lowest cost of 6 m/s/year is that for the satellite with $\alpha(0) = 90°$, which is slightly higher than the 5 m/s/year figure predicted by Eq. (10.54a). The highest cost of 41 m/s/year is incurred by the satellite with $\alpha(0) = 0°$. The result for this case matches perfectly with its analytical estimate, since the out-of-plane cost dominates the total cost. The dotted line in Fig. 10.19 shows the effect of fuel balancing; it is in fact a superposition of seven lines, one for each satellite. The average cost per satellite under the action of the fuel balancing control is 25.8 m/s/year. The total cost for formation maintenance without fuel balancing is 185 m/s/year as compared to 181 m/s/year, obtained with fuel balancing. A decrease, albeit small, in the total cost has been achieved, in addition to the main goal of homogenization of the inter-satellite fuel requirements.

The final set of results pertains to formation maintenances by the application of the two-impulse control scheme during every 10th orbit of the chief. The results for $\alpha(0) = 0$ and $\dot{\alpha} = 0$ are considered first. As can be seen from Fig. 10.20 and Fig. 10.13, the excursions away from the reference PCO are more pronounced when the impulses are applied during every tenth orbit. Notice also that the extreme deviations occur at four locations; these are functions of $\alpha(0)$ of the two-impulse scheme. Figure 10.21 shows that the variation of the error in the radius of the controlled PCO remains bounded within ±41 m. The secular growth is eliminated periodically by the control. The impulse

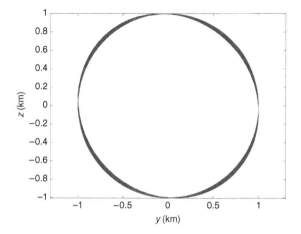

FIGURE 10.20 PCO resulting from the two-impulse control scheme applied during every tenth orbit of the chief, $\alpha(0) = 0$ and $\dot{\alpha} = 0$ (1-year simulation).

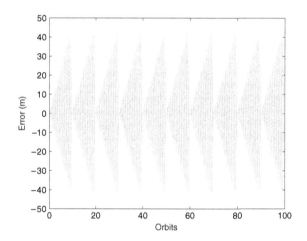

FIGURE 10.21 Evolution of the error in the radius of the PCO for $\alpha(0) = 0$ with control applied during every tenth orbit of the chief, $\dot{\alpha} = 0$ (100 orbits shown).

requirements are shown in Fig. 10.22 for a 100-orbit segment of time. The circles and the stars in this figure indicate, respectively, the magnitudes of the first and second impulses. A comparison of Fig. 10.22 and Fig. 10.14 shows that the impulse magnitudes have been scaled by a factor of 10 for the relaxed mode of operation. However, the total cost for this case is still approximately 41 m/s/year, unchanged from the corresponding case of control applied during every orbit. Delaying impulse application to a time when the orbit error is sufficiently large results in practically realizable thrust levels but it does not

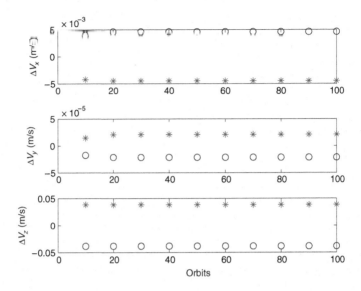

FIGURE 10.22 Impulse components for $\alpha(0) = 0$ and $\dot{\alpha} = 0$, two-impulse control applied during every tenth orbit of the chief, 100 orbits shown.

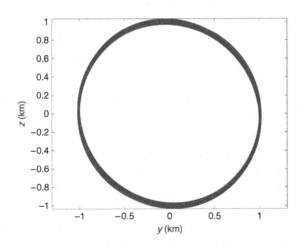

FIGURE 10.23 PCO resulting from the application of the two-impulse control scheme during every tenth orbit of the chief, $\alpha(0) = 0$ and $\dot{\alpha} = -2.723°/$day (1-year simulation).

save fuel because a secular perturbation is being dealt with. Figure 10.23 shows the relative orbits for $\alpha(0) = 0$ and $\dot{\alpha} = -2.723°/$day. It is seen that the resulting levels of excursions away from the reference are higher than those observed in Fig. 10.20. Figure 10.24 shows that the error in the radius of the PCO is bounded by ±49 m. A longer-horizon trace of the error is shown in Fig. 10.25, which indicates a stable, limit cycle behavior.

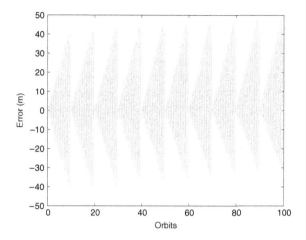

FIGURE 10.24 Evolution of the error in the radius of the PCO for $\alpha(0) = 0$ with control applied during every tenth orbit of the chief, $\dot\alpha = -2.723°$/day (100 orbits shown).

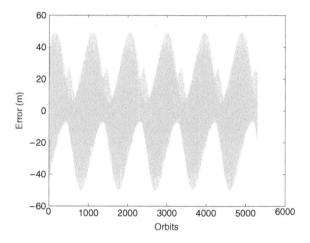

FIGURE 10.25 Evolution of the error in the radius of the PCO for $\alpha(0) = 0$ with control applied during every tenth orbit of the chief, $\dot\alpha = -2.723°$/day (1-year simulation).

SUMMARY

This chapter presented an overview of various control methods – continuous and discrete for formation establishment, maintenance, reconfiguration, and fuel balancing. Examples of the applications of the CLF and LQR methods were presented in conjunction with the feedback of errors based on mean and averaged elements. The application of an optimization technique for computing multi-impulse reconfiguration maneuvers was presented. Radial component of the thrust was inhibited for the example problems for which the formation can be controlled with tangential and cross-track controls. Application of the open-loop optimal multi-impulse maneuver strategy is impractical for long-duration formation maintenance problems. Hence, a simple two-impulse-per orbit scheme was also treated for which the impulses can be calculated analytically, effectively producing a feedback control approach. In all the cases presented, the benefit of controlling the mean rather than the osculating orbital elements was highlighted. Furthermore, many of the theoretical estimates for thrust or impulse magnitudes derived in the previous chapters were shown to be achievable under ideal control application to the nonlinear equations of motion. Simulations with higher-fidelity models and the effects of noise or the results of using estimated states were not treated in this chapter. These topics are addressed in the following chapters.

Chapter | eleven

Implementation of Δv Commands

It will have outlet, brave and not so brave.
weapons of war and implements of peace
Are but the points at which it finds release.

Robert Frost (1874–1963)

Historically, sensor noise has been a key factor in the analysis of system performance for formation flying spacecraft [39,172]. Since formation flying missions require coordination between multiple spacecraft, knowledge of the relative states must be as accurate as possible. This knowledge is typically used when planning trajectories to meet specific criteria, such as minimizing fuel use or maintaining a desired formation geometry. The planning process depends on knowledge of the initial conditions of the spacecraft [173], and degrades with increasing error. Specifically, Tillerson and How [38] showed that velocity estimation error was the primary cause of poor performance, and that an error of just 2 mm/s can result in errors of about 30 m after just one orbit.[1]

During the late 1990's and early 2000's the best navigation filters, using Carrier-Phase Differential GPS (CDGPS) signals, achieved velocity accuracy on the order of 0.5 mm/s [174]. Since then, the state-of-the-art has improved significantly, with Leung and Montenbruck [175] demonstrating a filter that can estimate relative position and velocity to within 1.5 mm and 5 µm/s, respectively. While those numbers represent a best-case performance for real-world operation, it is nonetheless important to understand how other sources of noise can affect formation flying. The PRISMA formation flying demonstration mission aims to validate sensor and actuator technologies for formation flight and rendezvous and docking. Consisting of a two-spacecraft formation deployed in a sun-synchronous 700 km orbit, detailed simulations of the estimation and

---

[1] This error estimate is consistent with the along-track drift term in Eq. (5.18). For $x(0) = 0$, the drift becomes $y_{\text{drift}}(t) = -3\dot{y}(0)t$. For an initial velocity error of 2 mm/s and a 90-minute orbit, the resulting drift is $y_{\text{drift}} = 32.4$ m.

Spacecraft Formation Flying; ISBN: 9780750685337

control processes have indicated that a driver of system performance is thruster performance [176]. Specifically, the minimum impulse bit of 0.7 mm/s is only adequate for controlling the relative mean along-track formation to within 90 m when using the impulsive control scheme presented in Ref. [3], indicating that incorrect implementation of a thruster burn is a significant source of error.

Thruster burns are commonly specified by the spacecraft's desired change in velocity (Δv), and the propulsion subsystem is then responsible for applying this Δv to the spacecraft. Mission-critical maneuvers require precise implementation of thruster burns. For example, for the Cassini Saturn Orbit Insertion ($\Delta v \approx 625$ m/s), an algorithm that measured the energy change of the spacecraft was used. This energy change was monitored autonomously by Cassini during the burn, using real-time measurements from onboard accelerometers [177]. The insertion maneuver was successfully terminated when the desired energy change was reached, and Cassini became the first spacecraft to orbit Saturn. Similarly, the smart impactor of the Deep Impact mission successfully targeted the comet Tempel 1 on July 4, 2005. Precision implementation of trajectory correction maneuvers (TCMs), made possible by the attitude determination and control (ADCS) software and accelerometers, is credited as a primary reason for mission success [178]. The TCM-1 maneuver had a command Δv of 28.568 m/s, and the control system delivered 28.561 m/s – a performance within 0.03% accuracy.

This chapter discusses the impact of Δv implementation error on the performance of spacecraft formation flying, rendezvous and docking. It also explores how accelerometers can be used to improve performance by providing accurate measurements of the applied Δv.

11.1 PLAN IMPLEMENTATION

If knowledge of the initial state of the spacecraft is sufficiently accurate to determine a good plan, the next key step is to accurately implement the plan. A good plan is one that meets performance objectives (formation geometry, fuel management, drift-free). If such a plan is improperly implemented, the performance objectives might not be met. At best, this probably means replanning and trying again, resulting in a loss of time and fuel. At worst, it could lead to the loss of a spacecraft or an entire formation. A spacecraft executes a Δv command as a sequence of one or more thruster firings. The on and off times of the thrusters can be pre-computed beforehand or calculated on-the-fly. Considering the simple case of a single thruster pointing in the direction of motion, one option is to calculate the burn duration based on an idealized thruster model, and then execute the burn. This open-loop strategy is

$$t_b = \frac{m \Delta v_d}{T_d} \qquad (11.1)$$

where t_b is the burn duration, m is the mass of the spacecraft (assumed constant throughout the burn), Δv_d is the desired velocity change, and T_d is the expected force that is provided by the thruster. This approach is easy to implement, but it is usually a poor idea. Although data on the expected thrust would typically be available, this data is likely based on laboratory testing under specific or idealized conditions. As the operating conditions of the thruster change, so too will its performance. This makes it difficult to predict the performance of the thruster or the open-loop system. The actual delivered thrust can instead be modeled as

$$T_a = T_d(1 + \delta) + w \qquad (11.2)$$

where T_a is the actual thrust delivered, δ is a small number representing a bias on the expected thrust, and w is a small random variable that accounts for any additional variations in the thrust. These additional variations are not limited to just unpredictable fluctuations in the engine thrust, but could also account for a spacecraft that is spinning slowly while thrusting (provided the time scale of the spin is small compared to the burn duration). Using the strategy in Eq. (11.1), and imposing the actuator model in Eq. (11.2), the actual Δv implemented using the open-loop strategy can be written as

$$\Delta v = \left(\frac{T_d}{m}\right) t_b + \left(\frac{T_d \delta + w}{m}\right) t_b \qquad (11.3)$$

which is just Δv_d plus an error term. It is clear that the resulting error from this implementation will be proportional to δ as well as the length of the burn. For high Δv maneuvers, this error could be quite large, and is therefore unacceptably risky.

11.1.1 Using accelerometers to improve Δv implementation

A straightforward way to improve the performance of the actuator is to design a feedback control loop that uses sensors to measure thruster performance. Rather than relying on GPS or CDGPS to measure the Δv changes *after* the burn, this subsection explores using other sensors as a means to measure and track the Δv *as it is applied*. This approach uses direct feedback and the addition of an *axial accelerometer* along the thrust direction to achieve this objective. To simplify the following analysis, it is assumed that the thruster can deliver a continuous range of thrusts. Figure 11.1 depicts a block diagram for the closed-loop thruster control system. For clarity, the state is chosen as

$$\mathbf{x} = \begin{bmatrix} V^* \\ V_a \\ \hat{V} \\ e \end{bmatrix} \qquad (11.4)$$

The commanded velocity is V^*, the actual velocity of the spacecraft is V_a, and the estimated velocity of the spacecraft is $\hat{V}$. The final element of the state vector, e, is the integral of the estimated velocity error. It will be used later by

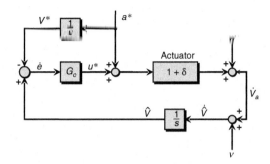

FIGURE 11.1 Closed-loop control system for a single thruster.

the controller. Although each of these velocities is actually a velocity *change*, the Δ is omitted for notational simplicity. The velocity command enters the system through the control input a^*, such that

$$\dot{V}^* = a^* \tag{11.5}$$

The actual acceleration experienced by the spacecraft is influenced by a number of factors. First, the acceleration command a^* is sent directly to the *thruster*, but is distorted by the *actuator* ("actuator" and "thruster" are used interchangeably) dynamics. Additionally, a corrective acceleration u^* is applied by the controller, but once more, the actuator dynamics distort u^* so that $u = (1 + \delta)u^*$. This corrective acceleration should be thought of as a throttling of the thruster. For example, if $\delta > 0$, then u^* will be negative, and the thrust command will be reduced to compensate for the unexpected amplification of the signal that happens in the actuator. Finally, random thruster process noise w is added.

$$\dot{V}_a = (a^* + u^*)(1 + \delta) + w \tag{11.6}$$

The accelerometer is mounted on the spacecraft and measures the actual acceleration directly with noise v

$$\dot{\hat{V}} = \dot{V}_a + v \tag{11.7}$$

It is assumed that both w and v are uncorrelated Gaussian, white-noise processes

$$w \sim \mathcal{N}(0, \sigma_w) \text{ and } v \sim \mathcal{N}(0, \sigma_v) \tag{11.8}$$

As noted above, e is the integral of the estimated velocity error.

$$e = \int_0^t (\hat{V} - V^*)dt \tag{11.9}$$

A proportional-integral (PI) controller $G_c(s)$ with gains k_p and k_i can be used to drive the estimated velocity error to zero, and is therefore a reasonable choice. The state of the integrator is e, as shown in Eq. (11.9). The control law for the PI controller may be written as

$$u^* = -k_p(\hat{V} - V^*) - k_i e \qquad (11.10)$$

Substituting Eq. (11.10) into Eq. (11.6) and Eq. (11.7) yields the following state relationships:

$$\dot{V}^* = a^* \qquad (11.11a)$$
$$\dot{V}_a = [-k_p(\hat{V} - V^*) - k_i e](1 + \delta) + a^*(1 + \delta) + w \qquad (11.11b)$$
$$\dot{\hat{V}} = [-k_p(\hat{V} - V^*) - k_i e](1 + \delta) + a^*(1 + \delta) + w + v \qquad (11.11c)$$
$$\dot{e} = \hat{V} - V^* \qquad (11.11d)$$

Or, in matrix form

$$
\begin{bmatrix} \dot{V}^* \\ \dot{V}_a \\ \dot{\hat{V}} \\ \dot{e} \end{bmatrix}
=
\begin{bmatrix}
0 & 0 & 0 & 0 \\
(1+\delta)k_p & 0 & -(1+\delta)k_p & -(1+\delta)k_i \\
(1+\delta)k_p & 0 & -(1+\delta)k_p & -(1+\delta)k_i \\
-1 & 0 & 1 & 0
\end{bmatrix}
\begin{bmatrix} V^* \\ V_a \\ \hat{V} \\ e \end{bmatrix}
$$
$$
+
\begin{bmatrix}
1 & 0 & 0 \\
(1+\delta) & 1 & 0 \\
(1+\delta) & 1 & 1 \\
0 & 0 & 0
\end{bmatrix}
\begin{bmatrix} a^* \\ w \\ v \end{bmatrix}
\qquad (11.12)
$$

Equations (11.11) and (11.12) provide a convenient form for simulating the system response to a range of acceleration commands. To gain more insight into the steady state behavior of this controller, begin with the generic system

$$\dot{x} = Ax + B_u u + B_w w$$
$$y = Cx + v \qquad (11.13)$$

where x is as in Eq. (11.4). In this formulation, the control inputs u and the random perturbations w are separated. If the control law is chosen as linear state feedback, such that

$$u = -Kx \qquad (11.14)$$

then Eq. (11.13) may be written as

$$\dot{x} = (A - B_u K)x + B_w w \Rightarrow$$
$$\dot{x} = A_{cl}x + B_w w \qquad (11.15)$$

The dynamics in Eq. (11.14) are that of a system driven by random process noise. For a linear time-invariant (LTI) system, if the process noise w is stationary, then the mean square value of the state, as $t \to \infty$, satisfies the Lyapunov equation

$$0 = A_{cl} X_{ss} + X_{ss} A_{cl}^T + B_w R_{ww} B_w^T \tag{11.16}$$

Given positive definite R_{ww}, Eq. (11.16) has a positive definite solution for the state covariance matrix X_{ss} if A_{cl} is stable [179]. Although Eq. (11.12) is convenient for simulating all of the parameters of interest, the A_{cl} matrix has two eigenvalues of 0, and therefore solution of Eq. (11.16) is not possible. By introducing $\epsilon = \hat{V} - V^*$ and $n = (w + v)$, a lower order system is obtained:

$$\begin{bmatrix} \dot{\epsilon} \\ \dot{e} \end{bmatrix} = \begin{bmatrix} -(1+\delta)k_p & -(1+\delta)k_i \\ 1 & 0 \end{bmatrix} \begin{bmatrix} \epsilon \\ e \end{bmatrix} + \begin{bmatrix} \delta & 1 \\ 0 & 0 \end{bmatrix} \begin{bmatrix} a^* \\ n \end{bmatrix} \tag{11.17}$$

Although a^* is not actually a random process noise, this formulation is still useful; provided $k_p > 0$ and $k_i > 0$, then A_{cl} for this system will be stable and there is a unique solution to the Lyapunov equation. While there are numerical algorithms for solving the equation, explicitly solving for the 2×2 case is feasible. The spectral intensity matrix R_{ww} for this problem is

$$R_{ww} = \begin{bmatrix} \sigma_{a^*}^2 & 0 \\ 0 & \sigma_n^2 \end{bmatrix} = \begin{bmatrix} 0 & 0 \\ 0 & \sigma_n^2 \end{bmatrix} \tag{11.18}$$

Since $\sigma_{a^*}^2$ is the reference input, it is known exactly and therefore the upper left entry of R_{ww} is 0 in Eq. (11.18). The steady-state covariance matrix for the state is

$$X_{ss} = \begin{bmatrix} \sigma_\epsilon^2 & \rho_{\epsilon e} \sigma_\epsilon \sigma_e \\ \rho_{e\epsilon} \sigma_e \sigma_\epsilon & \sigma_e^2 \end{bmatrix} = \begin{bmatrix} x_{11} & x_{12} \\ x_{21} & x_{22} \end{bmatrix} \tag{11.19}$$

Here, σ and ρ represent standard deviations and correlation coefficients, respectively. If the entries of A_{cl} are numbered in the same manner as Eq. (11.19), then the solution of the Lyapunov equation is straightforward. Performing the matrix multiplications in Eq. (11.16) yields a system of four equations for the four unknowns in X_{ss}:

$$2A_{11}x_{11} + A_{12}x_{12} + x_{21} = \sigma_n^2 \tag{11.20}$$

$$A_{21}x_{11} + (A_{11} + A_{22})x_{12} + A_{12}x_{22} = 0 \tag{11.21}$$

$$A_{21}x_{11} + (A_{11} + A_{22})x_{21} + A_{12}x_{22} = 0 \tag{11.22}$$

$$A_{21}x_{12} + A_{21}x_{21} + 2A_{22}x_{22} = 0 \tag{11.23}$$

Simultaneous solution of these equations gives the steady state performance of the closed-loop control system driven by the random input **w**:

$$X_{ss} = \begin{bmatrix} \dfrac{(\sigma_w + \sigma_v)^2}{2(1+\delta)k_p} & 0 \\ 0 & \dfrac{(\sigma_w + \sigma_v)^2}{2(1+\delta)k_p k_i} \end{bmatrix} \tag{11.24}$$

The diagonal elements x_{11} and x_{22} in Eq. (11.24) are the expected variances in the state variables ϵ and e, respectively. As might be expected, if either the process noise or sensor noise increases, the steady state behavior of the estimate degrades.

However, an important distinction is that Eq. (11.24) describes how well the *estimated velocity* will converge on the command velocity. Of perhaps more importance to overall performance is how well this estimated velocity tracks the actual imparted velocity, V_a. From Fig. 11.1,

$$\dot{\hat{V}} - \dot{V}_a = v \Rightarrow \hat{V} - V_a = \int_0^t v\,dt \tag{11.25}$$

As seen in Eq. (11.25), the estimation error during the burn grows as the integral of the random sensor noise v. While the expected value of the error is zero for any burn duration, the variance of the estimate will increase for longer burns. This means that for long burns, although the expected implementation error remains zero, the uncertainty grows. Fortunately, the sensor noise for a good accelerometer will be quite small, and the benefit of the added information far outweighs any potential pitfalls. For example, the Mars Reconnaissance Orbiter (MRO) accelerometers had a measurement noise level of just 0.005 mm/s^2 [180]. Furthermore, for long burns, it might be possible to periodically obtain velocity measurements from other sensors to reduce the estimate error.

Increasing the gains k_p and k_i in the controller will cause the estimated velocity to converge faster to the command velocity, but it does not affect the long-term tracking of the actual velocity; the Δv accuracy is purely limited by the sensor noise. As mentioned previously, provided the accelerometer is of reasonably good quality, then $v \ll \left(\frac{T_d}{m}\right)\delta$ and the error for the closed-loop system will grow much more slowly than for the open-loop system.

11.1.2 Discrete example

The analysis in Section 11.1.1 is for a continuous system, but in reality, a control system would be implemented digitally. This section discusses the extension of the previous results to the digital domain. The steady state performance of the system can be directly obtained from Eq. (11.24) by scaling the process and sensor noise. The relationship between *continuous noise* and *discrete noise* models is discussed in [181]. For this case, they are

$$\sigma_w = \sigma_{Dw}\sqrt{T} \tag{11.26a}$$

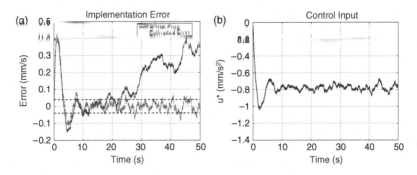

FIGURE 11.2 Results of a discrete simulation of the closed-loop algorithm.

$$\sigma_v = \sigma_{Dv}\sqrt{T} \tag{11.26b}$$

$$\sigma_n = \sigma_{Dv}\sqrt{T} \tag{11.26c}$$

where $\sigma_{D(*)}$ represents the standard deviation of a digital measurement or process, and T is the sample period. Substituting Eq. (11.26) into Eq. (11.24) gives the mean square performance of the digital estimate:

$$\sigma_\epsilon^2 = \frac{(\sigma_{Dw} + \sigma_{Dv})^2 T}{2(1 + \delta)k_p} \tag{11.27}$$

Now, the steady state convergence can be improved by increasing the sampling frequency (and reducing T). Figure 11.2 shows the results of a simulation of the closed-loop control system. The parameters for the simulation were set as follows: $a^* = 4$ cm/s^2, $V^* = 200$ cm/s, $\delta = 0.02$, $\sigma_{Dw} = 0.08$ cm/s, $\sigma_{Dv} = 0.1$ cm/s^2, $T = 0.001$ s, $k_d = 1$, and $k_i = 1$.

For a shorter burn, the gains could be set higher to obtain a faster convergence of the estimate, but they were left low in this case so that the corrective action of the controller is clearly visible. Since $\delta > 0$, the actual thrust is initially higher than the commanded thrust. Therefore, near the beginning of the burn, the accelerometers detect the variation and the estimated error starts to increase. The controller then applies a differential thrust in the negative direction to counteract δ. Because a^* is 4 cm/s^2 and $\delta = 0.02$, this differential thrust should be about -0.8 mm/s^2. The control history shown in Fig. 11.2(b) confirms this prediction.

For the values used in this simulation, Eq. (11.27) predicts that $\sigma_\epsilon \approx 0.04$ mm/s. This region is marked by the dashed line in Fig. 11.2(a) and closely matches the actual behavior. The data also shows how the actual error tends to drift from the estimated error as a consequence of Eq. (11.25). The overall performance of the closed-loop system results in an error of 0.36 mm/s, for a burn of 2000 mm/s, or less than 0.02% error. For the open-loop controller, the error would have been δ, or 2%. The closed-loop system delivers 100 times better performance.

11.2 IMPACT ON AUTONOMOUS RENDEZVOUS AND DOCKING

Autonomous rendezvous and docking is an area of ongoing research and promises to enable both space exploration and on-orbit assembly and servicing. The effect of process noise on an autonomous rendezvous scenario was simulated for two satellites initially separated by about 150 meters. One satellite, the chaser, must maneuver to the location of the second satellite, the target, over the course of one orbit. Only the chaser satellite fires its thrusters, and the orbit is circular at an altitude of 335 km (the approximate altitude of the International Space Station).

For a circular orbit, the time-invariant relative dynamics of two point-mass spacecraft separated by a short distance are described by the CW equations, discussed in Chapter 5; however, as we showed in Chapter 9, the CW equations do not provide accurate modeling of relative dynamics when the target and chaser spacecraft are arbitrarily shaped due to rotation-translation coupling. If the latter effect is neglected, the resulting perturbation must be absorbed by control accelerations.

In the presence of a control acceleration input, $\mathbf{u} = [u_x, u_y, u_z]^T$, the CW equations in the absence of rotation-translation coupling are given by

$$
\begin{bmatrix} \dot{x} \\ \dot{y} \\ \dot{z} \\ \ddot{x} \\ \ddot{y} \\ \ddot{z} \end{bmatrix} = \begin{bmatrix} 0 & 0 & 0 & 1 & 0 & 0 \\ 0 & 0 & 0 & 0 & 1 & 0 \\ 0 & 0 & 0 & 0 & 0 & 1 \\ 3n^2 & 0 & 0 & 0 & 2n & 0 \\ 0 & 0 & 0 & -2n & 0 & 0 \\ 0 & 0 & -n^2 & 0 & 0 & 0 \end{bmatrix} \begin{bmatrix} x \\ y \\ z \\ \dot{x} \\ \dot{y} \\ \dot{z} \end{bmatrix} + \begin{bmatrix} 0 & 0 & 0 \\ 0 & 0 & 0 \\ 0 & 0 & 0 \\ 1 & 0 & 0 \\ 0 & 1 & 0 \\ 0 & 0 & 1 \end{bmatrix} \begin{bmatrix} u_x \\ u_y \\ u_z \end{bmatrix} \quad (11.28)
$$

where, as usual, $n = \sqrt{\mu/a^3}$ is the mean motion on the reference orbit and a is the semimajor axis of the reference orbit.

The linear nature of Eq. (11.28) enables the use of *convex optimization* techniques to calculate a fuel-optimized plan [40]. The chaser's objective is to fire its thrusters and move to the origin (target satellite) over one orbit period. The orbit period is discretized into 1000 segments, with control inputs allowed during every segment. The only objectives for the rendezvous problem are to minimize fuel use and reach the target, and these objectives yield the simple optimization

$$
\min_{\mathbf{u}} \|\mathbf{u}\|_1 \quad \text{subject to} \quad \mathbf{x}_f = \mathbf{x}_d \quad (11.29)
$$

$\mathbf{x}_f$ is the actual state of the chaser at the end of the orbit, $\mathbf{x}_d$ is its desired state, and $\mathbf{u}$ is the sequence of control inputs applied at each step of the plan. The orbital dynamics enter through the constraint $\mathbf{x}_f = \mathbf{x}_d$, and the initial condition and control inputs are chosen to satisfy it.

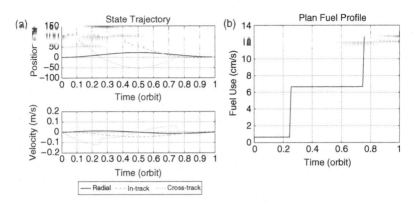

FIGURE 11.3 The ideal rendezvous trajectory and plan.

The minimization problem in Eq. (11.29) is a *linear programming* (LP) problem. It can be solved efficiently using one of many available commercial or free solvers (e. g., linprog in MATLAB® or COIN-LP[2]).

Ideally, the control system would add no process noise, and the plan would be executed perfectly. Figure 11.3 shows the ideal trajectory. At the end of the orbit, the chaser has reached the target (Fig. 11.3(a)). The total fuel consumed is 13.34 cm/s, with control applied at four different points in the orbit; two burns at the beginning and end of about 6.4 mm/s each, and two more burns of 6 cm/s around a quarter and three quarters of the way through the orbit. As this is the ideal performance case, all real control systems will perform worse, either by using more fuel or by failing to satisfy the terminal constraint.

Two parameters were independently varied for the simulations: Process noise and replan frequency. Process noise is modeled in the same manner as in Eq. (11.2), with a random δ for each burn which acts as a percentage error on the magnitude of each thruster firing. This way, longer burns with an inaccurate actuator lead to larger implementation errors. The random δ has a mean value $\bar{\delta}$ and a standard deviation σ_δ. If $\bar{\delta} \neq 0$ then the actuator is biased. As the process noise varies from one simulation to the next, σ_δ is the parameter that changes. Reducing $\bar{\delta}$ or σ_δ is equivalent to improving the actuator. Tested values for σ_δ ranged anywhere from 0.01% to 10% error. The replan frequency f_p is how many times the chaser satellite re-solves the optimization during a single orbit rendezvous mission. When the optimization is re-solved, the initial condition of the chaser satellite is adjusted to its current position. Errors in the initial conditions due to sensor limitations are not considered here; it is assumed that the relative positions and velocities are exactly known.

For the ideal case (with no process noise), replanning is not needed because the control inputs are applied precisely, but when process noise is introduced, errors in the implementation of the plan will cause the chaser satellite to deviate from its expected position. If left unchecked, this error will propagate all the

---

[2]See http://www.coin-or.org/index.html, accessed March 26, 2009.

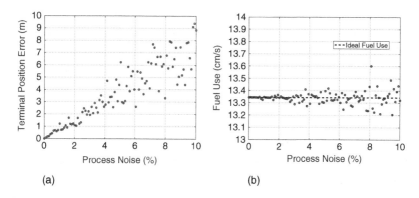

FIGURE 11.4 The effect of varying process noise on rendezvous performance. (a) Effect on terminal position. (b) Effect on fuel use.

way to the end of the trajectory. For a rendezvous mission, if this error is large enough, both the chaser and target satellites could be put at risk. Replanning provides a way to compensate for process error by detecting deviations and modifying the rendezvous trajectory accordingly.

In general, reducing process noise results in improved performance, as does increasing the replan frequency. Figure 11.4(a) shows how varying the process noise influences the performance of the rendezvous. The position error is measured as the rectilinear distance from the chaser to the target satellite at the end of the maneuver. For this set of simulations, the actuator was unbiased ($\bar{\delta} = 0$) and only the magnitude of σ_δ was varied. The control plan was created at the start of the maneuver and executed from start to finish, without replanning. Since no replanning was done, the process noise should have no effect on average fuel use because the deviations caused by incorrect Δv implementation are ignored; the planner does not expend fuel to try to correct them later. Figure 11.4(b) confirms this. The increased dispersion for larger σ_δ reflects the increased randomness of the thrusting, but there is no trend in the average value. For smaller σ_δ, the fuel use converges to the ideal case.

If instead, process noise is held constant while varying the replanning frequency, the performance changes as shown in Fig. 11.5. The process noise σ_δ was held fixed at 5% while replanning frequencies from 1 to 20 times per orbit were investigated. Additionally, in these tests, a bias of $\bar{\delta} = 10\%$ was introduced. As expected, the terminal position error is improved by increasing the replanning rate because replanning allows the effects of inaccurate thrusting to be caught and corrected. This enables the chaser to still reach the target satellite at the desired time. Still, as a result of the positive bias in the actuator, the fuel use remains above the ideal value. The true cause of the problem is not in the plan, but in the actuator that poorly implements it. Continuing to use the open-loop strategy while simply increasing the replanning frequency does nothing to address this root cause. The key distinction is that increasing the planning rate is a *reactive* solution; the errors must happen before they can be observed and corrected by the planner. For maneuvers

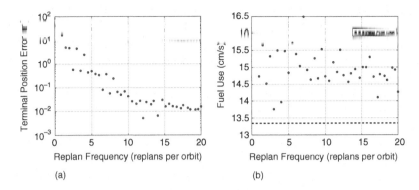

FIGURE 11.5 The effect of replanning frequency on rendezvous performance, with $\bar{\delta} = 10\%$ and $\sigma_\delta = 5\%$. (a) Effect on terminal position. (b) Effect on fuel use.

where a specific trajectory must be tracked closely or constraints avoided safely, replanning alone may not be enough to achieve the necessary performance. Additionally, the ability of a satellite to replan its trajectory is limited by several considerations. Accurate position and velocity estimation is required for both the chaser and the target satellites; repeatedly planning trajectories based on bad estimates is not only inefficient, but it could even result in a completely incorrect result. Moreover, the planning strategy or available computing power might place an upper bound on how often a trajectory can be replanned. Finally, thruster impingement constraints or science goals might limit the number and timing of possible firings.

On the other hand, using accelerometers to monitor the burn performance and more accurately implement Δv commands is a *proactive* solution; errors can be prevented from even happening. Provided the accelerometers are properly calibrated, the control system in Fig. 11.1 eliminates bias and significantly reduces the equivalent process noise.

11.2.1 Impact on formation reconfiguration

This section investigates the performance degradation of a formation reconfiguration maneuver when process noise is added to the system. It utilizes an optimization-based planner. A formation of five spacecraft are in an elliptical orbit defined by

$$\text{œ}_{ref}(t_0) = [\,4.69549 \quad 0.471 \quad 1.10497 \quad 4.24115 \quad 3.7350\pi\,]^T$$

where the first element of œ_{ref}, the semimajor axis, is in units of Earth radii, the eccentricity is unitless, and the inclination, ascending node, argument of perigee, and mean anomaly are in radians.

They begin in an along-track formation, separated by 50 m (Spacecraft 1 at 100 m along-track, Spacecraft 2 at 50 m and so on, so that Spacecraft 5 is at -100 m), with the chief at the center of the formation. The desired configuration is a box in a frame relative to the reference orbit, with the chief

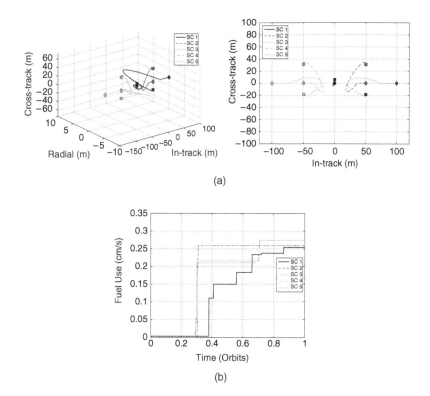

(a)

(b)

FIGURE 11.6 Nominal five spacecraft along-track to passive aperture reconfiguration maneuver. (a) Absolute frame (reference orbit fixed at origin). (b) Fuel use for the fleet.

at the origin and the following relative LVLH states: $[x, y, z, \dot{x}, \dot{y}, \dot{z}]^T$, with position components in meters and velocity components in meters/second):

$$\mathbf{x}_{d_2} = \begin{bmatrix} 0 & 50 & 25 & 0 & 0 & 0 \end{bmatrix}^T$$

$$\mathbf{x}_{d_3} = \begin{bmatrix} 0 & 50 & -25 & 0 & 0 & 0 \end{bmatrix}^T$$

$$\mathbf{x}_{d_4} = \begin{bmatrix} 0 & -50 & 25 & 0 & 0 & 0 \end{bmatrix}^T$$ (11.30)

$$\mathbf{x}_{d_5} = \begin{bmatrix} 0 & -50 & -25 & 0 & 0 & 0 \end{bmatrix}^T$$

This formation emulates an interferometer with four telescopes (the deputies) and a combiner located at the center (the chief), which attains a data-gathering geometry once per orbital period. An optimization-based formation reconfiguration plan for a full-orbit planning horizon was generated using the method in Ref. [182] and is shown in Fig. 11.6. Fuel consumption for each spacecraft (Fig. 11.6(b)) is 2.54 mm/s, 2.59 mm/s, 2.74 mm/s, 2.59 mm/s, and 2.74 mm/s, exhibiting good balance. The total fuel use is 13.20 mm/s. Spacecraft 3 and spacecraft 5 have farther to move to reach their desired positions, and so the formation center shifts 6.06 m in the positive out-of-plane direction in Fig. 11.6(a) to help equalize the fuel use.

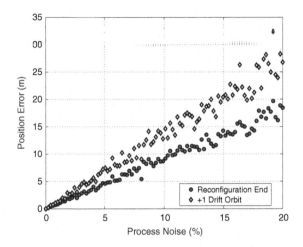

FIGURE 11.7 The effect of varying process noise on a formation reconfiguration.

A series of simulations was run with the formation implementing the plan with varying levels of process noise (0–20%). At the end of the maneuver, the orbit was allowed to continue to propagate for an extra orbit during which no control inputs were allowed (the drift orbit). The position errors for each spacecraft, measured as the rectilinear distance from the actual state to the desired state, were summed at both times. Each process noise level was simulated 20 times and the averaged results are plotted in Fig. 11.7. The circular points mark the position errors at the immediate conclusion of the plan, and the diamonds mark the position errors after the drift orbit. As expected, as the process noise increases, the formation deteriorates. However, the performance of the drift orbit also worsens; if errors are not quickly corrected, they compound. In each case, the position error grew during the drift orbit. The nominal plan does not visibly drift because the planner favors drift free orbits. Drift-free orbits are very sensitive to the initial conditions and this helps explain the rapid deterioration of formations with high implementation error.

SUMMARY

As sensing technology and the ability to design optimal trajectories for formations of spacecraft continue to improve, the need for control systems that accurately implement Δv is becoming more pronounced. This chapter developed a model of Δv implementation error and used it to investigate the dangers of poor thruster performance. For a rendezvous and docking scenario similar to what might be attempted on the International Space Station, small implementation errors of a few percent were found to result in terminal errors on the order of meters. For formation reconfigurations, the dangers are twofold: The desired relative geometry is not initially met, and the formation deteriorates more rapidly if it is allowed to drift. A feedback control system, using accelerometers to directly monitor thruster performance, was developed and analytical expressions describing its expected performance were obtained. For a thruster with 2% deviation from the expected thrust and realistic performance parameters, the closed-loop system improved the accuracy of the implemented Δv by two orders of magnitude. Although simply replanning to correct errors is a possibility, allowable thrust windows are often governed by science goals or other constraints; plans must be implemented correctly without the assumption that mistakes can just be corrected later.

Chapter twelve

Relative Measurements and Navigation

I roamed the infinite sky,
and soared in the ideal world,
and floated through the firmament.
But here I am, prisoner of measurement.

Khalil Gibran (1883–1931)

In this chapter, we discuss an application of both the EKF and UKF algorithms (discussed in Chapter 3) to the relative orbital navigation problem. The equations that relate Carrier-Phase Differential Global Positioning System (CDGPS) measurements to the relative state are the basis for the measurement models.

The relative state used in this chapter includes the quantities of interest for navigation and control, relative position and velocity of the vehicle, as well as several other quantities that are included in the augmented state so the filter will function properly. These other quantities, associated with the use of CDGPS, include the *clock offset*, the *clock drift rate*, and a *carrier-phase bias* for each GPS satellite that is tracked. The state vector used for relative navigation between two vehicles in this chapter is

$$
\mathbf{x}_k =
\begin{bmatrix}
\boldsymbol{\rho}_{ij}(t_k) \\
\Delta b_{ij}(t_k) \\
\dot{\boldsymbol{\rho}}_{ij}(t_k) \\
\Delta \dot{b}_{ij}(t_k) \\
\Delta \beta_{ij}^1 \\
\vdots \\
\Delta \beta_{ij}^N
\end{bmatrix}
=
\begin{bmatrix}
\textit{relative position vector} \\
\textit{clock offset} \\
\textit{relative velocity vector} \\
\textit{clock drift} \\
\textit{carrier phase bias, channel 1} \\
\vdots \\
\textit{carrier phase bias, channel N}
\end{bmatrix}
\tag{12.1}
$$

where the relative vectors are expressed in the ECEF frame $\mathcal{F}$, discussed in Section 2.1. The relative position and velocity dynamics are defined by relations

287

similar to the ones discussed in Chapter 4. The dynamical equations for the ~~absolute states for the deputy vehicles are shown in Section 12.1.~~

There are several methods for computing the relative position and velocity. The absolute states of the chief and deputy could be found and differenced. This has proven inadequate for close formations [43,183]. Another strategy, used here, calculates the relative state of the deputy vehicle, referenced to the position of the chief vehicle.

The absolute state is used in the process of obtaining GPS measurements and in functions external to navigation and control, so it is also estimated. There are many estimation techniques available to form an absolute state solution [42, 184,185]. The dynamics used for the absolute state estimation are presented here and are also used in the development of the relative dynamics equations.

12.1 DYNAMICAL MODELING

The two-body equations of motion in the ECI frame, Eq. (2.1), endowed with a thrust acceleration vector, $\mathbf{u}$, are given by

$$\ddot{\mathbf{r}} = -\frac{\mu \mathbf{r}}{\|\mathbf{r}\|^3} + \mathbf{u} \tag{12.2}$$

where the ECI position vector $\mathbf{r} = [\,X\ \ Y\ \ Z\,]^T$ has a magnitude of r.

The relative position and velocity terms in the state vector are governed by relative orbital dynamics equations, which can be derived from the absolute orbital dynamics equations, as discussed in Chapter 4. The relative position vector, $\boldsymbol{\rho}_{ij}$, is defined as the difference between the absolute position vectors of vehicles i and j,

$$\boldsymbol{\rho}_{ij} = \mathbf{r}_j - \mathbf{r}_i \tag{12.3}$$

It follows that the relative acceleration is defined by the difference of the two absolute accelerations, with thrust accelerations accounted for with the term $\Delta \mathbf{u}_{ij}$,

$$\ddot{\boldsymbol{\rho}}_{ij} = \ddot{\mathbf{r}}_j - \ddot{\mathbf{r}}_i + \Delta \mathbf{u}_{ij} \tag{12.4}$$

By using the two-body relation (12.2), one can obtain an expression for the relative dynamics in the ECI frame. However, the GPS constellation and the GPS navigation message use the ECEF reference frame, so ECEF is a natural reference frame for navigation filters using GPS measurements. To transform from the ECI frame to the ECEF frame, a correction term accounting for the Coriolis effect is added to the dynamics equation:

$$\ddot{\boldsymbol{\rho}}_{ij} = \frac{\mu}{\|\mathbf{r}_i\|^3} \left[\mathbf{r}_i - \frac{\|\mathbf{r}_i\|^3 (\mathbf{r}_i + \boldsymbol{\rho}_{ij})}{\sqrt{(\|\mathbf{r}_i\|^2 + 2\mathbf{r}_i \cdot \boldsymbol{\rho}_{ij} + \|\boldsymbol{\rho}_{ij}\|^2)^3}} \right]$$
$$+ \Delta \mathbf{u}_{ij} + \mathbf{C}_{ECEF} \tag{12.5}$$

where the differential perturbations have been added, and

$$C_{ECEF} = 2\boldsymbol{\Omega}_e \times \dot{\boldsymbol{\rho}}_{ij} + \boldsymbol{\Omega}_e \times (\boldsymbol{\Omega}_e \times \boldsymbol{\rho}_{ij}) \tag{12.6}$$

where $\boldsymbol{\Omega}_e \equiv {}^{\mathscr{I}}\boldsymbol{\Omega}^{\mathscr{F}} = [0, 0, \Omega_e]^T$ is the angular velocity vector of the ECEF frame relative to the ECI frame, so that $\Omega_e = 7.292 \times 10^{-5}$ rad/s is the rotation rate of the Earth about its axis. Other, more sophisticated, methods for performing the ECI to ECEF rotation are described in Ref. [12], but the propagation time in the Kalman filter is typically very short, so Eq. (12.6) is usually sufficient.

The Jacobian F of the nonlinear relative dynamics in Eq. (12.5) is required for the linear state propagation scheme and for the covariance propagation. The linear dynamics matrix for the position and velocity states is given as

$$F = \begin{bmatrix} F_{pp} & F_{pv} \\ F_{vp} & F_{vv} \end{bmatrix} \tag{12.7}$$

where the position and velocity partitions are

$$F_{pp} = 0_{3\times3} \tag{12.8}$$
$$F_{pv} = I_3 \tag{12.9}$$

$$F_{vp} = \frac{\mu}{r^3} \begin{bmatrix} -1 + 3\dfrac{X^2}{r^2} - \dfrac{r^3}{\mu}\Omega_e^2 & 3\dfrac{XY}{r^2} & 3\dfrac{XZ}{r^2} \\ 3\dfrac{YX}{r^2} & -1 + 3\dfrac{Y^2}{r^2} - \dfrac{r^3}{\mu}\Omega_e^2 & 3\dfrac{YZ}{r^2} \\ 3\dfrac{ZX}{r^2} & 3\dfrac{ZY}{r^2} & -1 + 3\dfrac{Z^2}{r^2} \end{bmatrix} \tag{12.10}$$

$$F_{vv} = \begin{bmatrix} 0 & 2\Omega_e & 0 \\ -2\Omega_e & 0 & 0 \\ 0 & 0 & 0 \end{bmatrix} \tag{12.11}$$

CDGPS navigation techniques require that time be known with a high degree of accuracy. The clocks on the local receivers are relatively low-quality and unstable, so a clock offset and a clock drift rate must be estimated by including each in the Kalman filter state definition. The dynamics of the clock offset from GPS time, b, and the clock drift rate, $\dot{b}$, are modeled as

$$\begin{bmatrix} \dot{b} \\ \ddot{b} \end{bmatrix} = \begin{bmatrix} 0 & 1 \\ 0 & 0 \end{bmatrix} \begin{bmatrix} b \\ \dot{b} \end{bmatrix} + \begin{bmatrix} 0 \\ 1 \end{bmatrix} w_b \tag{12.12}$$

When the single differences are performed between two vehicles, the relative clock dynamics retain only the differential white noise term,

$$\begin{bmatrix} \Delta\dot{b}_{ij} \\ \Delta\ddot{b}_{ij} \end{bmatrix} = \begin{bmatrix} 0 & 1 \\ 0 & 0 \end{bmatrix} \begin{bmatrix} \Delta b_{ij} \\ \Delta\dot{b}_{ij} \end{bmatrix} + \begin{bmatrix} 0 \\ 1 \end{bmatrix} w_{\Delta b} \tag{12.13}$$

Thus, in the state propagation step, the clock model will contribute nothing to the state transition matrix but will introduce terms in the noise covariance model.

Each GPS carrier phase measurement includes a bias term. This bias is treated as a constant that must be estimated. A description of the carrier bias is found in Section 12.2. The differential biases are modeled as constants,

$$\Delta \dot{\beta}_{ij}^m = 0 \qquad (12.14)$$

To summarize, the full relative state is defined as

$$\mathbf{x}_k = \begin{bmatrix} \boldsymbol{\rho}_{ij}(t_k) \\ \Delta b_{ij}(t_k) \\ \dot{\boldsymbol{\rho}}_{ij}(t_k) \\ \Delta \dot{b}_{ij}(t_k) \\ \Delta \beta_{ij}^1 \\ \vdots \\ \Delta \beta_{ij}^N \end{bmatrix} \qquad (12.15)$$

The carrier biases, $\Delta \beta_{ij}^m$, are constant and are ignored during the state propagation step. A truncated vector, which excludes the biases $\Delta \beta_{ij}^1, \ldots, \Delta \beta_{ij}^N$, is propagated. The linearized dynamics of the position, velocity, and clock states are modeled as

$$\dot{\mathbf{x}}(t) = \begin{bmatrix} F_{pp} & 0 & F_{pv} & 0 \\ 0 & 0 & 0 & 1 \\ F_{vp} & 0 & F_{vv} & 0 \\ 0 & 0 & 0 & 0 \end{bmatrix} \mathbf{x}(t) + \mathbf{w} + \mathbf{u} \qquad (12.16)$$

where

$$\mathbf{w} = \begin{bmatrix} \mathbf{0}_{1\times4} & \mathbf{w}_\rho & w_{\Delta b} \end{bmatrix}^T \qquad (12.17)$$

$$\mathbf{u} = \begin{bmatrix} \mathbf{0}_{1\times4} & \Delta \mathbf{u}_{ij} & 0 \end{bmatrix}^T \qquad (12.18)$$

The dynamics and process noise models are then discretized as discussed in Section 3.8.

12.2 MEASUREMENT UPDATE: CARRIER-PHASE DIFFERENTIAL GPS

The code-based pseudorange is used to calculate the absolute state. The code phase, χ, is

$$\chi_i^m = \|\mathbf{r}^{m_i} - \mathbf{r}_i\| + b_i + B^{m_i} + I_i^m + \nu_\chi \qquad (12.19)$$

where $\|\mathbf{r}^{m_i} - \mathbf{r}_i\|$ represents the true range between where the vehicle i is at the measurement time and the GPS satellite m at the transmission time. Offset errors in the clock of vehicle i and the GPS satellite m are captured in the terms b_i and B^{m_i}. The unmodeled (and unknowable) phenomena that affect the code phase measurement are included in the noise term, v_χ. The term I_i^m models the delay imposed on the signal by the ionosphere. This term is modeled as

$$I_i^m = \frac{82.1 \times TEC}{F_c^2 \times \sqrt{\sin^2 \gamma_i^m + 0.076} + \sin \gamma_i^m} \tag{12.20}$$

where TEC is the total electron count in the atmosphere, a varying quantity influenced by, among other things, local solar illumination and sunspot activity. The signal frequency F_c, and the elevation angle of the GPS satellite m with respect to vehicle i, γ_i^m, both influence the path delay caused by the ionosphere.

The carrier phase pseudorange, similarly relating range, clock states, and ionospheric delay, is

$$\phi_i^m = \|\mathbf{r}^{m_i} - \mathbf{r}_i\| + b_i + B^{m_i} + \beta_i^m - I_i^m + v_\phi \tag{12.21}$$

A carrier phase noise term, v_ϕ in Eq. (12.21), replaces the code noise from Eq. (12.19). The difference in the effect of wave delay seen by the carrier and group delay seen by the code is reflected in the carrier pseudorange in Eq. (12.21) that has an ionospheric delay term that is opposite in sign.

The additional term β_i^m introduced in the carrier pseudorange is a carrier phase bias. The bias is required to deal with an integer ambiguity in the phase measurement. The distance between the GPS satellite and the vehicle can be expressed as the sum of the carrier phase ϕ, and an integer multiple k of the carrier wavelength λ,

$$d = \phi + k\lambda$$

where $\lambda \approx 19.2$ cm. The distance viewed as being measured in units of carrier wavelengths, the fractional part of the distance, which is the carrier phase measurement, is known very accurately. The part of the distance that is covered by the integer multiple of wavelengths cannot be determined immediately from the information in the carrier phase measurement. Fortunately, there are a number of techniques available to determine this integer number. We will use a passive technique called *kinematic positioning*. As the GPS constellation and the spacecraft move relative to each other, the range measurements will change, but the bias remains constant [43]. With measurements collected over time, the biases are then observable and can be estimated. While this technique results in a longer startup time, it is quite simple and the biases do not change after the initial startup period. When new GPS satellites enter the antenna's field of view, the biases in their measurements can be determined very quickly. Another advantage to this approach is that because the bias estimates are not necessarily required to be integers, the bias estimate can include constant errors, such as those potentially introduced by an antenna line bias or the correlator inside the receiver.

The line-of-sight (LOS) vector is a unit vector whose origin is vehicle i and points towards GPS satellite m,

$$\mathbf{los}_i^m = \frac{\mathbf{r}^{m_i} - \mathbf{r}_i}{\|\mathbf{r}^{m_i} - \mathbf{r}^i\|} \tag{12.22}$$

The vectors in the LOS equation refer to the positions of vehicle i at the time of measurement and the GPS satellite m at the time of signal transmission. The measurement matrix H includes the LOS vectors for each GPS satellite tracked,

$$H_{\mathbf{LOS}} = \begin{bmatrix} \mathbf{los}_i^1 \\ \vdots \\ \mathbf{los}_i^N \end{bmatrix} \tag{12.23}$$

The Geometric Dilution of Precision (GDOP), indicates the distribution of satellites,

$$\text{GDOP} = \sqrt{\text{trace}\,[(H_{\mathbf{LOS}}^T H_{\mathbf{LOS}})^{-1}]} \tag{12.24}$$

A low GDOP indicates good GPS satellite coverage, which means that measurements are available in all directions, providing good observability of the state. Conversely, a large GDOP indicates poor coverage and may result in degraded estimates.

When two vehicles in close proximity track the same GPS satellites, the measurement for GPS satellite m taken by vehicle i will see many of the same errors as the measurement taken by vehicle j. If these measurements are differenced, then the errors cancel to a large degree. This is the crux of the advantage of CDGPS. The carrier differential phase is defined as

$$\Delta\phi_{ij}^m = \phi_j^m - \phi_i^m \tag{12.25}$$

where ϕ_i^m and ϕ_j^m are the raw carrier phases from GPS satellite m measured by vehicles i and j. This difference is formed for each GPS satellite commonly tracked by both vehicles. Substituting the Eq. (12.21) into this difference yields an expression for the carrier differential phase measurement,

$$\Delta\phi_{ij}^m = \|\mathbf{r}^{m_i} - \mathbf{r}_i\| - \|\mathbf{r}^{m_j} - \mathbf{r}_j\| + \Delta\beta_{ij}^m$$
$$+ \Delta b_{ij} + \Delta B_{ij}^m + \Delta I_{ij}^m + v_{\Delta\phi} \tag{12.26}$$

The carrier differential phase can be expressed explicitly as a function of the relative state, as defined in Eq. (12.3),

$$
\Delta\phi_{ij}^m = \|\mathbf{r}^{m_i} - \mathbf{r}_i\| - \|\mathbf{r}^{m_j} - (\mathbf{r}_i + \boldsymbol{\rho}_{ij})\| + \Delta\beta_{ij}^m
$$
$$
+ \Delta b_{ij} + \Delta B_{ij}^m + \Delta I_{ij}^m + v_{\Delta\phi} \tag{12.27}
$$

As in the equations for relative orbital mechanics, the relative carrier phase measurement equation retains the absolute state of the reference vehicle. Also, the error terms introduced for the raw carrier phase measurements have become differential terms. If the vehicles are close, it is reasonable to assume that the terms modeling the GPS satellite clock error and the ionospheric delay cancel,

$$
\Delta\phi_{ij}^m = \|\mathbf{r}^{m_i} - \mathbf{r}_i\| - \|\mathbf{r}^{m_j} - (\mathbf{r}_i + \boldsymbol{\rho}_{ij})\| + \Delta\beta_{ij}^m + \Delta b_{ij} + v_{\Delta\phi} \tag{12.28}
$$

In summary, the measurement vector, $\mathbf{y}$, contains all the measurements that are used in the Kalman filter. The GPS receiver used herein tracks up to 12 GPS satellites, so the measurement vector

$$
\mathbf{y}_k = \begin{bmatrix} \Delta\phi_{ij}^1(t_k) \\ \vdots \\ \Delta\phi_{ij}^N(t_k) \end{bmatrix} \tag{12.29}
$$

may include as many as $N = 12$ differential carrier phase measurements. Given an estimate of the relative state, the nonlinear measurement equation is

$$
\hat{\mathbf{y}}_k = \begin{bmatrix} \|\hat{\mathbf{r}}^{1_i} - \hat{\mathbf{r}}_i\| - \|\hat{\mathbf{r}}^{1_j} - (\hat{\mathbf{r}}_i + \boldsymbol{\rho}_{ij})\| + \Delta b_{ij} + \Delta\beta_{ij}^1 + \Delta\hat{B}_{ij}^1 \\ \vdots \\ \|\hat{\mathbf{r}}^{N_i} - \hat{\mathbf{r}}_i\| - \|\hat{\mathbf{r}}^{N_j} - (\hat{\mathbf{r}}_i + \boldsymbol{\rho}_{ij})\| + \Delta b_{ij} + \Delta\beta_{ij}^N + \Delta\hat{B}_{ij}^N \end{bmatrix}
$$
$$
+ \begin{bmatrix} \Delta\hat{I}_{ij}^1(\hat{\mathbf{r}}_i, \hat{\boldsymbol{\rho}}_{ij}, \hat{\mathbf{r}}^m) + v_{\Delta\phi} \\ \vdots \\ \Delta\hat{I}_{ij}^N(\hat{\mathbf{r}}_i, \hat{\boldsymbol{\rho}}_{ij}, \hat{\mathbf{r}}^m) + v_{\Delta\phi} \end{bmatrix} = \hat{h}_k(\hat{\mathbf{x}}_k^-) + v \tag{12.30}
$$

and the associated Jacobian is

$$
H_k = \begin{bmatrix} H_{\mathrm{LOS}(N\times3)} & \mathbf{1}_{N\times1} & \mathbf{0}_{N\times3} & \mathbf{0}_{N\times1} & I_N \end{bmatrix} \tag{12.31}
$$

12.3 COMPARISON OF EKF AND UKF FOR RELATIVE NAVIGATION

The purpose of this section is to investigate whether the UKF may improve the relative navigation for formation flying spacecraft, particularly when vehicle separations exceed 1 km. Tests are performed with simulated measurements to show that there is little performance difference between the two forms of the UKF discussed in Chapter 3 for this application. These simulations also show

Table 12.1 Comparison of the standard additive and square root forms of the UKF

	Position (m)	Velocity (m)
Mean, $\mu_{UKF-S} - \mu_{UKF-A}$	3.8277e-005	4.5264e-008
Standard deviation, $\sigma_{UKF-S} - \sigma_{UKF-A}$	1.2474e-005	3.0500e-008

that the additive form of the UKF, described in Subsection 3.9.2, requires much less computation.

We performed a set of simulations to compare the performance of the UKF to the performance of the EKF. Simulations used data generated with orbits calculated in FreeFlyer^{TM} and using stored GPS receiver measurements compared to truth data from experiments at NASA Goddard Space Flight Center (GSFC). The results of these simulations confirm that the UKF outperforms EKF when the separation between vehicles is greater than 1 km or the discrete time step of the filter is greater than 10–15 seconds.

The UKF-S has the potential for improved performance and numerical stability, but these improvements are not guaranteed, so both the Additive and Square Root form of the UKF (UKF-A and UKF-S) were implemented and compared. Since the UKF was designed to better handle the nonlinearities, a 10 km baseline (BL) example was used for this comparison. The estimator was given simulated measurements that were created from stored absolute trajectories. White noise was applied to the dynamics and the measurements. The performance was seen to vary when the filter was run for different randomly generated measurement noise sequences. Twenty different random measurement noise profiles were generated and stored, and the simulation was run for each profile. The average of the results from 20 profiles provided a better basis for evaluating the filter than the results from any single noise profile. Because nonlinearities in the dynamics may have more of an effect when the filter has a longer time step, the UKF-A and UKF-S were compared for time steps between 5 and 60 seconds.

Figure 12.1 shows the differences in the means and standard deviations for the position and velocity estimates from the UKF-A and UKF-S. The results for the individual runs are shown in the background in gray, and the mean over all the noise profiles is shown in a strong black line. The differences for any single run are very small, and, when averaged over the 20 noise profiles, the difference is negligible. Also, the mean differences do not grow or shrink significantly as the time step is increased. The results, summarized in Table 12.1, indicate that there is no performance advantage to using the UKF-S in relative navigation filters. Because there is no significant performance difference, and the UKF-A is easier to implement, it was the one used for comparisons of the UKF and EKF.

Since the UKF was developed to better handle nonlinear dynamics and measurements, it is expected that the UKF will perform better than the EKF only when the nonlinearities become significant. The errors associated with

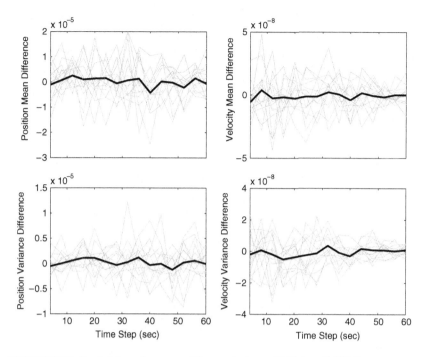

FIGURE 12.1 Negligible performance difference between UKF-A and UKF-S.

nonlinearities are expected to increase as the baseline distance between the two vehicles grows and as the filter time step increases.

Several sets of simulations were created to explore the performance differences in the EKF and UKF. To observe the effects of both increased baseline and time step, the simulations were conducted for baselines of 100 m, 1 km, and 10 km for time steps between 5 seconds and 1 minute. Comparisons were repeated in both a simulated environment and a more realistic environment created using stored data from GSFC. The two environments have different advantages. The total simulation environment provides control over every variable, from noise levels, to perturbations, to satellite coverage. However, simulation results are usually required to be corroborated by data from real hardware. For this reason, simulations based on FreeFlyer^{TM} trajectories and simulations based on recorded hardware experiments at GSFC are used. A summary of these environments is given below.

- FreeFlyer^{TM}-**based simulations** The original MATLAB$^{®}$ simulation trajectories were created with a simple propagator. Perturbations and other real-world effects were coarsely simulated by adding white noise into the dynamics propagation. This is exactly the dynamics model used in the EKF and the UKF. The simulation is more realistic if it uses truth trajectories created with a dynamics model that has a much higher fidelity than the model used in the filter. The FreeFlyer^{TM} commercial orbital dynamics simulation

Table 12.2 Summary of the simulations and results

Baseline	FreeFlyer^{TM} data	GSFC data
100 m	Fig. 12.2	Fig. 12.5
1 km	Fig. 12.3	Fig. 12.6
10 km	Fig. 12.4	Fig. 12.7

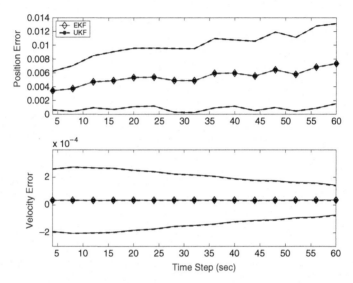

FIGURE 12.2 FreeFlyer^{TM}, BL $=$ 100 m.

software can create very high fidelity trajectories that include per-turbation forces including higher-order gravity terms, solar radiation pressure, third body gravity effects, and aerodynamic forces [186]. For the EKF/UKF comparison, FreeFlyer^{TM} was used to create the truth trajectories from which the simulated measurements were derived. Each simulation was repeated for 20 stored noise profiles and the results were averaged.

- **GSFC stored-data simulations** The GSFC Formation Flying Testbed has a Spirent simulator that models vehicle motion and the GPS satellite constellation and creates an RF signal that mimics the input of the vehicle antenna in space. The Spirent trajectories can be stored as truth data and the output of the GPS receiver is stored to provide future measurement inputs to the filter. The truth and measurements can be post-processed to evaluate various estimators.

A summary of the simulations used to compare the EKF and UKF is shown in Table 12.2. The results of the comparisons are shown in Figs. 12.2–12.7. Each figure has subplots for position and velocity. The mean of the estimate error and bounds for the standard deviation of the error are shown

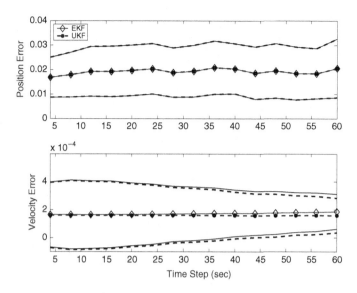

FIGURE 12.3 FreeFlyer^{TM}, BL = 1 km.

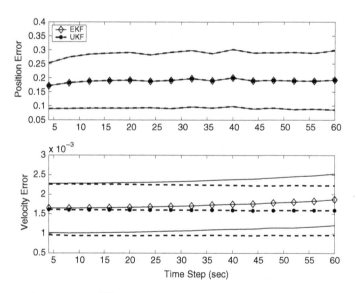

FIGURE 12.4 FreeFlyer^{TM}, BL = 10 km.

for both the EKF and UKF. The errors are shown against an axis, indicating the discrete time step of the filter. These six plots contain a large amount of information, and will be used to demonstrate several trends. The set of figures will be discussed (i) as individual entities; (ii) across a set of baseline distances, evaluating FreeFlyer^{TM} and GSFC results separately; and (iii) between FreeFlyer^{TM} and GSFC results, evaluating each baseline distance separately. Prior to initiating a detailed discussion of the results, a summary of

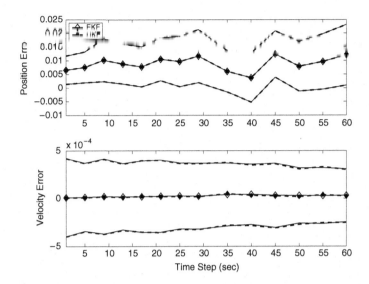

FIGURE 12.5 GSFC, BL = 100 m.

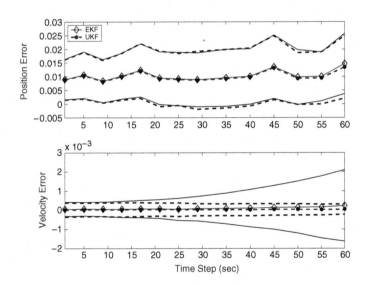

FIGURE 12.6 GSFC, BL = 1 km.

the questions addressed is listed below.

1. **Examining a single figure**
   - How do the errors for the EKF and UKF results compare?
   - Do the means grow as the time step is increased?
   - Do the standard deviations grow as the time step is increased?
   - Does the time step increase have a greater effect on position or velocity estimates?

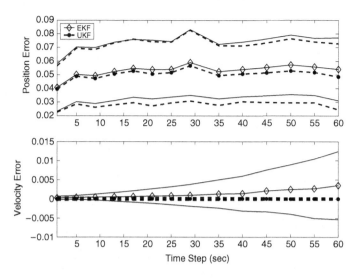

FIGURE 12.7 GSFC, BL = 10 km.

2. **Comparing plots across 3 baselines for same environment**
   - How do the mean values change across the 3 baselines?
   - Do the standard deviations increase across the 3 baselines?
   - If the standard deviation increases with time step for one baseline, does it in the others? If not, is there a reasonable explanation?
   - Does changing the baseline affect the position and velocity differently?

3. **Comparing FreeFlyer™ GSFC plots for same baselines**
   The results from the FreeFlyer™ and GSFC simulations are not expected to agree numerically. The FreeFlyer™ based simulations provide much more control over the truth trajectories and perturbation effects. The measurements are created by adding white noise to the output of equations. This model was also used in the estimator. At GSFC, the real Orion™ hardware with the many associated uncertainties, was used to create the measurements. For example, thermal variation has been known to affect the measurements. Clock uncertainties, poor electrical connections, or unexpected interference might also contribute errors. In general, the simulated measurements produce better estimates. However, trends seen in simulation are also found in hardware tests, so both are used in the comparisons.
   - When the trends in the FreeFlyer™ and GSFC simulations are similar, how do the FreeFlyer™ results reinforce interpretation for GSFC performance?
   - When the trends are different, do simulation differences provide a reasonable explanation?

- Finally, do the FreeFlyer™ and GSFC provide the same answer for the question of when does the UKF performs better?

Several expectations about the performance of the EKF and UKF and about the results of FreeFlyer™ and GSFC simulations are also useful to consider before proceeding.

- The UKF should perform better when the dynamics model breaks down. The dynamics model breaks down when the time step or baseline distance increases significantly.
- Dynamics models in both the EKF and UKF are more similar to the model that created the FreeFlyer™ trajectory. Since the velocity estimate depends on the quality of the velocity model, FreeFlyer™ simulations may perform better for velocity estimates.
- The filter measurement models closely match those used to create measurements from the FreeFlyer™ simulated trajectory. The hardware used to record GSFC measurements may or may not reflect the measurement model in the filter. This suggests FreeFlyer™ simulations may perform better.

Though results are shown for both position and velocity estimates, the velocity performance is much more important. The error in the velocity estimate has more effect on the knowledge of semimajor axis error than does the error in position. Since the semimajor axis error influences closed-loop control performance, and the navigation system exists primarily to aid in the control system, the velocity performance will be used as the final discriminator between the EKF and UKF.

12.3.1 Comparison for single baseline, as time step increases

Each of the six figures shows the mean and the 1σ bounds around the mean for position and velocity errors. The individual figures are useful in evaluating whether there is an advantage in using the UKF for a particular scenario. For example, at 100 m, the position and velocity means are nearly identical for the UKF and EKF. The standard deviation increases very slightly with the time step. In general, the nonlinearities at 100 m, even at a longer time step, are not significant enough to warrant using a UKF. At 1 km and 10 km, the velocity means are higher for the EKF, and the velocity standard deviations diverge as the time step increases. In these cases, the UKF would be a better choice.

12.3.2 Comparison as baselines increase

Figures 12.2–12.7 show that as the baseline is increased, the mean values for position and velocity errors also increase. This degradation of estimation accuracy concurs with results reported in Ref. [43]. When the nonlinear equations were linearized, the higher-order terms of the series expansion are truncated.

The linearization error increases with the distance between vehicles. The truncation is accounted for in the filter by including it in the process noise term. The process noise must therefore be increased as the baseline (and the corresponding truncation error) is increased. The results of this increase in process noise are increases in the position and velocity errors. Similarly, as the baseline grows, the standard deviation trends change from being tapered or constant as the time step increases, to diverging with time step increases.

Though the errors grow with the baseline size, the UKF still offers advantages over the EKF. This is seen in the velocity performance of the UKF, particularly in Figs. 12.6 and 12.7. Interesting differences in the velocity standard deviation of the FreeFlyer^{TM} and GSFC simulations arise and are discussed in the following subsubsection.

12.3.3 Comparison for FreeFlyer^{TM} and GSFC simulations

As stated before, the results from FreeFlyer^{TM} and GSFC simulations are not expected to agree numerically. The 100 m baseline case was unremarkable: the EKF and UKF produced nearly identical results, with position means and standard deviations both showing slight increases in the FreeFlyer^{TM} and GSFC simulations. The velocity means were constant for both simulation environments. The velocity standard deviation bounds became smaller, though this effect was more subtle for the GSFC simulations.

With 1 km baselines, the position trends are similar in the FreeFlyer^{TM} and GSFC simulations. The mean values increase slightly with time step, and the standard deviation bounds flare slightly at the largest time steps. There is no significant difference in EKF and UKF performance in position estimation. As in the 100 m baseline simulations, the mean of velocity errors show little growth as the time step increases. The velocity mean of the EKF is slightly higher than the UKF mean for the largest time steps.

An interesting difference is seen in the standard deviations of velocity error produced by the FreeFlyer^{TM} and GSFC simulations, at a 1 km baseline. In the FreeFlyer^{TM} simulations, the standard deviations for both the EKF and the UKF decrease as the time step increases. In the GSFC simulations, the UKF velocity standard deviation shows a slight decrease as well. However, this is nearly obscured by the dramatic *increase* in the EKF velocity standard deviation. The question is why the EKF velocity diverges in GSFC simulations and not in the FreeFlyer^{TM} simulations. The dynamics model in the filter is a simple two-body model with J_2 perturbations. This is closer to the model used to create the FreeFlyer^{TM} trajectory than it is to the model that governs the GSFC dynamics. FreeFlyer^{TM} corresponds to a case that is in between "perfectly modeled dynamics" and "fully realistic dynamics". Since the velocity estimates strongly depend on the quality of the dynamics model, it is reasonable to expect that the GSFC simulations will show poor performance for combinations of the separation and time step that are shorter than for the FreeFlyer^{TM} simulations. In addition, the "measurements" used in the FreeFlyer^{TM} simulations were created with the measurement equation, while those used in the GSFC simulations were recorded with real hardware. Accordingly, FreeFlyer^{TM} simulations also fall somewhere between "perfectly modeled measurements" and "fully realistic measurements".

This difference in velocity standard deviation behavior in FreeFlyer^{TM} and GSFC simulations continues in the 10 km baseline examples. In the GSFC simulations, in Fig. 12.7, the EKF velocity mean becomes much larger than the UKF velocity mean at large time steps exceeding even the UKF 1σ bounds. The EKF is so much worse in this case that the UKF mean and standard deviation lines appear on top of each other – the EKF velocity standard deviation bounds diverge fairly explosively. Overall, the errors in the EKF velocity at 10 km eclipse the UKF errors. In comparison, while the FreeFlyer^{TM} simulations show the UKF errors are smaller than the EKF errors, the difference is not as dramatic. This is attributed to the differences in the FreeFlyer^{TM} and GSFC simulation setups.

Also of significance, in the 1 km FreeFlyer^{TM} simulations, the velocity standard deviations decreased for the larger time steps, in Fig. 12.3. Conversely, they appear nearly constant in the 10 km FreeFlyer^{TM} simulations, in Fig. 12.4. This suggests the FreeFlyer^{TM} simulations will follow a trend seen in the 100 m and 1 km GSFC simulations, where the velocity standard deviation *decreased* for the larger time steps in the former, but *increased* for larger time steps in the latter. The trend appears more gradually in the FreeFlyer^{TM} simulations than it did in the GSFC simulations, where the EKF velocity standard deviations failed dramatically when the distance was increased from 100 m to 1 km. Still, it appears that the dynamics model is beginning to fail even in the FreeFlyer^{TM} simulations. This trend is affirmed for the FreeFlyer^{TM} model in a final simulation with extreme nonlinearities, in Subsection 12.3.4, where the EKF mean and standard deviations diverge with the sharpness seen in the GSFC simulations.

Overall, the UKF performs better than the EKF when the baseline distance and time step are both increased. This assessment is based on the smaller means and standard deviation bounds of the velocity errors, which is the parameter that most strongly influences closed-loop control performance. The advantage of the UKF is especially apparent in GSFC simulations, whose dynamics and measurements are more realistic than the FreeFlyer^{TM} simulations.

12.3.4 A final example

The previous discussion presented examples that showed the UKF outperforming the EKF as the baselines and time steps increased. This confirmed the hypothesis that the UKF, designed to better handle nonlinearities in the dynamics and measurement models, would become advantageous when the system nonlinearities are accentuated.

A final example with an extremely long baseline of 100 km is presented in Fig. 12.8. The nonlinearities are especially insidious here, as the CW equations fail rapidly at this large separation. Also, at this distance, the ionospheric effects

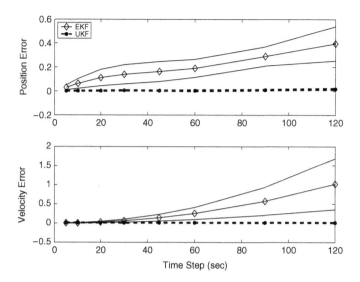

FIGURE 12.8 FreeFlyer^TM, baseline 100 km.

would begin to dominate [187]. At very small time steps, the EKF and UKF performance is comparable. This is because the estimate is corrected very frequently with new measurements. At higher time steps, the filter depends on the dynamics model to propagate between states, but when the dynamics model is poor, the propagated state error grows rapidly. If this error is not corrected quickly, then the filter will diverge. The UKF uses the nonlinear dynamics and measurement equations and employs a much better method of propagating the state error covariance. As a result, the performance is more consistent at longer baselines and longer time steps, and the UKF does not diverge, which is in stark contrast to the EKF results in Fig. 12.8.

This example underscores the potential for the UKF in situations when nonlinearities, including those caused by long time steps or large separations, are particularly important. Long time separations and eccentric orbits are likely to be required of some future missions and could reasonably benefit from the UKF.

SUMMARY

We discussed relative measurements and relative navigation using CDGPS, and compared the estimation of spacecraft relative states using EKF and UKF. Overall, the UKF performs better than the EKF for larger separations and longer time steps. This assessment is based on the smaller mean values and standard deviation bounds of the velocity errors, which is the parameter that most strongly influences closed-loop control performance. The advantage of the UKF is especially apparent in the GSFC simulations, which represent a more realistic setup (in particular the measurements) than the FreeFlyer^TM simulations.

Chapter | thirteen

High-Fidelity Formation Flying Simulation

I breathed enough to learn the trick,
And now, removed from air,
I simulate the breath so well,
That one, to be quite sure.

Emily Dickinson (1830–1886)

In this chapter, we will illustrate most of the previously-introduced material by performing a series of nonlinear simulations of portions of a formation flying reference mission using the FreeFlyer^{TM} orbit simulator [186]. The purpose is to investigate the effects of perturbations and control techniques on a realistic satellite formation through all planned stages of operation. A commercial orbit propagator, FreeFlyer^{TM}, was used to propagate the orbits in an absolute frame for both satellites. Propagation included the effects of many realistic perturbations, including drag, lift, solar radiation pressure, and J_2. The propagator interacts with the controller through a MATLAB$^{\circledR}$ interface. During each propagation step, the control algorithm is queried. If the controller is currently implementing a thrusting plan, the thrusts corresponding to the current position in the plan are converted into appropriate orbital element offsets and returned to the propagator. After each plan is completely implemented, a new plan is created.

The simulation presented in the following sections involves a three-satellite formation. The reference orbit, represented by a virtual satellite with properties similar to the average of the fleet, has a semimajor axis of 6900 km, inclination 45°, and eccentricity 0.003. Realistic perturbations (drag, J_2, solar radiation pressure, Sun/Moon effects) were included in all simulations.

Each satellite is modeled using specifications for the MELCO formation flying mission [188]. Each satellite has a mass of 900 kg and a ballistic coefficient of 0.4. The satellite thrusters are restricted to provide a maximum of 2 N of force over a 10.8-second time step.

The MELCO formation flying mission consists of two different formation shapes: (i) along-track separation formations and (ii) along-track/cross-track **305**

Spacecraft Formation Flying; ISBN: 9780750685337

passive-aperture formations (triangular). Over the course of a mission, the for-mation is designed to achieve four different configurations in the following order: (1) along-track formation with 1 km (or optimum) of separation between spacecraft (represents deployment configuration); (2) passive-aperture forma-tion with a 50-m baseline; (3) passive-aperture formation with a 500-m base-line; and (4) passive-aperture formation with a 5-km baseline.

Lawden's time-varying equations, described in Subsection 5.6.2, are used to determine the desired state for each spacecraft; however, the CW equations, described in Section 5.1, are used in solving an LP problem, to be discussed shortly. This is consistent with the observation in Ref. [26] that slightly eccentric orbits ($0.0005 < e < 0.01$) require eccentricity-invariant initial conditions, but not time-varying dynamics. Ten minutes of each orbit are reserved for observations: During this time position constraints are enforced, but no thrusting is permitted.

For each type of formation at each baseline, one or more 18-day simulations were conducted to determine the average fuel usage. A representative mission accuracy requirement is that each formation size must use an *error box* that is 10% of the baseline.

13.1 SIMULATION CONTROLLER CONFIGURATION

A *planning controller* is used in the simulations in this section. Each space-craft designs its control individually and coordination between spacecraft is accomplished through the design of the formation desired states at the time the formation is initialized. Error box constraints of the type discussed in Ref. [26] are added at every 6 time steps to ensure that satellites achieve their perfor-mance objectives, while reducing the computational burden of imposing them at every time step. In addition, fuel inputs are permitted every 6 time steps for station-keeping (in which spacecraft are tasked to remain in formation) and every time step for formation maneuvers (those periods during which the for-mation is being switched from one configuration to another). Mission require-ments are chosen to ensure that a ten-minute time window at apogee should be reserved during formation-keeping maneuvers in which no thruster inputs are permitted.

An always-feasible formulation was used for the simulation controller, guar-anteeing that a plan is always returned, even if no feasible solution exists which satisfies the error box constraints. The always-feasible formulation is based on the *constraint-enlarging approach* presented in Ref. [26], but with a modifica-tion allowing several degrees of constraint violation. If no feasible solution is possible for the desired error box, the approach in Ref. [26] enlarges the error box as little as possible until a feasible solution is found. However, the result-ing plan may end with the spacecraft outside the nominal box, which would guarantee that the next plan would use as much fuel as necessary to return the spacecraft to the error box on the first step of the plan or would also require an expansion of the nominal box. The always-feasible formulation used in this section also enlarges the error box until a feasible plan is found, but includes another soft constraint, which prefers that the planned trajectory still ends inside the nominal error box. This additional constraint is also implemented using an additional high cost-penalty variable that is used to ensure that it is only relaxed in the event that no feasible solution exists that enlarges the error box and ends inside the nominal error box. The modified always-feasible formulation allows the controller to prefer plans that make future optimizations initially feasible.

13.1.1 Parameters examined

An initial simulation of the mission described above was conducted. The results of that simulation indicated that levels of CDGPS noise expected were excessive for the station-keeping in a 50-m passive aperture formation with 5 m × 5 m × 5 m error box, the most tightly constricting phase of the mission. The controller was unable to keep the satellite constrained to its error box under those conditions and experienced errors that grew over time. In order to investigate this phenomenon and find a stable configuration for the controller that met as many mission requirements as possible, a number of variations of the basic controller setup have been implemented and tested in realistic simulations. Section 13.2 presents the results of those simulations and evaluates the effectiveness of the parameter variations in terms of effectiveness at preventing error box violations and average fuel use. The control parameters considered are as follows.

THRUSTING DURING OBSERVATIONS

Mission specifications are chosen to ensure that observations cannot occur when any spacecraft in the formation is thrusting. As a result, the basic configuration prevents thrusting during a 10 minute period at the apogee of every orbit that the formation is in the passive aperture configuration. However, the mission reference orbit is 95 minutes, so the effect of the observation thrusting prohibition is to reduce the overall control authority of any plan by more than 10%. Although this requirement is a hard constraint for the mission, it is included as a parameter in the simulation study so that its effects on formation flying mission performance can be judged.

ERROR BOXES RELAXED WHEN NOT OBSERVING

One concept for a formation flying mission could specify that a formation should be in a passive aperture during periods when the formation is taking observations and transition to a widely separated mode during periods when observations are not needed. A maneuver generation analysis indicates that there is generally no fuel advantage to formation-keeping in an along-track formation versus a passive aperture formation and that the cost of maneuvering into and out-of holding configurations is unnecessary.

Given that the tight error box requirements are derived from the needs of the distributed observation instrument, a reasonable modification of the mission requirements is to enlarge the error boxes during the 85 minutes of each orbit in which no observations are taking place and only enforce tight performance constraints when they are strictly needed. This constraint is implemented by doubling the error box size during the time steps outside the 10 minute period reserved for observations at apogee.

DYNAMICALLY MOTIVATED ERROR BOX SHAPES

Because the 50 m passive aperture is the most fuel-intensive of the formation configurations, a 5 × 10 × 5 meter (radial/along-track/cross-track) error box is examined as an alternative to the 10% requirement for that portion of the mission. The 5 × 5 × 5 meter error box is sufficiently small that the navigation errors [38] strongly influence the closed-loop behavior. Enlarging the along-track dimension allows a slightly more natural relative elliptical motion in the radial/cross-track plane (typically a 1 × 2 ellipse). This observation is derived from the form of Hill's equations, in which the coefficients of the harmonic terms for the along-track axis are exactly twice the value of the coefficients of the harmonics terms for the radial axis.

TERMINAL-INVARIANCE CONDITION

A plan that is optimized to guarantee that a spacecraft remains inside an error box over some future horizon will accomplish that goal, but will not provide any guarantees for the future behavior of the spacecraft. Occasionally, a situation may arise where the spacecraft would approach its constraint boundary in the near future after the end of a plan. This situation is prevented from resulting in a constraint violation by the creation and implementation of a new plan. However, if the new plan is forced to react quickly to a potential constraint violation in the near term, the only feasible solution may require a great deal of fuel. In an effort to reduce the need for vehicles to make short term "emergency" corrections, an invariant-set terminal condition was examined. In this case, the condition guarantees that after a plan ends, the spacecraft will naturally (i.e., with no thrust inputs) remain inside its box for a full orbit and return to its state at the time the plan ended. Within the time span that the dynamics can accurately propagate the states, this terminal constraint guarantees perpetual collision avoidance in the absence of state knowledge uncertainty. This constraint is imposed using the nominal estimate of the spacecraft state, because a robust implementation is generally infeasible.

13.2 SIMULATION RESULTS

Multiple simulations were performed to study the formation flying mission and the effects of the control system parameters introduced in Section 13.1. These simulations are described in Table 13.1 and summarized in Fig. 13.1.

In Figs. 13.2–13.19, the vast majority of the trajectories remain inside the error boxes. Several instances of error box constraint violations occurred as a result of the always-feasible formulation, but were quickly corrected. Several general trends are visible: In most cases the 50 m passive aperture requires the most fuel to maintain and the 500 m passive aperture requires the least fuel to maintain. Also, enlarging the along-track dimension of the error box tended to reduce fuel use a great deal in comparable simulation configurations (i.e., Sim 1 & Sim 4 and Sim 3 & Sim 6).

Table 13.1 Fuel use results for formation flying simulations: Fuel costs for station-keeping (SK) are given in mm/s/orbit/satellite and fuel costs for maneuvers (Mvr) are given in m/s for the entire formation

	50 m PA: 5 × 5 × 5 m error box			50 m PA: 5 × 10 × 5 m error box		
	Sim 1	Sim 2	Sim 3	Sim 4	Sim 5	Sim 6
No thrust observation	No	Yes	Yes	No	Yes	Yes
No observe large EB	No	Yes	No	No	No	No
Mnvr. end invariance	No	No	Yes	No	No	Yes
SK 50 m	45.96	106.51	53.73	8.39	33.47	12.01
SK 500 m	0.06	0.09	1.54	0.06	0.04	1.51
SK 5 km	5.83	27.47	1.69	6.8	29.47	1.72
Mvr #1	1.06	0.56	0.65	0.71	0.57	0.57
Mvr #2	17.94	36.57	19.97	5.78	13.51	6.8
Mvr #3	51.37	65.73	48.59	39.4	48.15	35.42

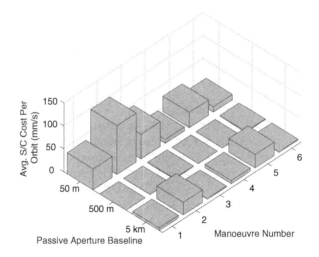

FIGURE 13.1 Summary of station-keeping results from Table 13.1.

Interestingly, the least restrictive error box (500 m for a 5 km formation baseline) did not result in the lowest fuel use. This phenomenon is likely due to the reduced likelihood of encountering constraints in a single orbit planning horizon for the larger error box. As a result, it is possible for a spacecraft to enter into a large, high-speed relative orbit that may end near an error box constraint. In those situations, future trajectory optimizations would need to take immediate corrective action to avoid constraint boundaries. This

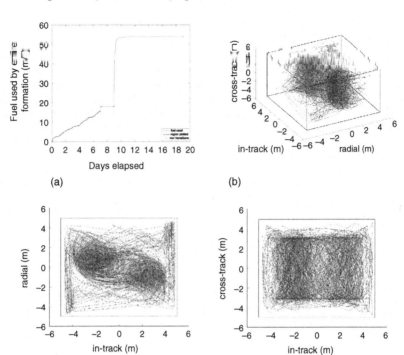

FIGURE 13.2 Simulation #1: Station-keeping in a passive aperture formation (50 m). (a) Fuel use. (b) 3D LVLH trajectory. (c) LVLH trajectory: Along-track/radial. (d) LVLH trajectory: Along-track/cross-track.

hypothesis is supported by significant reductions in fuel use for the 5 km passive aperture between Sim 1 & Sim 3 and Sim 4 & Sim 6, even with fewer thrusting times available in Sim 3 and Sim 6 due to constraints against thrusting during observations. In both cases, the only difference between the simulations is the inclusion of the terminal invariance constraint and a restriction against thrusting during observations. The restriction against thrusting tends to increase the fuel use, as in the Sim 4 & Sim 5 pair. In Sim 4, thrusting is permitted at all times in the orbit, whereas in Sim 5 10% of the orbit is reserved for observations. In the 50 m and 5 km configurations, the 10% restriction increased fuel use by more than an order of magnitude for those simulations. It is likely that the difference between the fuel use numbers for the 500 m configuration is insignificant due to random elements in the simulations.

The addition of the terminal invariance constraint in Sim 6 lowers the fuel consumption for the 5 km configuration, but raises it for the 50 m and 500 m baselines compared to Sim 4. At the 5 km level, the reduction is greater than a factor of 3. Figures 13.13 and 13.19 show the error box motion for all three satellites in the formation for station-keeping in the 5 km formation for Sims 4 and 6, respectively. While Fig. 13.13 shows a considerable amount of motion throughout the entire error box, Fig. 13.19 shows that the spacecraft is actively

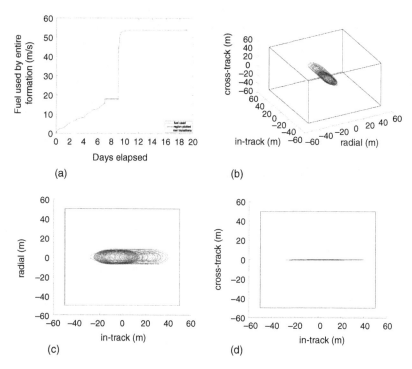

FIGURE 13.3 Simulation #1: Station-keeping in a passive aperture formation (500 m). (a) Fuel use. (b) 3D LVLH trajectory. (c) LVLH trajectory: Along-track/radial. (d) LVLH trajectory: Along-track/cross-track.

controlled in the center of the error box. These results are counterintuitive for several reasons: (i) the addition of constraints to an optimization would typically be associated with increased fuel cost; and (ii) expending fuel to keep a spacecraft in a small box would usually be more expensive than allowing the same spacecraft to stay in a large box. The fact that the fuel costs were reduced is not in conflict with the fact that individual optimizations should have increased cost; the fuel costs measured are the steady-state closed-loop levels of consumption, as opposed to expected maneuver costs. Also, note that the fuel use in Sim 4 remained high for a period after the maneuver to the 5 km configuration before settling into a steady state, whereas the Sim 5 fuel use settled almost immediately. This is likely because the terminal invariance condition forces the trajectories to enter closed ellipses. In Sim 4, where no invariance condition was specified, the spacecraft did not start in a closed ellipse, but did eventually enter one and did not exit for the duration of the simulation. This is because even without requiring a closed-ellipse, the optimization recognizes that there is no need to expend fuel to change a trajectory, which will result in no constraint violations. An almost identical pattern is visible between Sim 1 (which does not impose terminal invariance) and Sim 3 (which uses terminal invariance).

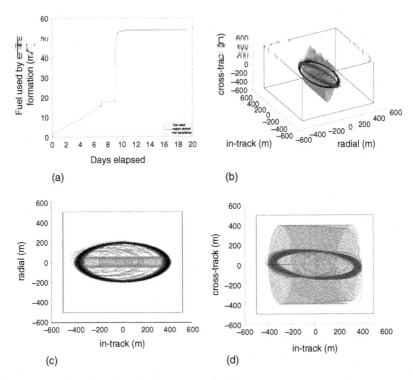

FIGURE 13.4 Simulation #1: Station-keeping in a passive aperture formation (5000 m). (a) Fuel use. (b) 3D LVLH trajectory. (c) LVLH trajectory: Along-track/radial. (d) LVLH trajectory: Along-track/cross-track.

Terminal invariance raises the fuel cost between Sims 1 & 3 and Sims 4 & 6 for the 50 m and 500 m configurations. At the 50 m level, this effect is complicated by the fact that the increase is small and Sims 3 and 6 both also restrict thrusting at apogee. Sims 1 and 4 were found to not be feasible with thrust restrictions. Sim 3 was feasible most of the time, but Fig. 13.8 shows that there were a number of instances in which error box violations occurred. It should be noted that Sim 3 was the only one of the simulations that used the original MELCO mission control specifications and succeeded in remaining stable. All of the other simulations used modifications that enlarged the error boxes at some or all times. Sim 6 used a 5×10×5 meter error box for the 50 m configuration and did not require any error box violations. At the 500 m level, the addition of the invariance constraint causes an almost two orders of magnitude fuel consumption increase. It appears that this is because almost no fuel is used without invariance as a constraint, but the resulting trajectories are naturally invariant (see Figs. 13.3, 13.6, 13.12 and 13.15) with slight semimajor axis mismatches, which cause the ellipses to travel inside the error box. The trajectories with invariance (see Figs. 13.18 and 13.19) take on similar shapes but expend fuel to cancel any real or perceived drift introduced through navigation errors with each successive optimization.

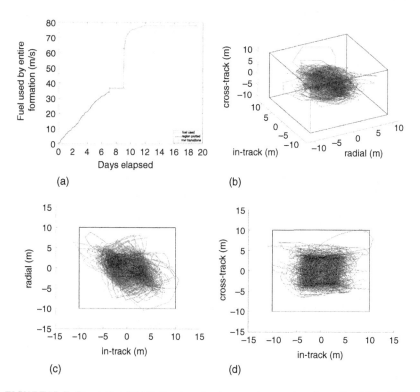

FIGURE 13.5 Simulation #2: Station-keeping in a passive aperture formation (50 m). (a) Fuel use. (b) 3D LVLH trajectory. (c) LVLH trajectory: Along-track/radial. (d) LVLH trajectory: Along-track/cross-track.

Overall, it appears that the most consistent combination of successful constraint satisfaction and low fuel use came from the simulations using the invariance constraints (Sims 3 and 6). These simulations used the observation thrusting restriction and the specified error box size for Sim 3 and a slightly larger error box for the 50 m configuration in Sim 6. Thus, the 500 m and 5 km configurations were the same in Sims 3 and 6 and, as would be expected, they have nearly identical fuel use. The only difference is at the 50 m level, where a 5 m increase in the along-track error box size decreases the fuel use by more than a factor of 4.

The fuel numbers for the maneuvers between formation types are included for completeness, but cannot be used to draw conclusions because of the stochastic nature of the simulations. Many additional simulations would need to be run and averages examined. This fact does not reduce the validity of the conclusions regarding the fuel use for station-keeping, because in those situations the fuel use tends to reach a steady state, as is evidenced by the linear (fixed slope) rates of fuel use in the Δv plots in this section.

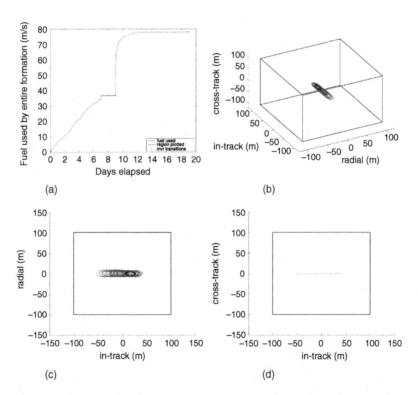

(a)

(b)

(c)

(d)

FIGURE 13.6 Simulation #2: Station-keeping in a passive aperture formation (500 m). (a) Fuel use. (b) 3D LVLH trajectory. (c) LVLH trajectory: Along-track/radial. (d) LVLH trajectory: Along-track/cross-track.

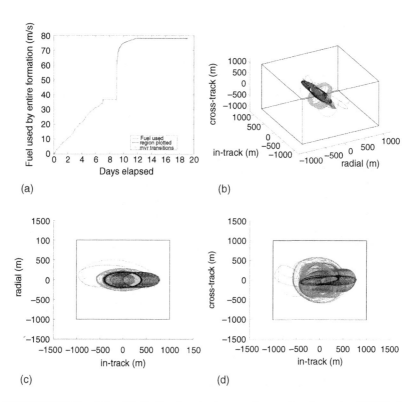

FIGURE 13.7 Simulation #2: Station-keeping in a passive aperture formation (5000 m). (a) Fuel use. (b) 3D LVLH trajectory. (c) LVLH trajectory: Along-track/radial. (d) LVLH trajectory: Along-track/cross-track.

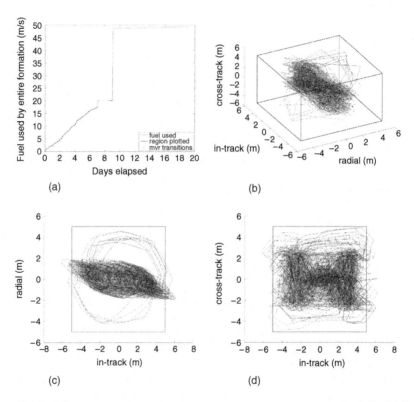

FIGURE 13.8 Simulation #3: Station-keeping in a passive aperture formation (50 m). (a) Fuel use. (b) 3D LVLH trajectory. (c) LVLH trajectory: Along-track/radial. (d) LVLH trajectory: Along-track/cross-track.

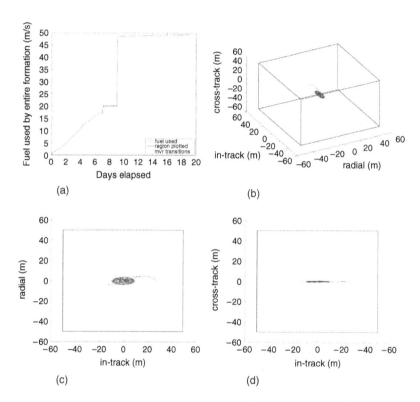

FIGURE 13.9 Simulation #3: Station-keeping in a passive aperture formation (500 m). (a) Fuel use. (b) 3D LVLH trajectory. (c) LVLH trajectory: Along-track/radial. (d) LVLH trajectory: Along-track/cross-track.

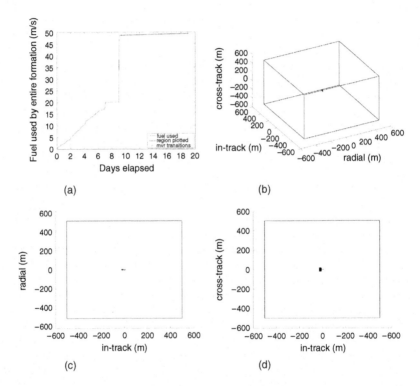

FIGURE 13.10 Simulation #3: Station-keeping in a passive aperture formation (5000 m). (a) Fuel use. (b) 3D LVLH trajectory. (c) LVLH trajectory: Along-track/radial. (d) LVLH trajectory: Along-track/cross-track.

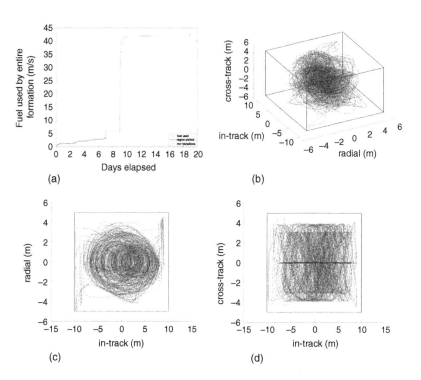

FIGURE 13.11 Simulation #4: Station-keeping in a passive aperture formation (50 m). (a) Fuel use. (b) 3D LVLH trajectory. (c) LVLH trajectory: Along-track/radial. (d) LVLH trajectory: Along-track/cross-track.

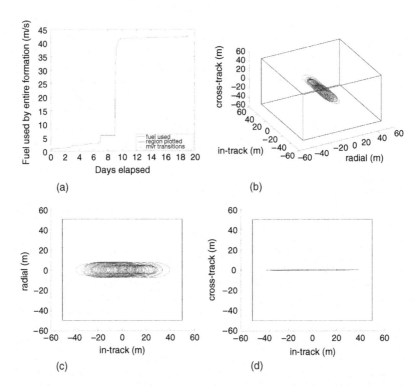

FIGURE 13.12 Simulation #4: Station-keeping in a passive aperture formation (500 m). (a) Fuel use. (b) 3D LVLH trajectory. (c) LVLH trajectory: Along-track/radial. (d) LVLH trajectory: Along-track/cross-track.

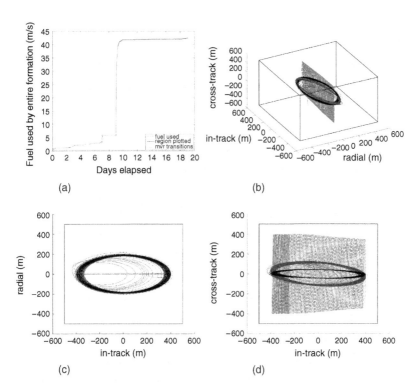

FIGURE 13.13 Simulation #4: Station-keeping in a passive aperture formation (5000 m). (a) Fuel use. (b) 3D LVLH trajectory. (c) LVLH trajectory: Along-track/radial. (d) LVLH trajectory: Along-track/cross-track.

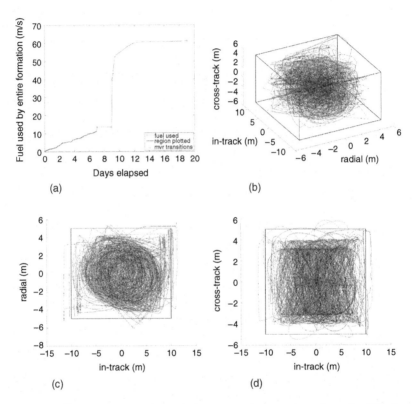

FIGURE 13.14 Simulation #5: Station-keeping in a passive aperture formation (50 m). (a) Fuel use. (b) 3D LVLH trajectory. (c) LVLH trajectory: Along-track/radial. (d) LVLH trajectory: Along-track/cross-track.

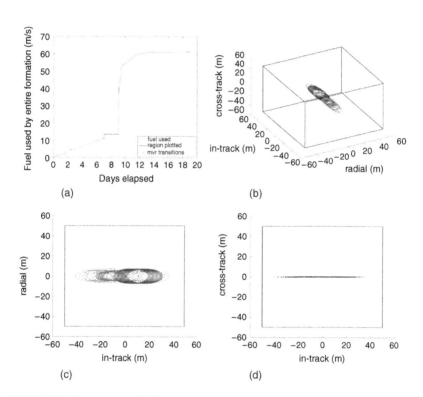

FIGURE 13.15 Simulation #5: Station-keeping in a passive aperture formation (500 m). (a) Fuel use. (b) 3D LVLH trajectory. (c) LVLH trajectory: Along-track/radial. (d) LVLH trajectory: Along-track/cross-track.

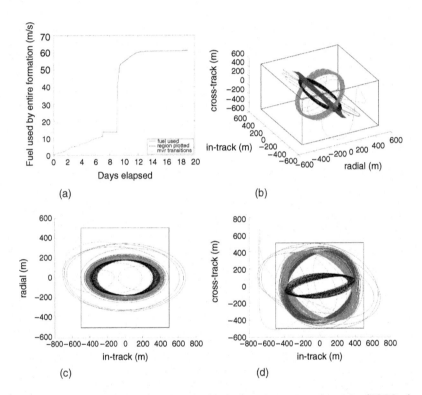

FIGURE 13.16 Simulation #5: Station-keeping in a passive aperture formation (5000 m). (a) Fuel use. (b) 3D LVLH trajectory. (c) LVLH trajectory: Along-track/radial. (d) LVLH trajectory: Along-track/cross-track.

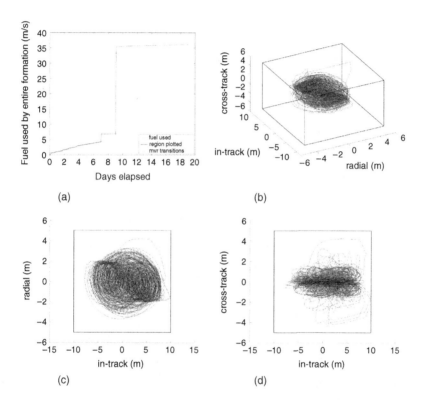

FIGURE 13.17 Simulation #6: Station-keeping in a passive aperture formation (50 m). (a) Fuel use. (b) 3D LVLH trajectory. (c) LVLH trajectory: Along-track/radial. (d) LVLH trajectory: Along-track/cross-track.

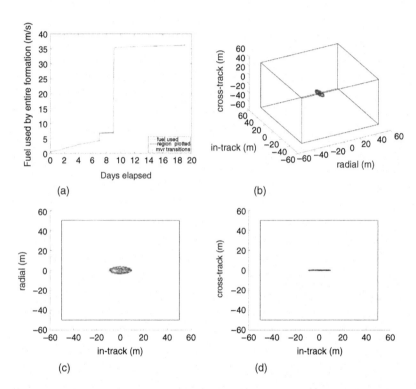

FIGURE 13.18 Simulation #6: Station-keeping in a passive aperture formation (500 m). (a) Fuel use. (b) 3D LVLH trajectory. (c) LVLH trajectory: Along-track/radial. (d) LVLH trajectory: Along-track/cross-track.

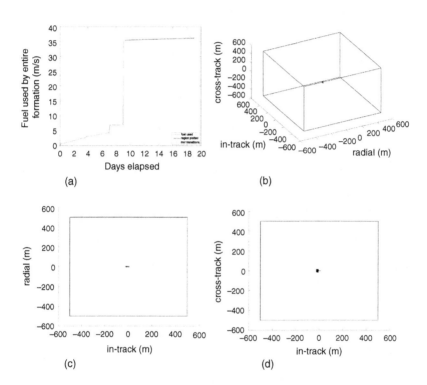

FIGURE 13.19 Simulation #6: Station-keeping in a passive aperture formation (5000 m). (a) Fuel use. (b) 3D LVLH trajectory. (c) LVLH trajectory: Along-track/radial. (d) LVLH trajectory: Along-track/cross-track.

SUMMARY

Multiple high-fidelity simulations of a reference formation flying mission were performed successfully. These simulations demonstrated, with the most rigorous tools available, that a realistic LEO reference mission can be feasibly controlled and made to meet all of its performance constraints. The simulations indicated that the fuel cost of reserving passive observation time can be significant. Also, it appears that the fewest constraint violations and most consistently low fuel usage occur when a terminal-invariance constraint is added to the control formulation. It was demonstrated that small increases in error box size can result in large fuel savings.

Chapter | fourteen

Summary and Future Prospects

What surety is there
That we will meet again,
On other worlds some
Future time undated.

Maya Angelou (1928–)

Given the advantages that can be gained by flying spacecraft in formations and all the associated research into the topic, a natural question to ask would be: Why are there no spacecraft formations on orbit currently?

Are we confident in our models to maintain 10–100 m separations? Can we foresee a collision and take evasive action before it is too late? With increased miniaturization and the advent of nano-satellites, the problem of navigating the jungle will become more complex. A scientific mission can be carried out for long durations if the satellites do not thrust while collecting images. For thrust-free operations, we must find good natural solutions, which allow very long-term drift-free formation flying. Even though the satellites of the future will become smaller and cheaper to build, the launch costs are still too high to make formation flying affordable. These are the main research issues; progress must be made in all fronts before we develop enough confidence to go on with real missions. But there is no doubt regarding the payoffs.

14.1 RISK REDUCTION

No formations have ever been flown and a formation is by definition multiple spacecraft in close proximity. The leap in complexity and cost from controlling a single spacecraft to multiple spacecraft incurs a large quantity of risk. In particular, verifying and validating the control system algorithms is expected to be a key barrier and cost driver. Communications reliability will also need to be considered from both the perspective of the control system performance in off-nominal communication conditions and the feasibility of the control requirements on the communications system.

Spacecraft Formation Flying; ISBN: 9780750685337

14.2 FUEL REQUIREMENTS

Unlike a monolithic spacecraft, the component spacecraft in a formation are expected to require regular on-orbit corrections to prevent drift within the formation. These regular corrections are expected to use more fuel per satellite than that required for maintaining the orbit of a satellite which is not part of a formation. Furthermore, when any one satellite runs out of fuel, the entire formation flying mission will be over. One can argue, however, that the remaining satellites will still be able to perform the mission, albeit with a degradation in performance.

Thus, either a high degree of confidence must be placed in a fuel-minimizing control system or a great deal of extra fuel will be required for each satellite on orbit. Raising the redundant fuel margin for each satellite will raise the mission cost considerably.

14.3 MISSION OPERATIONS

Although many control approaches for spacecraft formations have been proposed in the literature, few papers discuss the methods and costs involved in operating a spacecraft formation. A balance will need to be found between guidance conducted by human-monitored ground systems and fully autonomous control on-orbit. Furthermore, the culture between spacecraft autonomy and spacecraft automatic control must be bridged to collect the best ideas from both areas of research. Although the presentation in this book has focused on control systems, a large quantity of research exists in the areas of spacecraft fault detection, automated command sequence generation, and future software architectures [189–192]. The future of implementing, verifying, and validating higher level control systems on board satellites will depend critically on successful integration with the spacecraft software community.

Appendices

A THE TRANSFORMATION MATRIX $\Sigma(T)$

This appendix presents the elements of the transformation matrix $\Sigma(t) = [A(t) + A_2 B(t)]$, where $A_2 = 3 J_2 R_e^2$. All the orbital elements are the elements of the chief; for brevity, the subscript "0" has been dropped. The following quantities will be needed:

$$r = p/(1 + q_1 \cos\theta + q_2 \sin\theta)$$
$$p = a\left(1 - q_1^2 - q_2^2\right)$$
$$v_t = \sqrt{\frac{\mu}{p}}\,(1 + q_1 \cos\theta + q_2 \sin\theta)$$
$$v_r = \sqrt{\frac{\mu}{p}}\,(q_1 \sin\theta - q_2 \cos\theta) \qquad (A.1)$$
$$v_{rt} = \frac{v_r}{v_t}$$
$$n = \sqrt{\frac{\mu}{a^3}}$$

$$\Sigma_{11} = \frac{r}{a}$$
$$\Sigma_{12} = v_{rt}r$$
$$\Sigma_{13} = 0$$
$$\Sigma_{14} = -\frac{r}{p}\,(2aq_1 + r\cos\theta) \qquad (A.2)$$
$$\Sigma_{15} = -\frac{r}{p}\,(2aq_2 + r\sin\theta)$$
$$\Sigma_{16} = 0$$

$$\Sigma_{21} = -\frac{1}{2}\frac{v_r}{a}$$
$$\Sigma_{22} = \left(1 - \frac{r}{p}\right)v_t$$
$$\Sigma_{23} = 0$$
$$\Sigma_{24} = \left[v_{rt}\left(\frac{a}{p}\right)q_1 + \left(\frac{r}{p}\right)\sin\theta\right]v_t \qquad (A.3)$$
$$\Sigma_{25} = \left[v_{rt}\left(\frac{a}{p}\right)q_2 - \left(\frac{r}{p}\right)\cos\theta\right]v_t$$
$$\Sigma_{26} = 0$$

$$\Sigma_{31} = 0$$
$$\Sigma_{32} = r$$
$$\Sigma_{33} = 0, \Sigma_{34} = 0, \Sigma_{35} = 0 \tag{A.1}$$
$$\Sigma_{36} = r \cos i$$

$$\Sigma_{41} = -\frac{3}{2}\frac{v_t}{a}$$

$$\Sigma_{42} = -v_r$$

$$\Sigma_{43} = -\left(\frac{A_2}{pr}\right)\left(\sin i \cos i \sin^2 \theta\right) v_t$$

$$\Sigma_{44} = \left(\frac{r}{p}\right)\left[3\left(\frac{a}{r}\right)q_1 + 2\cos\theta\right] v_t \tag{A.5}$$

$$\Sigma_{45} = \left(\frac{r}{p}\right)\left[3\left(\frac{a}{r}\right)q_2 + 2\sin\theta\right] v_t$$

$$\Sigma_{46} = v_r \cos i + \left(\frac{A_2}{pr}\right)\left(\sin^2 i \cos i \sin\theta \cos\theta\right) v_t$$

$$\Sigma_{51} =, \Sigma_{52} = 0$$
$$\Sigma_{53} = r \sin\theta$$
$$\Sigma_{54} = 0, \Sigma_{55} = 0 \tag{A.6}$$
$$\Sigma_{56} = -r \sin i \cos\theta$$

$$\Sigma_{61} = 0$$

$$\Sigma_{62} = \left(\frac{A_2}{pr}\right)\left(\sin i \cos i \sin\theta\right) v_t$$

$$\Sigma_{63} = v_t \cos\theta + v_r \sin\theta \tag{A.7}$$

$$\Sigma_{64} = 0, \Sigma_{65} = 0$$

$$\Sigma_{66} = v_t \left(\sin\theta - v_{rt}\cos\theta\right)\sin i + \left(\frac{A_2}{pr}\right)\left(\sin i \cos^2 i \sin\theta\right) v_t$$

B THE TRANSFORMATION MATRIX $\Sigma(T)^{-1}$

This appendix presents the elements of the transformation matrix $\Sigma(t)^{-1} = [A(t) + A_2 B(t)]^{-1}$.

$$\Sigma_{11}^{-1} = -\left(\frac{2a}{r}\right)\left[3\left(\frac{a}{r}\right)\left(1 - \frac{p}{r}\right) - 2\left(1 + \left(\frac{ap}{r^2}\right)v_{rt}^2\right)\right]$$

$$\Sigma_{12}^{-1} = \frac{2p}{v_t}\left(\frac{a}{r}\right)^2 v_{rt}$$

$$\Sigma_{13}^{-1} = \left(\frac{2a}{r}\right)\left[2\left(\frac{a}{r}\right)\left(1 - \frac{p}{r}\right) - \left(1 + \left(\frac{ap}{r^2}\right)v_{rt}^2\right)\right] v_{rt} \tag{B.1}$$

$$\Sigma_{14}^{-1} = -\left(\frac{2a}{v_t}\right)\left[2\left(\frac{a}{r}\right)\left(1 - \frac{p}{r}\right) - \left(1 + \left(\frac{ap}{r^2}\right)v_{rt}^2\right)\right]$$

$$\Sigma_{15}^{-1} = -\left(\frac{A_2}{r^2}\right)\left(\frac{a}{p}\right)(2\sin i \cos i \sin\theta)\left[\left(1 - \frac{p}{r}\right) - \left(1 + \left(\frac{ap}{r^2}\right)v_{rt}^2\right)\right]$$

$$\Sigma_{16}^{-1} = 0$$

$$\Sigma_{21}^{-1} = 0, \ \Sigma_{22}^{-1} = 0$$

$$\Sigma_{23}^{-1} = \frac{1}{r} + \frac{A_2}{r^2}\left(\frac{\cos^2 i \sin^2\theta}{p}\right)$$

$$\Sigma_{24}^{-1} = 0 \tag{B.2}$$

$$\Sigma_{25}^{-1} = (\cos\theta + v_{rt}\sin\theta)\left(\frac{\cos i}{r\sin i}\right)$$

$$\Sigma_{26}^{-1} = -\left(\frac{\sin\theta}{v_t}\right)\left(\frac{\cos i}{\sin i}\right)$$

$$\Sigma_{31}^{-1} = 0, \ \Sigma_{32}^{-1} = 0$$

$$\Sigma_{33}^{-1} = -\left(\frac{A_2}{r^2}\right)\left(\frac{\sin i \cos i \sin\theta \cos\theta}{p}\right)$$

$$\Sigma_{34}^{-1} = 0 \tag{B.3}$$

$$\Sigma_{35}^{-1} = \frac{(\sin\theta - v_{rt}\cos\theta)}{r}$$

$$\Sigma_{36}^{-1} = \frac{\cos\theta}{v_t}$$

$$\Sigma_{41}^{-1} = \left(\frac{p}{r^2}\right)(3\cos\theta + 2v_{rt}\sin\theta)$$

$$\Sigma_{42}^{-1} = \left(\frac{p}{rv_t}\right)\sin\theta$$

$$\Sigma_{43}^{-1} = -\left(\frac{1}{r}\right)\left[\left(\frac{p}{r} - 1\right)\sin\theta + \left(\frac{p}{r}\right)v_{rt}(\cos\theta + v_{rt}\sin\theta)\right]$$
$$+ \left(\frac{A_2}{r^2}\right)\left(\frac{\cos^2 i \sin^2\theta}{r}\right)\left[\left(\frac{r}{p} - 1\right)\sin\theta + v_{rt}\cos\theta\right]$$

$$\Sigma_{44}^{-1} = \left(\frac{p}{rv_t}\right)(2\cos\theta + v_{rt}\sin\theta) \tag{B.4}$$

$$\Sigma_{45}^{-1} = (\cos\theta + v_{rt}\sin\theta)\left[\left(1 - \frac{p}{r}\right)\sin\theta + \left(\frac{p}{r}\right)v_{rt}\cos\theta\right]\left(\frac{\cos i}{r\sin i}\right)$$
$$+ \left(\frac{A_2}{r^2}\right)\left(\frac{\sin i \cos i \sin\theta}{r}\right)(2\cos\theta + v_{rt}\sin\theta)$$

$$\Sigma_{46}^{-1} = -\left(\frac{\sin\theta}{v_t}\right)\left[\left(\frac{p}{r}\right)v_{rt}\cos\theta + \left(1 - \frac{p}{r}\right)\sin\theta\right]\left(\frac{\cos i}{\sin i}\right)$$

$$\Sigma_{51}^{-1} = \left(\frac{p}{r^2}\right)(3\sin\theta - 2v_{rt}\cos\theta)$$

$$\Sigma_{52}^{-1} = -\left(\frac{p}{rv_t}\right)\cos\theta$$

$$\Sigma_{53}^{-1} = \left(\frac{1}{r}\right)\left[\left(\frac{p}{r} - 1\right)\cos\theta + \left(\frac{p}{r}\right)v_{rt}(v_{rt}\cos\theta - \sin\theta)\right]$$

$$+ \left(\frac{A_2}{r^2}\right)\left(\frac{\cos^2 i \sin^2\theta}{r}\right)\left[\left(1 - \frac{r}{p}\right)\cos\theta + v_{rt}\sin\theta\right]$$

(B.5)

$$\Sigma_{54}^{-1} = \left(\frac{p}{rv_t}\right)(2\sin\theta - v_{rt}\cos\theta)$$

$$\Sigma_{55}^{-1} = \left(\frac{1}{r}\right)(\cos\theta + v_{rt}\sin\theta)\left[\left(\frac{p}{r} - 1\right)\cos\theta + \left(\frac{p}{r}\right)v_{rt}\sin\theta\right]\left(\frac{\cos i}{\sin i}\right)$$

$$+ \left(\frac{A_2}{r^2}\right)\left(\frac{\sin i \cos i \sin\theta}{r}\right)(2\sin\theta - v_{rt}\cos\theta)$$

$$\Sigma_{56}^{-1} = \left(\frac{\sin\theta}{rv_t}\right)\left[\left(1 - \frac{p}{r}\right)\cos\theta - \frac{p}{r}v_{rt}\sin\theta\right]\left(\frac{\cos i}{\sin i}\right)$$

$$\Sigma_{61}^{-1} = 0, \ \Sigma_{62}^{-1} = 0$$

$$\Sigma_{63}^{-1} = -\left(\frac{A_2}{r^2}\right)\left(\frac{\cos i \sin^2\theta}{p}\right)$$

$$\Sigma_{64}^{-1} = 0$$

(B.6)

$$\Sigma_{65}^{-1} = -\left(\frac{1}{r\sin i}\right)(\cos\theta + v_{rt}\sin\theta)$$

$$\Sigma_{66}^{-1} = \frac{\sin\theta}{v_t \sin i}$$

C THE MATRIX $\bar{B}(T)$

$$\bar{B}_{1j} = \bar{B}_{3j} = \bar{B}_{5j} = 0, \quad j = 1 - 6 \tag{C.1}$$

$$\bar{B}_{21} = \left(\frac{5nrv_{rt}}{8ap^2}\right)(5\cos^2 i - 1)$$

$$\bar{B}_{22} = \left(\frac{nr}{4p^2}\right)(1 - 5\cos^2 i)\left[\left(1 - \frac{r}{p}\right) + 2v_{rt}^2\right]$$

$$\bar{B}_{23} = \left(\frac{5nrv_{rt}}{2p^2}\right)(\sin i \cos i)$$

$$\bar{B}_{24} = \left(\frac{nr}{4p^2}\right)(5\cos^2 i - 1)\left[2v_{rt}\left(\frac{r}{p}\cos\theta - \frac{a}{p}q_1\right) - \left(\frac{r}{p}\right)\sin\theta\right]$$

(C.2)

$$\bar{B}_{25} = \left(\frac{nr}{4p^2}\right)(5\cos^2 i - 1)\left[2v_{rt}\left(\frac{r}{p}\sin\theta - \frac{a}{p}q_2\right) + \left(\frac{r}{p}\right)\cos\theta\right]$$

$$\bar{B}_{26} = 0$$

$$\bar{B}_{41} = \left(\frac{7}{4}\frac{nr}{ap^2}\right)\cos^2 i$$

$$\bar{B}_{42} = \left(\frac{1}{4}\frac{nr v_{rt}}{p^2}\right)\left(1 - 5\cos^2 i\right)$$

$$\bar{B}_{43} = \left(\frac{1}{2}\frac{nr}{p^2}\right)\sin i \cos i$$

$$\bar{B}_{44} = -\left(\frac{2nraq_1}{p^3}\right)\cos^2 i \qquad (\text{C.3})$$

$$\bar{B}_{45} = -\left(\frac{2nraq_2}{p^3}\right)\cos^2 i$$

$$\bar{B}_{46} = \left(\frac{1}{4}\frac{nr v_{rt} \cos i}{p^2}\right)\left(1 - 5\cos^2 i\right)$$

$$\bar{B}_{61} = -\left(\frac{7}{4}\frac{nr}{ap^2}\right)\cos\theta \sin i \cos i$$

$$\bar{B}_{62} = 0$$

$$\bar{B}_{63} = \left(\frac{1}{4}\frac{nr}{p^2}\right)\left[v_{rt}\sin\theta\left(1 - 5\cos^2 i\right) - 2\cos\theta \sin^2 i\right]$$

$$\bar{B}_{64} = \left(\frac{2nraq_1}{p^3}\right)\cos\theta \sin i \cos i \qquad (\text{C.4})$$

$$\bar{B}_{65} = \left(\frac{2nraq_2}{p^3}\right)\cos\theta \sin i \cos i$$

$$\bar{B}_{66} = \left(\frac{1}{4}\frac{nr v_{rt}}{p^2}\right)\cos\theta \sin i \left(5\cos^2 i - 1\right)$$

D THE STATE TRANSITION MATRIX FOR RELATIVE MEAN ELEMENTS

In this appendix, all the variables are mean variables, that is, those that result from the averaged Hamiltonian. The subscript "0" indicates the value at the initial time, t_0, except for the quantities G_{γ_0}, $\gamma = \{\theta, q_1, q_2\}$, for which the definition is given at the end of the Appendix. We use this notation convention to be consistent with the notation in Ref. [75].

$$\bar{\phi}_{\text{œ}11} = 1 \qquad (\text{D.1})$$
$$\bar{\phi}_{\text{œ}1j} = 0, \quad j = 2,\ldots,6$$

$$\bar{\phi}_{\text{œ}21} = -\frac{(t-t_0)}{G_\theta}\left[\left(\frac{3}{2}\frac{n}{a}\right) + \left(\frac{7A_2}{8p^2}\right)\left(\frac{n}{a}\right)\left[\eta\left(3\cos^2 i - 1\right)\right.\right.$$
$$\left.\left. + K\left(5\cos^2 i - 1\right)\right]\right]$$

$$\bar{\phi}_{\text{œ}22} = -\frac{G_{\theta_0}}{G_\theta}$$

$$\bar{\phi}_{œ23} = -\frac{(t-t_0)}{G_0}\left[\left(\frac{A_2}{?p^2}\right)n\,(\sin i\cos i)\,(3\eta + 5K)\right]$$

$$\bar{\phi}_{œ24} = -\frac{1}{G_\theta}\left(G_{q_{10}} + G_{q_1}\cos(\Delta\omega) + G_{q_2}\sin(\Delta\omega)\right) + \frac{(t-t_0)}{G_\theta}$$
$$\times\left(\frac{A_2}{4p^2}\right)\left(\frac{anq_{10}}{p}\right)\left[3\eta\left(3\cos^2 i - 1\right) + 4K\left(5\cos^2 i - 1\right)\right] \quad (D.2)$$

$$\bar{\phi}_{œ25} = -\frac{1}{G_\theta}\left(G_{q_{20}} - G_{q_1}\sin(\Delta\omega) + G_{q_2}\cos(\Delta\omega)\right) + \frac{(t-t_0)}{G_\theta}$$
$$\times\left(\frac{A_2}{4p^2}\right)\left(\frac{anq_{20}}{p}\right)\left[3\eta\left(3\cos^2 i - 1\right) + 4K\left(5\cos^2 i - 1\right)\right]$$

$$\bar{\phi}_{œ26} = 0$$

$$\bar{\phi}_{œ33} = 1 \quad (D.3)$$
$$\bar{\phi}_{œ3j} = 0, \quad j = 1,\ldots,6$$

$$\bar{\phi}_{œ41} = \left(\frac{7A_2}{8p^2}\right)\left(\frac{n}{a}\right)(q_{10}\sin(\Delta\omega) + q_{20}\cos(\Delta\omega))\left(5\cos^2 i - 1\right)(t-t_0)$$

$$\bar{\phi}_{œ42} = 0$$

$$\bar{\phi}_{œ43} = \left(\frac{5}{2}\frac{A_2}{p^2}\right)n\,(q_{10}\sin(\Delta\omega) + q_{20}\cos(\Delta\omega))\,(\sin i\cos i)\,(t-t_0)$$

$$\bar{\phi}_{œ44} = \cos(\Delta\omega) - \left(\frac{A_2}{p^2}\right)\left(\frac{nq_{10}}{\eta^2}\right)(q_{10}\sin(\Delta\omega) + q_{20}\cos(\Delta\omega))$$
$$\times\left(5\cos^2 i - 1\right)(t-t_0) \quad (D.4)$$

$$\bar{\phi}_{œ45} = -\sin(\Delta\omega) - \left(\frac{A_2}{p^2}\right)\left(\frac{nq_{20}}{\eta^2}\right)(q_{10}\sin(\Delta\omega) + q_{20}\cos(\Delta\omega))$$
$$\times\left(5\cos^2 i - 1\right)(t-t_0)$$

$$\bar{\phi}_{œ46} = 0$$

$$\bar{\phi}_{œ51} = -\left(\frac{7A_2}{8p^2}\right)\left(\frac{n}{a}\right)(q_{10}\cos(\Delta\omega) - q_{20}\sin(\Delta\omega))\left(5\cos^2 i - 1\right)(t-t_0)$$

$$\bar{\phi}_{œ52} = 0$$

$$\bar{\phi}_{œ53} = -\left(\frac{5}{2}\frac{A_2}{p^2}\right)n\,(q_{10}\cos(\Delta\omega) - q_{20}\sin(\Delta\omega))\,(\sin i\cos i)\,(t-t_0)$$

$$\bar{\phi}_{œ54} = \sin(\Delta\omega) + \left(\frac{A_2}{p^2}\right)\left(\frac{nq_{10}}{\eta^2}\right)(q_{10}\cos(\Delta\omega)$$
$$- q_{20}\sin(\Delta\omega))\left(5\cos^2 i - 1\right)(t-t_0) \quad (D.5)$$

$$\bar{\phi}_{œ55} = \cos(\Delta\omega) + \left(\frac{A_2}{p^2}\right)\left(\frac{nq_{20}}{\eta^2}\right)(q_{10}\cos(\Delta\omega)$$
$$- q_{20}\sin(\Delta\omega))\left(5\cos^2 i - 1\right)(t-t_0)$$

$$\bar{\phi}_{œ56} = 0$$

$$\bar{\phi}_{œ61} = \left(\frac{7}{4}\frac{A_2}{p^2}\right)\left(\frac{n\cos i}{a}\right)(t - t_0)$$

$$\bar{\phi}_{œ62} = 0$$

$$\bar{\phi}_{œ63} = \left(\frac{A_2}{2p^2}\right)(n\sin i)(t - t_0)$$

$$\bar{\phi}_{œ64} = -\left(\frac{2A_2}{p^2}\right)\left(\frac{nq_{10}\cos i}{\eta^2}\right)(t - t_0) \qquad\qquad\text{(D.6)}$$

$$\bar{\phi}_{œ65} = -\left(\frac{2A_2}{p^2}\right)\left(\frac{nq_{20}\cos i}{\eta^2}\right)(t - t_0)$$

$$\bar{\phi}_{œ66} = 1$$

where

$$\Delta\omega = \dot{\omega}^{(s)}(t - t_0)$$

$$\dot{\omega}^{(s)} = 0.75 J_2 n\left(\frac{R_e}{p}\right)^2\left(5\cos^2 i - 1\right) \qquad\qquad\text{(D.7)}$$

$$K = 1 + G_{q1}[q_{10}\sin(\Delta\omega) + q_{20}\cos(\Delta\omega)]$$
$$\quad - G_{q2}[q_{10}\cos(\Delta\omega) - q_{20}\sin(\Delta\omega)]$$

$$G = \lambda - \lambda(t_0) = \dot{\lambda}^{(s)}(t - t_0)$$

$$\dot{\lambda}^{(s)} = n\left[1 + 0.75 J_2\left(\frac{R_e}{a}\right)^2\left(\frac{1}{\eta^4}\right)\left[\eta\left(3\cos^2 i - 1\right) + \left(5\cos^2 i - 1\right)\right]\right]$$

$$G_\gamma = \frac{\partial G}{\partial \gamma}$$

where γ denotes any variable, and the superscript $(\cdot)^{(s)}$ denotes the secular value.

$$G_\theta = \frac{rn}{v_t} \qquad\qquad\text{(D.8)}$$

$$G_{\theta_0} = -\frac{r_0 n_0}{v_{t0}}$$

$$G_{q1} = \frac{q_2}{\eta\,(1 + \eta)} + \frac{q_1 v_{rt}}{\eta} - \eta\left(\frac{r}{p}\right)^2\left(1 + \frac{a}{r}\right)(q_2 + \sin\theta)$$

$$G_{q10} = -\frac{q_{20}}{\eta_0\,(1 + \eta_0)} - \frac{q_{10} v_{rt0}}{\eta_0} + \eta_0\left(\frac{r_0}{p_0}\right)^2\left(1 + \frac{a_0}{r_0}\right)(q_{20} + \sin\theta_0)$$

$$\qquad\qquad\qquad\qquad\qquad\qquad\qquad\qquad\qquad\qquad\text{(D.9)}$$

$$G_{q2} = -\frac{q_1}{\eta\,(1 + \eta)} + \frac{q_2 v_{rt}}{\eta} - \eta\left(\frac{r}{p}\right)^2\left(1 + \frac{a}{r}\right)(q_1 + \cos\theta)$$

$$G_{q20} = \frac{q_{10}}{\eta_0\,(1 + \eta_0)} - \frac{q_{20} v_{rt0}}{\eta_0} - \eta_0\left(\frac{r_0}{p_0}\right)^2\left(1 + \frac{a_0}{r_0}\right)(q_{10} + \cos\theta_0)$$

E TRANSFORMATION FROM MEAN TO OSCULATING ELEMENTS

$$a^{(lp)} = 0 \tag{E.1}$$

$$\lambda^{(lp)} = \left[\frac{q_1 q_2 \sin^2 i}{8a^2 \eta^2 (1 + \eta)} \right] \left(1 - 10\Theta \cos^2 i \right) + \left(\frac{q_1 q_2}{16a^2 \eta^4} \right)$$
$$\times \left(3 - 55 \cos^2 i - 280\Theta \cos^4 i - 400\Theta^2 \cos^6 i \right) \tag{E.2}$$

$$\theta^{(lp)} = \lambda^{(lp)} - \left(\frac{\sin^2 i}{16a^2 \eta^4} \right) \left(1 - 10\Theta \cos^2 i \right)$$
$$\times \left\{ q_1 q_2 \left[3 + \frac{2\eta^2}{(1 + \eta)} \right] + 2 \left(q_1 \sin \theta + q_2 \cos \theta \right) + \frac{\varepsilon_1 \sin 2\theta}{2} \right\} \tag{E.3}$$

$$i^{(lp)} = \left(\frac{\sin 2i}{32a^2 \eta^4} \right) \left(1 - 10\Theta \cos^2 i \right) \left(q_1^2 - q_2^2 \right) \tag{E.4}$$

$$q_1^{(lp)} = - \left(\frac{q_1 \sin^2 i}{16a^2 \eta^2} \right) \left(1 - 10\Theta \cos^2 i \right)$$
$$- \left(\frac{q_1 q_2^2}{16a^2 \eta^4} \right) \left(3 - 55 \cos^2 i - 280\Theta \cos^4 i - 400\Theta^2 \cos^6 i \right) \tag{E.5}$$

$$q_2^{(lp)} = \left(\frac{q_2 \sin^2 i}{16a^2 \eta^2} \right) \left(1 - 10\Theta \cos^2 i \right)$$
$$+ \left(\frac{q_1^2 q_2}{16a^2 \eta^4} \right) \left(3 - 55 \cos^2 i - 280\Theta \cos^4 i - 400\Theta^2 \cos^6 i \right) \tag{E.6}$$

$$\Omega^{(lp)} = \left(\frac{q_1 q_2 \cos i}{8a^2 \eta^4} \right) \left(11 + 80\Theta \cos^2 i + 200\Theta^2 \cos^4 i \right) \tag{E.7}$$

$$a^{(sp1)} = \left[\frac{(1 - 3 \cos^2 i)}{2a\eta^6} \right] \left[(1 + \varepsilon_2)^3 - \eta^3 \right] \tag{E.8}$$

$$\lambda^{(sp1)} = \left[\frac{\varepsilon_3 (1 - 3 \cos^2 i)}{4a^2 \eta^4 (1 + \eta)} \right] \left[(1 + \varepsilon_2)^2 + (1 + \varepsilon_2) + \eta^2 \right]$$
$$+ \left[\frac{3 (1 - 5 \cos^2 i)}{4a^2 \eta^4} \right] (\theta - \lambda + \varepsilon_3) \tag{E.9}$$

$$\theta^{(sp1)} = \lambda^{(sp1)} - \left[\frac{\varepsilon_3 \left(1 - 3\cos^2 i\right)}{4a^2\eta^4(1+\eta)} \right] \left[(1+\varepsilon_2)^2 + \eta(1+\eta)\right] \tag{E.10}$$

$$i^{(sp1)} = 0 \tag{E.11}$$

$$
\begin{aligned}
q_1^{(sp1)} &= \left[\frac{\left(1 - 3\cos^2 i\right)}{4a^2\eta^4(1+\eta)} \right] \left\{ \left[(1+\varepsilon_2)^2 + \eta^2\right][q_1 + (1+\eta)\cos\theta] \right. \\
&\quad + \left. (1+\varepsilon_2)\left[(1+\eta)\cos\theta + q_1(\eta - \varepsilon_2)\right] \right\} \\
&\quad - \left[\frac{3q_2 \left(1 - 5\cos^2 i\right)}{4a^2\eta^4} \right] (\theta - \lambda + \varepsilon_3)
\end{aligned}
\tag{E.12}
$$

$$
\begin{aligned}
q_2^{(sp1)} &= \left[\frac{\left(1 - 3\cos^2 i\right)}{4a^2\eta^4(1+\eta)} \right] \left\{ \left[(1+\varepsilon_2)^2 + \eta^2\right][q_2 + (1+\eta)\sin\theta] \right. \\
&\quad + \left. (1+\varepsilon_2)\left[(1+\eta)\sin\theta + q_2(\eta - \varepsilon_2)\right] \right\} \\
&\quad + \left[\frac{3q_1 \left(1 - 5\cos^2 i\right)}{4a^2\eta^4} \right] (\theta - \lambda + \varepsilon_3)
\end{aligned}
\tag{E.13}
$$

$$\Omega^{(sp1)} = \left(\frac{3\cos i}{2a^2\eta^4} \right) [(\theta - \lambda) + \varepsilon_3] \tag{E.14}$$

$$a^{(sp2)} = -\left(\frac{3\sin^2 i}{2a\eta^6} \right) (1+\varepsilon_2)^3 \cos 2\theta \tag{E.15}$$

$$
\begin{aligned}
\lambda^{(sp2)} &= -\left[\frac{3\varepsilon_3 \sin^2 i \cos 2\theta}{4a^2\eta^4(1+\eta)} \right] (1+\varepsilon_2)(2+\varepsilon_2) - \left[\frac{\sin^2 i}{8a^2\eta^2(1+\eta)} \right] \\
&\quad \times [3\,(q_1 \sin\theta + q_2 \cos\theta) + (q_1 \sin 3\theta - q_2 \cos 3\theta)] \\
&\quad - \left[\frac{\left(3 - 5\cos^2 i\right)}{8a^2\eta^4} \right] [3\,(q_1 \sin\theta + q_2 \cos\theta) + 3\sin 2\theta \\
&\quad + (q_1 \sin 3\theta - q_2 \cos 3\theta)]
\end{aligned}
\tag{E.16}
$$

$$
\theta^{(sp2)} = \lambda^{(sp2)} - \left[\frac{\sin^2 i}{32a^2\eta^4(1+\eta)} \right] \\
\times \left\{
\begin{aligned}
&36q_1q_2 - 4\left(3\eta^2 + 5\eta - 1\right)(q_1 \sin\theta + q_2 \cos\theta) \\
&+ 12\varepsilon_2 q_1 q_2 - 32(1+\eta)\sin 2\theta \\
&- (\eta^2 + 12\eta + 39)(q_1 \sin 3\theta - q_2 \cos 3\theta) \\
&+ 36q_1q_2 \cos 4\theta - 18\left(q_1^2 - q_2^2\right)\sin 4\theta \\
&- 3\left(q_1^2 - q_2^2\right)q_1 \sin 5\theta + 3\left(3q_1^2 - q_2^2\right)q_2 \cos 5\theta
\end{aligned}
\right\}
\tag{E.17}
$$

$$i^{(sp2)} = -\left(\frac{\sin 2i}{8a^2\eta^4}\right)[3\,(q_1\cos\theta - q_2\sin\theta)$$
$$+ 3\sin 2\theta + (q_1\cos 3\theta + q_2\sin 3\theta)] \tag{E.18}$$

$$q_1^{(sp2)} = \left[\frac{q_2\left(3 - 5\cos^2 i\right)}{8a^2\eta^4}\right][3\,(q_1\sin\theta + q_2\cos\theta) + 3\sin 2\theta$$
$$+ (q_1\sin 3\theta - q_2\cos 3\theta)]$$
$$+ \left(\frac{\sin^2 i}{8a^2\eta^4}\right)\left[3\left(\eta^2 - q_1^2\right)\cos\theta + 3q_1q_2\sin\theta\right.$$
$$\left. - \left(\eta^2 + 3q_1^2\right)\cos 3\theta - 3q_1q_2\sin 3\theta\right] - \left(\frac{3\sin^2 i\cos 2\theta}{16a^2\eta^4}\right)$$
$$\times \left[\begin{array}{l} 10q_1 + \left(8 + 3q_1^2 + q_2^2\right)\cos\theta + 2q_1q_2\sin\theta \\ + 6\,(q_1\cos 2\theta + q_2\sin 2\theta) \\ + \left(q_1^2 - q_2^2\right)\cos 3\theta + 2q_1q_2\sin 3\theta \end{array}\right] \tag{E.19}$$

$$q_2^{(sp2)} = -\left[\frac{q_1\left(3 - 5\cos^2 i\right)}{8a^2\eta^4}\right][3\,(q_1\sin\theta + q_2\cos\theta)$$
$$+ 3\sin 2\theta + (q_1\sin 3\theta - q_2\cos 3\theta)]$$
$$- \left(\frac{\sin^2 i}{8a^2\eta^4}\right)\left[3\left(\eta^2 - q_2^2\right)\sin\theta + 3q_1q_2\cos\theta\right.$$
$$\left. + \left(\eta^2 + 3q_2^2\right)\sin 3\theta + 3q_1q_2\cos 3\theta\right] - \left(\frac{3\sin^2 i\cos 2\theta}{16a^2\eta^4}\right)$$
$$\times \left[\begin{array}{l} 10q_2 + \left(8 + q_1^2 + 3q_2^2\right)\sin\theta + 2q_1q_2\cos\theta \\ + 6\,(q_1\sin 2\theta - q_2\cos 2\theta) \\ + \left(q_1^2 - q_2^2\right)\sin 3\theta - 2q_1q_2\cos 3\theta \end{array}\right] \tag{E.20}$$

$$\Omega^{(sp2)} = -\left(\frac{\cos i}{4a^2\eta^4}\right)[3\,(q_1\sin\theta + q_2\cos\theta)$$
$$+ 3\sin 2\theta + (q_1\sin 3\theta - q_2\cos 3\theta)] \tag{E.21}$$

$$\lambda_{q_1} = \left(\frac{\partial\lambda}{\partial q_1}\right) = \frac{q_2}{\eta(1+\eta)} + \frac{q_1}{\eta}v_{rt} - \frac{\eta r(a+r)}{p^2}\,(q_2 + \sin\theta) \tag{E.22}$$

$$\lambda_{q_2} = \left(\frac{\partial\lambda}{\partial q_2}\right) = -\frac{q_1}{\eta(1+\eta)} + \frac{q_2}{\eta}v_{rt} + \frac{\eta r(a+r)}{p^2}\,(q_1 + \cos\theta) \tag{E.23}$$

$$\Theta = \frac{1}{\left(1 - 5\cos^2 i\right)} \tag{E.24}$$

$$\varepsilon_1 = \sqrt{q_1^2 + q_2^2} \tag{E.25}$$

$$\varepsilon_2 = q_1 \cos\theta + q_2 \sin\theta \tag{E.26}$$

$$\varepsilon_3 = q_1 \sin\theta - q_2 \cos\theta \tag{E.27}$$

F JACOBIAN FOR MEAN TO OSCULATING ELEMENTS

This appendix contains the Jacobian for the mean to osculating transformation. The variables $\varepsilon_1, \varepsilon_2, \varepsilon_3$ were defined in Eqs. (E.25)–(E.27). The Jacobian D is defined as

$$D = \frac{\partial \text{œ}}{\partial \overline{\text{œ}}} = I - J_2 R_e^2 \left(D^{(lp)} + D^{(sp1)} + D^{(sp2)} \right) \tag{F.1}$$

$$D_{11}^{(lp)} = -\left(\frac{1}{a}\right) a^{(lp)}, \quad D_{12}^{(lp)} = D_{13}^{(lp)} = D_{14}^{(lp)} = D_{15}^{(lp)} = D_{16}^{(lp)} = 0 \tag{F.2}$$

$$D_{21}^{(lp)} = -\left(\frac{2}{a}\right)\theta^{(lp)}$$

$$D_{22}^{(lp)} = -\left(\frac{\sin^2 i}{16a^2\eta^4}\right)\left(1 - 10\Theta\cos^2 i\right)[2\,(q_1\cos\theta - q_2\sin\theta) + \varepsilon_1\cos 2\theta]$$

$$D_{23}^{(lp)} = \left(\frac{\sin 2i}{16a^2\eta^4}\right)\Big\{5q_1q_2\left(11 + 112\Theta\cos^2 i\right.$$
$$+ 520\Theta^2\cos^4 i + 800\Theta^3\cos^6 i\Big)$$
$$- [2q_1q_2 + (2+\varepsilon_2)(q_1\sin\theta + q_2\cos\theta)]$$
$$\times \left[\left(1 - 10\Theta\cos^2 i\right) + 10\Theta\sin^2 i\left(1 + 50\Theta\cos^2 i\right)\right]\Big\}$$

$$D_{24}^{(lp)} = \left(\frac{1}{16a^2\eta^6}\right)\Big\{\left(\eta^2 + 4q_1^2\right)$$
$$\times \left[q_2\left(3 - 55\cos^2 i - 280\Theta\cos^4 i - 400\Theta^2\cos^6 i\right)\right.$$
$$\left. - \sin^2 i\left(1 - 10\Theta\cos^2 i\right)(3q_2 + 2\sin\theta)\right]$$
$$- 2\sin^2 i\left(1 - 10\Theta\cos^2 i\right)[4q_2 + \sin\theta(1+\varepsilon_1)]q_1\cos\theta\Big\}$$

$$D_{25}^{(lp)} = \left(\frac{1}{16a^2\eta^6}\right)\Big\{\left(\eta^2 + 4q_2^2\right)$$
$$\times \left[q_1\left(3 - 55\cos^2 i - 280\Theta\cos^4 i - 400\Theta^2\cos^6 i\right)\right.$$
$$\left. - \sin^2 i\left(1 - 10\Theta\cos^2 i\right)(3q_1 + 2\cos\theta)\right]$$
$$- 2\sin^2 i\left(1 - 10\Theta\cos^2 i\right)[4q_1 + \cos\theta(1+\varepsilon_1)]q_2\sin\theta\Big\}$$

$$D_{26}^{(lp)} = 0 \tag{F.3}$$

$$D_{31}^{(lp)} = -\left(\frac{2}{a}\right) i^{(lp)}, \quad D_{32}^{(lp)} = 0$$

$$D_{33}^{(lp)} = \left(\frac{q_1^2 - q_2^2}{16a^2\eta^4}\right) \left[\cos 2i \left(1 - 10\Theta \cos^2 i\right) + 5\Theta \sin^2 2i \left(1 + 5\Theta \cos^2 i\right)\right]$$

$$D_{34}^{(lp)} = \left(\frac{q_1 \sin 2i}{16a^2\eta^6}\right) \left(1 - 10\Theta \cos^2 i\right) \left[\eta^2 + 2\left(q_1^2 - q_2^2\right)\right] \qquad \text{(F.4)}$$

$$D_{35}^{(lp)} = -\left(\frac{q_2 \sin 2i}{16a^2\eta^6}\right) \left(1 - 10\Theta \cos^2 i\right) \left[\eta^2 - 2\left(q_1^2 - q_2^2\right)\right]$$

$$D_{36}^{(lp)} = 0$$

$$D_{41}^{(lp)} = -\left(\frac{2}{a}\right) q_1^{(lp)}, \quad D_{42}^{(lp)} = 0$$

$$D_{43}^{(lp)} = -\left(\frac{q_1 \sin 2i}{16a^2\eta^4}\right) \left[\eta^2 \left[\left(1 - 10\Theta \cos^2 i\right) + 10\Theta \sin^2 i \left(1 + 5\Theta \cos^2 i\right)\right]\right.$$
$$\left. + 5q_2^2 \left(11 + 112\Theta \cos^2 i + 520\Theta^2 \cos^4 i + 800\Theta^3 \cos^6 i\right)\right]$$

$$D_{44}^{(lp)} = -\left(\frac{1}{16a^2\eta^6}\right) \left[\eta^2 \sin^2 i \left(1 - 10\Theta \cos^2 i\right) \left(\eta^2 + 2q_1^2\right)\right. \qquad \text{(F.5)}$$
$$\left. + q_2^2 \left(3 - 55 \cos^2 i - 280\Theta \cos^4 i - 400\Theta^2 \cos^6 i\right) \left(\eta^2 + 4q_1^2\right)\right]$$

$$D_{45}^{(lp)} = -\left(\frac{q_1 q_2}{8a^2\eta^6}\right) \left[\eta^2 \sin^2 i \left(1 - 10\Theta \cos^2 i\right)\right.$$
$$\left. + \left(3 - 55 \cos^2 i - 280\Theta \cos^4 i - 400\Theta^2 \cos^6 i\right) \left(\eta^2 + 2q_2^2\right)\right]$$

$$D_{46}^{(lp)} = 0$$

$$D_{51}^{(lp)} = -\left(\frac{2}{a}\right) q_2^{(lp)}, \quad D_{52}^{(lp)} = 0$$

$$D_{53}^{(lp)} = \left(\frac{q_2 \sin 2i}{16a^2\eta^4}\right) \left[\eta^2 \left[\left(1 - 10\Theta \cos^2 i\right) + 10\Theta \sin^2 i \left(1 + 5\Theta \cos^2 i\right)\right]\right.$$
$$\left. + 5q_1^2 \left(11 + 112\Theta \cos^2 i + 520\Theta^2 \cos^4 i + 800\Theta^3 \cos^6 i\right)\right]$$

$$D_{54}^{(lp)} = \left(\frac{q_1 q_2}{8a^2\eta^6}\right) \left[\eta^2 \sin^2 i \left(1 - 10\Theta \cos^2 i\right)\right. \qquad \text{(F.6)}$$
$$\left. + \left(3 - 55 \cos^2 i - 280\Theta \cos^4 i - 400\Theta^2 \cos^6 i\right) \left(\eta^2 + 2q_1^2\right)\right]$$

$$D_{55}^{(lp)} = \left(\frac{1}{16a^2\eta^6}\right) \left[\eta^2 \sin^2 i \left(1 - 10\Theta \cos^2 i\right) \left(\eta^2 + 2q_2^2\right)\right.$$
$$\left. + q_1^2 \left(3 - 55 \cos^2 i - 280\Theta \cos^4 i - 400\Theta^2 \cos^6 i\right) \left(\eta^2 + 4q_2^2\right)\right]$$

$$D_{56}^{(lp)} = 0$$

$$D_{61}^{(lp)} = -\left(\frac{2}{a}\right)\Omega^{(lp)}, \quad D_{62}^{(lp)} = 0$$

$$D_{63}^{(lp)} = -\left(\frac{q_1 q_2 \sin i}{8a^2\eta^4}\right)\left[\left(11 + 80\Theta\cos^2 i + 200\Theta^2\cos^4 i\right)\right.$$

$$\left. +160\Theta\cos^2 i\left(1 + 5\Theta\cos^2 i\right)^2\right] \tag{F.7}$$

$$D_{64}^{(lp)} = \left(\frac{q_2\cos i}{8a^2\eta^6}\right)\left(11 + 80\Theta\cos^2 i + 200\Theta^2\cos^4 i\right)\left(\eta^2 + 4q_1^2\right)$$

$$D_{65}^{(lp)} = \left(\frac{q_1\cos i}{8a^2\eta^6}\right)\left(11 + 80\Theta\cos^2 i + 200\Theta^2\cos^4 i\right)\left(\eta^2 + 4q_2^2\right)$$

$$D_{66}^{(lp)} = 0$$

$$D_{11}^{(sp1)} = -\left(\frac{1}{a}\right)a^{(sp1)}$$

$$D_{12}^{(sp1)} = -\left(\frac{3\varepsilon_3}{2a\eta^6}\right)\left(1 - 3\cos^2 i\right)(1 + \varepsilon_2)^2$$

$$D_{13}^{(sp1)} = \left(\frac{3\sin 2i}{2a\eta^6}\right)\left[(1 + \varepsilon_2)^3 - \eta^3\right]$$

$$D_{14}^{(sp1)} = \left[\frac{3\left(1 - 3\cos^2 i\right)}{2a\eta^8}\right]$$

$$\times\left[2q_1(1 + \varepsilon_2)^3 + \eta^2(1 + \varepsilon_2)^2\cos\theta - \eta^3 q_1\right] \tag{F.8}$$

$$D_{15}^{(sp1)} = \left[\frac{3\left(1 - 3\cos^2 i\right)}{2a\eta^8}\right]$$

$$\times\left[2q_2(1 + \varepsilon_2)^3 + \eta^2(1 + \varepsilon_2)^2\sin\theta - \eta^3 q_2\right]$$

$$D_{16}^{(sp1)} = 0$$

$$D_{21}^{(sp1)} = -\left(\frac{2}{a}\right)\theta^{(sp1)}$$

$$D_{22}^{(sp1)} = \left[\frac{(1 - 3\cos^2 i)}{4a^2\eta^4(1 + \eta)}\right]\left[\varepsilon_2(1 + \varepsilon_2 - \eta) - \varepsilon_3^2\right]$$

$$+\left[\frac{3\left(1 - 5\cos^2 i\right)}{4a^2\eta^4(1 + \varepsilon_2)^2}\right]\left[(1 + \varepsilon_2)^3 - \eta^3\right]$$

$$D_{23}^{(sp1)} = \left[\frac{3\varepsilon_3\sin 2i}{4a^2\eta^4(1 + \eta)}\right]\left[(1 + \varepsilon_2) + (5 + 4\eta)\right] + \left(\frac{15\sin 2i}{4a^2\eta^4}\right)(\theta - \lambda) \tag{F.9}$$

$$D_{24}^{(sp1)} = \left[\frac{\left(1 - 3\cos^2 i\right)}{4a^2\eta^6(1+\eta)^2}\right]\left\{\eta^2\left[\varepsilon_1\sin\theta + (1+\eta)(\varepsilon_2\sin\theta + \varepsilon_3\cos\theta)\right]\right.$$
$$\left. + q_1\varepsilon_3\left[4(\varepsilon_1 + \varepsilon_2) + \eta(2 + 5\varepsilon_2)\right]\right\}$$
$$+ \left[\frac{3\left(1 - 5\cos^2 i\right)}{4a^2\eta^6}\right]\left\{4q_1\left[(\theta - \lambda) + \varepsilon_3\right] + \eta^2\sin\theta\right\}$$
$$- \left[\frac{3\left(1 - 5\cos^2 i\right)}{4a^2\eta^4}\right]\left(\lambda_{q_1}\right)$$

$$D_{25}^{(sp1)} = -\left[\frac{\left(1 - 3\cos^2 i\right)}{4a^2\eta^6(1+\eta)^2}\right]\left\{\eta^2\left[\varepsilon_1\cos\theta + (1+\eta)(\varepsilon_2\cos\theta - \varepsilon_3\sin\theta)\right]\right.$$
$$\left. - q_2\varepsilon_3\left[4(\varepsilon_1 + \varepsilon_2) + \eta(2 + 5\varepsilon_2)\right]\right\}$$
$$+ \left[\frac{3\left(1 - 5\cos^2 i\right)}{4a^2\eta^6}\right]\left\{4q_2\left[(\theta - \lambda) + \varepsilon_3\right] - \eta^2\cos\theta\right\}$$
$$- \left[\frac{3\left(1 - 5\cos^2 i\right)}{4a^2\eta^4}\right]\left(\lambda_{q_2}\right)$$

$$D_{26}^{(sp1)} = 0$$

$$D_{31}^{(sp1)} = -\left(\frac{2}{a}\right)i^{(sp1)},$$

$$D_{32}^{(sp1)} = D_{33}^{(sp1)} = D_{34}^{(sp1)} = D_{35}^{(sp1)} = D_{36}^{(sp1)} = 0$$

(F.10)

$$D_{41}^{(sp1)} = -\left(\frac{2}{a}\right)q_1^{(sp1)}$$

$$D_{42}^{(sp1)} = -\left[\frac{\left(1 - 3\cos^2 i\right)}{4a^2\eta^4}\right]\left[(1 + \varepsilon_2)(2\sin\theta + \varepsilon_2\sin\theta + 2\varepsilon_3\cos\theta)\right.$$
$$\left. + \varepsilon_3(q_1 + \cos\theta) + \eta^2\sin\theta\right]$$
$$- \left[\frac{3q_2\left(1 - 5\cos^2 i\right)}{4a^2\eta^4(1+\varepsilon_2)^2}\right]\left[(1+\varepsilon_2)^3 - \eta^3\right]$$

$$D_{43}^{(sp1)} = \left[\frac{3q_1\sin 2i}{4a^2\eta^2(1+\eta)}\right] + \left(\frac{3\sin 2i}{4a^2\eta^4}\right)\left\{(1+\varepsilon_2)\left[q_1 + (2+\varepsilon_2)\cos\theta\right]\right.$$
$$\left. - 5q_2\varepsilon_3 + \eta^2\cos\theta\right\} - \left(\frac{15q_2\sin 2i}{4a^2\eta^4}\right)(\theta - \lambda)$$

$$D_{44}^{(sp1)} = \left[\frac{(1 - 3\cos^2 i)}{4a^2\eta^2(1 + \eta)}\right] + \left[\frac{(1 - 3\cos^2 i)}{8a^2\eta^6}\right]$$

$$\times \left\{\eta^2 \left[5 + 2(5q_1\cos\theta + 2q_2\sin\theta) + (3 + 2\varepsilon_2)\cos 2\theta\right]\right.$$

$$\left. + 2q_1\left[4(1 + \varepsilon_2)(2 + \varepsilon_2)\cos\theta + (3\eta + 4\varepsilon_2)q_1\right]\right\}$$

$$+ \left[\frac{(1 - 3\cos^2 i)\, q_1^2(4 + 5\eta)}{4a^2\eta^6(1 + \eta)^2}\right]$$

$$- \left[\frac{3q_2\left(1 - 5\cos^2 i\right)}{4a^2\eta^6}\right]\left(4q_1\varepsilon_3 + \eta^2\sin\theta\right)$$

$$- \left[\frac{3q_1q_2\left(1 - 5\cos^2 i\right)}{a^2\eta^6}\right](\theta - \lambda) + \left[\frac{3q_2\left(1 - 5\cos^2 i\right)}{4a^2\eta^4}\right](\lambda_{q_1}) \quad \text{(F.11)}$$

$$D_{45}^{(sp1)} = \left[\frac{(1 - 3\cos^2 i)}{8a^2\eta^6}\right]\left\{\eta^2\left[2(q_1\sin\theta + 2q_2\cos\theta) + (3 + 2\varepsilon_2)\sin 2\theta\right]\right.$$

$$\left. + 2q_2\left[4(1 + \varepsilon_2)(2 + \varepsilon_2)\cos\theta + (3\eta + 4\varepsilon_2)q_1\right]\right\}$$

$$+ \left[\frac{(1 - 3\cos^2 i)\, q_1q_2(4 + 5\eta)}{4a^2\eta^6(1 + \eta)^2}\right]$$

$$- \left[\frac{3\left(1 - 5\cos^2 i\right)}{4a^2\eta^6}\right]\left[\varepsilon_3\left(\eta^2 + 4q_2^2\right) - \eta^2 q_2\cos\theta\right]$$

$$- \left[\frac{3\left(1 - 5\cos^2 i\right)}{4a^2\eta^6}\right]\left[\left(\eta^2 + 4q_2^2\right)(\theta - \lambda)\right]$$

$$+ \left[\frac{3q_2\left(1 - 5\cos^2 i\right)}{4a^2\eta^4}\right](\lambda_{q_2})$$

$$D_{46}^{(sp1)} = 0$$

$$D_{51}^{(sp1)} = -\left(\frac{2}{a}\right)q_2^{(sp1)}$$

$$D_{52}^{(sp1)} = \left[\frac{(1 - 3\cos^2 i)}{4a^2\eta^4}\right]\left[(1 + \varepsilon_2)(2\cos\theta + \varepsilon_2\cos\theta - 2\varepsilon_3\sin\theta)\right.$$

$$\left. - \varepsilon_3(q_2 + \sin\theta) + \eta^2\cos\theta\right]$$

$$+ \left[\frac{3q_1\left(1 - 5\cos^2 i\right)}{4a^2\eta^4(1 + \varepsilon_2)^2}\right]\left[(1 + \varepsilon_2)^3 - \eta^3\right]$$

$$D_{53}^{(sp1)} = \left[\frac{3q_2\sin 2i}{4a^2\eta^2(1 + \eta)}\right]$$

$$+ \left(\frac{3\sin 2i}{4a^2\eta^4}\right)\left\{(1 + \varepsilon_2)\left[q_2 + (2 + \varepsilon_2)\sin\theta\right] + 5q_1\varepsilon_3 + \eta^2\sin\theta\right\}$$

$$- \left(\frac{15q_1\sin 2i}{4a^2\eta^4}\right)(\theta - \lambda) \quad \text{(F.12)}$$

$$D_{54}^{(sp1)} = \left[\frac{(1 - 3\cos^2 i)}{8a^2\eta^6}\right]\left\{\eta^2\left[2(2q_1\sin\theta + q_2\cos\theta) + (3 + 2\varepsilon_2)\sin 2\theta\right]\right.$$

$$\left. + 2q_1\left[4(1 + \varepsilon_2)(2 + \varepsilon_2)\sin\theta + (3\eta + 4\varepsilon_2)q_2\right]\right\}$$

$$+ \left[\frac{(1 - 3\cos^2 i)\, q_1 q_2(4 + 5\eta)}{4a^2\eta^6(1 + \eta)^2}\right]$$

$$+ \left[\frac{3\left(1 - 5\cos^2 i\right)}{4a^2\eta^6}\right]\left[\varepsilon_3\left(\eta^2 + 4q_1^2\right) + \eta^2 q_1\sin\theta\right]$$

$$+ \left[\frac{3\left(1 - 5\cos^2 i\right)}{4a^2\eta^6}\right]\left[\left(\eta^2 + 4q_1^2\right)(\theta - \lambda)\right]$$

$$- \left[\frac{3q_1\left(1 - 5\cos^2 i\right)}{4a^2\eta^4}\right](\lambda_{q_1})$$

$$D_{55}^{(sp1)} = \left[\frac{(1 - 3\cos^2 i)}{4a^2\eta^2(1 + \eta)}\right] + \left[\frac{(1 - 3\cos^2 i)}{8a^2\eta^6}\right]\left\{\eta^2[5 + 2(2q_1\cos\theta + 5q_2\sin\theta)\right.$$

$$\left. - (3 + 2\varepsilon_2)\cos 2\theta] + 2q_2\left[4(1 + \varepsilon_2)(2 + \varepsilon_2)\sin\theta + (3\eta + 4\varepsilon_2)q_2\right]\right\}$$

$$+ \left[\frac{(1 - 3\cos^2 i)\, q_2^2(4 + 5\eta)}{4a^2\eta^6(1 + \eta)^2}\right]$$

$$+ \left[\frac{3q_1\left(1 - 5\cos^2 i\right)}{4a^2\eta^6}\right]\left(4q_2\varepsilon_3 - \eta^2\cos\theta\right)$$

$$+ \left[\frac{3q_1 q_2\left(1 - 5\cos^2 i\right)}{a^2\eta^6}\right](\theta - \lambda) - \left[\frac{3q_1\left(1 - 5\cos^2 i\right)}{4a^2\eta^4}\right](\lambda_{q_2})$$

$$D_{56}^{(sp1)} = 0$$

$$D_{61}^{(sp1)} = -\left(\frac{2}{a}\right)\Omega^{(sp1)}$$

$$D_{62}^{(sp1)} = \left[\frac{3\cos i}{2a^2\eta^4(1 + \varepsilon_2)^2}\right]\left[(1 + \varepsilon_2)^3 - \eta^3\right]$$

$$D_{63}^{(sp1)} = -\left(\frac{3\varepsilon_3\sin i}{2a^2\eta^4}\right) - \left(\frac{3\sin i}{2a^2\eta^4}\right)(\theta - \lambda)$$

$$D_{64}^{(sp1)} = \left(\frac{3\cos i}{2a^2\eta^6}\right)\left(4q_1\varepsilon_3 + \eta^2\sin\theta\right) + \left(\frac{6q_1\cos i}{a^2\eta^6}\right)(\theta - \lambda)$$

$$- \left(\frac{3\cos i}{2a^2\eta^4}\right)(\lambda_{q_1})$$

(F.13)

$$D_{65}^{(sp1)} = \left(\frac{3\cos i}{2a^2\eta^6}\right)\left(4q_2\varepsilon_3 - \eta^2\cos\theta\right) + \left(\frac{6q_2\cos i}{a^2\eta^6}\right)(\theta - \lambda)$$

$$- \left(\frac{3\cos i}{2a^2\eta^4}\right)(\lambda_{q_2})$$

$$D_{66}^{(sp1)} = 0$$

$$D_{11}^{(sp2)} = -\left(\frac{1}{a}\right) a^{(sp2)}$$

$$D_{12}^{(sp2)} = \left(\frac{3\sin^2 i}{2a\eta^6}\right)(1+\varepsilon_2)^2 \left[2(1+\varepsilon_2)\sin 2\theta + 3\varepsilon_3 \cos 2\theta\right]$$

$$D_{13}^{(sp2)} = -\left(\frac{3\sin 2i \cos 2\theta}{2a\eta^6}\right)(1+\varepsilon_2)^3$$

$$D_{14}^{(sp2)} = -\left(\frac{9\sin^2 i \cos 2\theta}{2a\eta^8}\right)(1+\varepsilon_2)^2 \left[2q_1(1+\varepsilon_2) + \eta^2 \cos\theta\right]$$

$$D_{15}^{(sp2)} = -\left(\frac{9\sin^2 i \cos 2\theta}{2a\eta^8}\right)(1+\varepsilon_2)^2 \left[2q_2(1+\varepsilon_2) + \eta^2 \sin\theta\right]$$

$$D_{16}^{(sp2)} = 0$$

(F.14)

$$D_{21}^{(sp2)} = -\left(\frac{2}{a}\right) \theta^{(sp2)}$$

$$D_{22}^{(sp2)} = -\left(\frac{1}{8a^2\eta^4}\right)$$
$$\times \left\{\begin{array}{l} 3\left(3-5\cos^2 i\right)\left[(q_1\cos\theta - q_2\sin\theta) + 2\cos 2\theta \right.\\ + (q_1\cos 3\theta + q_2\sin 3\theta)] - \sin^2 i[5(q_1\cos\theta - q_2\sin\theta) \\ + 16\cos 2\theta + 9(q_1\cos 3\theta + q_2\sin 3\theta)] \end{array}\right\}$$

$$D_{23}^{(sp2)} = -\left(\frac{\sin 2i}{8a^2\eta^4}\right)[10(q_1\sin\theta + q_2\cos\theta)$$
$$+ 7\sin 2\theta + 2(q_1\sin 3\theta - q_2\cos 3\theta)]$$

(F.15)

$$D_{24}^{(sp2)} = -\left[\frac{(3-5\cos^2 i)}{8a^2\eta^6}\right]\left\{4q_1\left[3\sin 2\theta + q_2(3\cos\theta - \cos 3\theta)\right]\right.$$

$$+ \left(\eta^2 + 4q_1^2\right)(3\sin\theta + \sin 3\theta)\right\} - \left[\frac{\sin^2 i}{8a^2\eta^2(1+\eta)}\right]$$

$$\times (3\sin\theta + \sin 3\theta) - \left[\frac{\sin^2 i}{32a^2\eta^4(1+\eta)}\right]$$

$$\times \left\{\begin{array}{l} 36q_2 - 4(2+3\eta)\sin\theta - \left(39 + 12\eta + \eta^2\right)\sin 3\theta + 9\varepsilon_1\sin 5\theta \\ + 12q_2(2q_1\cos\theta + q_2\sin\theta) + 9q_1(q_1\sin 3\theta - q_2\cos 3\theta) \\ + 18(3q_1\sin 4\theta + 2q_2\cos 4\theta) - 3q_1(q_1\sin 5\theta - 11q_2\cos 5\theta) \\ + 24[(1+\varepsilon_2)(2+\varepsilon_2)\sin\theta + \varepsilon_3(3+2\varepsilon_2)\cos\theta]\cos 2\theta \end{array}\right\}$$

$$- \left[\frac{3\sin^2 i}{32a^2\eta^4(1+\eta)^2}\right][4\sin\theta - 6q_1\sin 4\theta - q_1(q_1\sin 5\theta + q_2\cos 5\theta)]$$

$$+ \left[\frac{q_1\sin^2 i}{8a^2\eta^6(1+\eta)}\right]\left[\begin{array}{l} 20(1+\eta)(q_1\sin\theta + q_2\cos\theta) + 32(1+\eta)\sin 2\theta \\ + 3(4+3\eta)(q_1\sin 3\theta - q_2\cos 3\theta) \end{array}\right]$$

$$-\left[\frac{q_1 \sin^2 i\,(4+5\eta)}{32a^2\eta^6(1+\eta)^2}\right]$$

$$\times \begin{cases} 24\,(q_1 \sin\theta + q_2 \cos\theta) + 24\varepsilon_3(1+\varepsilon_2)(2+\varepsilon_2)\cos 2\theta \\ -(27+3\eta)\,(q_1 \sin 3\theta - q_2 \cos 3\theta) - 18\sin 4\theta \\ -3\,(q_1 \sin 5\theta + q_2 \cos 5\theta) + 12q_2[(3+\varepsilon_2)q_1 \\ +3\,(q_1 \cos 4\theta + q_2 \sin 4\theta) + q_1\,(q_1 \cos 5\theta + q_2 \sin 5\theta)] \end{cases}$$

$$D_{25}^{(sp2)} = -\left[\frac{(3-5\cos^2 i)}{8a^2\eta^6}\right]\left\{4q_2\,[3\sin 2\theta + q_1\,(3\sin\theta + \sin 3\theta)]\right.$$

$$\left. + \left(\eta^2 + 4q_2^2\right)(3\cos\theta - \cos 3\theta)\right\} - \left[\frac{\sin^2 i}{8a^2\eta^2(1+\eta)}\right]$$

$$\times (3\cos\theta - \cos 3\theta) - \left[\frac{\sin^2 i}{32a^2\eta^4(1+\eta)}\right]$$

$$\times \begin{cases} 36q_1 - 4(2+3\eta)\cos\theta + \left(39+12\eta+\eta^2\right)\cos 3\theta + 9\varepsilon_1\cos 5\theta \\ +12q_1\,(q_1 \cos\theta + 2q_2 \sin\theta) + 9q_2\,(q_1 \sin 3\theta - q_2 \cos 3\theta) \\ +18\,(2q_1 \cos 4\theta + 7q_2 \sin 4\theta) + 3q_2\,(11q_1 \sin 5\theta - q_2 \cos 5\theta) \\ +24\,[\varepsilon_3(3+2\varepsilon_2)\sin\theta - (1+\varepsilon_2)(2+\varepsilon_2)\cos\theta]\cos 2\theta \end{cases}$$

$$-\left[\frac{3\sin^2 i}{32a^2\eta^4(1+\eta)^2}\right][4\cos\theta - 6q_2 \sin 4\theta - q_2\,(q_1 \sin 5\theta + q_2 \cos 5\theta)]$$

$$+\left[\frac{q_2 \sin^2 i}{8a^2\eta^6(1+\eta)}\right]\left[\begin{matrix}20(1+\eta)\,(q_1 \sin\theta + q_2 \cos\theta) + 32(1+\eta)\sin 2\theta \\ +3(4+3\eta)\,(q_1 \sin 3\theta - q_2 \cos 3\theta)\end{matrix}\right]$$

$$-\left[\frac{q_1 \sin^2 i\,(4+5\eta)}{32a^2\eta^6(1+\eta)^2}\right]$$

$$\times \begin{cases} 24\,(q_1 \sin\theta + q_2 \cos\theta) + 24\varepsilon_3(1+\varepsilon_2)(2+\varepsilon_2)\cos 2\theta \\ -(27+3\eta)\,(q_1 \sin 3\theta - q_2 \cos 3\theta) - 18\sin 4\theta \\ -3\,(q_1 \sin 5\theta + q_2 \cos 5\theta) + 12q_2[(3+\varepsilon_2)q_1 \\ +3\,(q_1 \cos 4\theta + q_2 \sin 4\theta) + q_1\,(q_1 \cos 5\theta + q_2 \sin 5\theta)] \end{cases}$$

$$D_{26}^{(sp2)} = 0$$

$$D_{31}^{(sp2)} = -\left(\frac{2}{a}\right)i^{(sp2)}$$

$$D_{32}^{(sp2)} = \left(\frac{3\sin 2i}{8a^2\eta^4}\right)[(q_1 \sin\theta + q_2 \cos\theta) + 2\sin 2\theta$$
$$+ (q_1 \sin 3\theta - q_2 \cos 3\theta)] \tag{F.16}$$

$$D_{33}^{(sp2)} = -\left(\frac{\cos 2i}{4a^2\eta^4}\right)[3\,(q_1 \cos\theta - q_2 \sin\theta) + 3\cos 2\theta$$
$$+ (q_1 \cos 3\theta + q_2 \sin 3\theta)]$$

$$D_{34}^{(sp2)} = -\left(\frac{\sin 2i}{8a^2\eta^6}\right)\left\{4q_1\left[3\cos 2\theta - q_2\left(3\sin\theta - \sin 3\theta\right)\right]\right.$$
$$\left. + \left(\eta^2 + 4q_1^2\right)\left(3\cos\theta + \cos 3\theta\right)\right\}$$

$$D_{35}^{(sp2)} = -\left(\frac{\sin 2i}{8a^2\eta^6}\right)\left\{4q_2\left[3\cos 2\theta + q_1\left(3\cos\theta + \cos 3\theta\right)\right]\right.$$
$$\left. - \left(\eta^2 + 4q_2^2\right)\left(3\sin\theta - \sin 3\theta\right)\right\}$$

$$D_{36}^{(sp2)} = 0$$

$$D_{41}^{(sp2)} = -\left(\frac{2}{a}\right)q_1^{(sp2)}$$

$$D_{42}^{(sp2)} = \left[\frac{3q_2\left(3 - 5\cos^2 i\right)}{8a^2\eta^4}\right]\left[(q_1\cos\theta - q_2\sin\theta)\right.$$
$$\left. + 2\cos 2\theta + (q_1\cos 3\theta + q_2\sin 3\theta)\right]$$
$$+ \left(\frac{3\sin^2 i}{32a^2\eta^4}\right)\left\{2\begin{bmatrix}2q_2\varepsilon_2 - 9q_2\left(q_1\cos 3\theta + q_2\sin 3\theta\right)\\ +12\left(q_1\sin 4\theta - q_2\cos 4\theta\right)\\ -5q_2\left(q_1\cos 5\theta + q_2\sin 5\theta\right)\end{bmatrix}\right.$$
$$\left. + \left[4\left(1 + 3q_1^2\right)\sin\theta + 40q_1\sin 2\theta\right.\right.$$
$$\left.\left. + (28 + 17\varepsilon_1)\sin 3\theta + 5\varepsilon_1\sin 5\theta\right]\right\}$$

$$D_{43}^{(sp2)} = -\left(\frac{\sin 2i}{32a^2\eta^4}\right)\left\{2\begin{bmatrix}36q_1\left(q_1\cos\theta - q_2\sin\theta\right)\\ +30\left(q_1\cos 2\theta - q_2\sin 2\theta\right)\\ -q_2\left(q_1\sin 3\theta - q_2\cos 3\theta\right)\\ +9\left(q_1\cos 4\theta + q_2\sin 4\theta\right)\\ +3q_2\left(q_1\sin 5\theta - q_2\cos 5\theta\right)\end{bmatrix}\right.$$
$$\left. + \left[6q_1\left(3 + 2q_1\cos\theta\right) + 12(1 - 4\varepsilon_1)\cos\theta\right.\right.$$
$$\left.\left. + (28 + 17\varepsilon_1)\cos 3\theta + 3\varepsilon_1\cos 5\theta\right]\right\}$$

$$D_{44}^{(sp2)} = \left[\frac{q_2\left(3 - 5\cos^2 i\right)}{8a^2\eta^6}\right]\left\{4q_1\left[3\sin 2\theta + q_2\left(3\cos\theta - \cos 3\theta\right)\right]\right.$$
$$\left. + \left(\eta^2 + 4q_1^2\right)\left(3\sin\theta + \sin 3\theta\right)\right\}$$
$$- \left(\frac{\sin^2 i}{8a^2\eta^4}\right)\left\{\begin{array}{l}[8q_1\cos 3\theta - 3q_2\left(\sin\theta - \sin 3\theta\right)]\\ +3\left[5 + \varepsilon_2 + 3\cos 2\theta + 3\left(q_1\cos 3\theta\right.\right.\\ \left.\left. + q_2\sin 3\theta\right)\right]\cos 2\theta\end{array}\right\}$$
$$- \left(\frac{3q_1\sin^2 i}{4a^2\eta^6}\right)\left\{\begin{array}{l}2q_1\left[(q_1\cos\theta - q_2\sin\theta)\right.\\ \left. + (q_1\cos 3\theta + q_2\sin 3\theta)\right]\\ + \begin{bmatrix}9\cos\theta - \cos 3\theta + 2q_1(5 + \varepsilon_2)\\ +6\left(q_1\cos 2\theta + q_2\sin 2\theta\right)\\ +2q_1\left(q_1\cos 3\theta + q_2\sin 3\theta\right)\end{bmatrix}\cos 2\theta\end{array}\right\}$$

$$D_{45}^{(sp2)} = \left[\frac{(3 - 5\cos^2 i)}{8a^2\eta^6}\right]\left\{\left(\eta^2 + 4q_2^2\right)[3\sin 2\theta + q_1(3\sin\theta + \sin 3\theta)]\right.$$

$$+ 2q_2\left(\eta^2 + 2q_2^2\right)(3\cos\theta - \cos 3\theta)\Big\}$$

$$+ \left(\frac{\sin^2 i}{16a^2\eta^4}\right)[6(q_1\sin\theta + 2q_2\cos\theta) - (9q_1\sin 3\theta + q_2\cos 3\theta)$$

$$- 9\sin 4\theta - 3(q_1\sin 5\theta + q_2\cos 5\theta)]$$

$$- \left(\frac{3q_2\sin^2 i}{8a^2\eta^6}\right)\left\{2q_1\begin{bmatrix}3 + 2(2q_1\cos\theta - q_2\sin\theta)\\+10\cos 2\theta\\+3(q_1\cos 3\theta + q_2\sin 3\theta)\\+(q_1\cos 5\theta + q_2\sin 5\theta)\end{bmatrix}\right.$$

$$+ [8\cos\theta + 9\cos 3\theta$$

$$+ 6(q_1\cos 4\theta + q_2\sin 4\theta) - \cos 5\theta]\Big\}$$

$$D_{46}^{(sp2)} = 0 \qquad\qquad\qquad\qquad\qquad\qquad\qquad\qquad\qquad\qquad \text{(F.17)}$$

$$D_{51}^{(sp2)} = -\left(\frac{2}{a}\right)q_2^{(sp2)}$$

$$D_{52}^{(sp2)} = -\left[\frac{3q_1(3 - 5\cos^2 i)}{8a^2\eta^4}\right][(q_1\cos\theta - q_2\sin\theta)$$

$$+ 2\cos 2\theta + (q_1\cos 3\theta + q_2\sin 3\theta)]$$

$$+ \left(\frac{3\sin^2 i}{32a^2\eta^4}\right)\left\{2\begin{bmatrix}2q_1\varepsilon_2 + 9q_1(q_1\cos 3\theta + q_2\sin 3\theta)\\-12(q_1\cos 4\theta + q_2\sin 4\theta)\\-5q_1(q_1\cos 5\theta + q_2\sin 5\theta)\end{bmatrix}\right.$$

$$+ \left[4\left(1 + 3q_2^2\right)\cos\theta + 40q_2\sin 2\theta\right.$$

$$\left.- (28 + 17\varepsilon_1)\cos 3\theta + 5\varepsilon_1\cos 5\theta\right]\Big\}$$

$$D_{53}^{(sp2)} = -\left(\frac{\sin 2i}{32a^2\eta^4}\right)\left\{2\begin{bmatrix}36q_1(q_1\sin\theta + q_2\cos\theta)\\+30(q_1\sin 2\theta + q_2\cos 2\theta)\\+q_1(q_1\sin 3\theta - q_2\cos 3\theta)\\+9(q_1\sin 4\theta - q_2\cos 4\theta)\\+3q_1(q_1\sin 5\theta - q_2\cos 5\theta)\end{bmatrix}\right.$$

$$- [6q_2(3 + 2q_2\sin\theta) + 12(1 + 2\varepsilon_1)\sin\theta$$

$$\left.- (28 + 17\varepsilon_1)\sin 3\theta + 3\varepsilon_1\sin 5\theta]\right\}$$

$$D_{54}^{(sp2)} = -\left[\frac{(3 - 5\cos^2 i)}{8a^2\eta^6}\right]\left\{\left(\eta^2 + 4q_1^2\right)[3\sin 2\theta + q_2(3\cos\theta - \cos 3\theta)]\right.$$

$$+ 2q_1\left(\eta^2 + 2q_1^2\right)(3\sin\theta + \sin 3\theta)\Big\}$$

$$- \left(\frac{\sin^2 i}{16a^2\eta^4}\right)[6(2q_1\sin\theta + q_2\cos\theta) + (q_1\sin 3\theta + 9q_2\cos 3\theta)$$

$$+ 9\sin 4\theta - 3(q_1\sin 5\theta + q_2\cos 5\theta)]$$

$$+\left(\frac{3q_1\sin^2 i}{8a^2\eta^6}\right)\left\{2q_2\begin{bmatrix}3-2\left(2q_1\cos\theta-2q_2\sin\theta\right)\\-10\cos 2\theta-3\left(q_1\cos 3\theta+q_2\sin 3\theta\right)\\+\left(q_1\cos 5\theta+q_2\sin 5\theta\right)\\+[8\sin\theta-9\sin 3\theta-6\left(q_1\sin 4\theta\right.\\\left.-q_2\cos 4\theta\right)-\sin 5\theta]\end{bmatrix}\right\}$$

$$D_{55}^{(sp2)}=-\left[\frac{q_1\left(3-5\cos^2 i\right)}{8a^2\eta^6}\right]\left\{4q_2\left[3\sin 2\theta+q_1\left(3\sin\theta+\sin 3\theta\right)\right]\right.$$
$$\left.+\left(\eta^2+4q_2^2\right)\left(3\cos\theta-\cos 3\theta\right)\right\}$$
$$-\left(\frac{\sin^2 i}{8a^2\eta^4}\right)\left\{\begin{matrix}[8q_2\sin 3\theta+3q_1\left(\cos\theta+\cos 3\theta\right)]\\+3\left[5+\varepsilon_2-3\cos 2\theta\right.\\-\left(q_1\cos 3\theta-q_2\sin 3\theta\right)]\cos 2\theta\end{matrix}\right\}$$
$$-\left(\frac{3\sin^2 i}{4a^2\eta^6}\right)\begin{bmatrix}9\sin\theta-\sin 3\theta+2q_2(5+\varepsilon_2)\\+6\left(q_1\sin 2\theta-q_2\cos 2\theta\right)\\+2q_1\left(q_1\sin 3\theta-q_2\cos 3\theta\right)\end{bmatrix}\left(q_2\cos 2\theta\right)$$

$$D_{56}^{(sp2)}=0 \tag{F.18}$$

$$D_{61}^{(sp2)}=-\left(\frac{2}{a}\right)\Omega^{(sp2)}$$

$$D_{62}^{(sp2)}=-\left(\frac{3\cos i}{4a^2\eta^4}\right)\left[\left(q_1\cos\theta-q_2\sin\theta\right)+2\cos 2\theta\right.$$
$$\left.+\left(q_1\cos 3\theta+q_2\sin 3\theta\right)\right]$$

$$D_{63}^{(sp2)}=\left(\frac{\sin i}{4a^2\eta^4}\right)\left[3\left(q_1\sin\theta+q_2\cos\theta\right)+3\sin 2\theta\right.$$
$$\left.+\left(q_1\sin 3\theta-q_2\cos 3\theta\right)\right]$$

$$D_{64}^{(sp2)}=-\left(\frac{\cos i}{4a^2\eta^6}\right)\left\{4q_1\left[3\sin 2\theta+q_2\left(3\cos\theta-\cos 3\theta\right)\right]\right. \tag{F.19}$$
$$\left.+\left(\eta^2+4q_1^2\right)\left(3\sin\theta+\sin 3\theta\right)\right\}$$

$$D_{65}^{(sp2)}=-\left(\frac{\cos i}{4a^2\eta^6}\right)\left\{4q_2\left[3\sin 2\theta+q_1\left(3\sin\theta+\sin 3\theta\right)\right]\right.$$
$$\left.+\left(\eta^2+4q_2^2\right)\left(3\cos\theta-\cos 3\theta\right)\right\}$$

$$D_{66}^{(sp2)}=0$$

G SMALL ECCENTRICITY THEORY

This appendix contains the equations for a theory that is valid for small eccentricities. First-order eccentricity terms are included for the terms that do not have J_2 as a factor. In the terms that include J_2 the eccentricity is set to zero, i.e., $e=q_1=q_2=0$. The Σ and Σ^{-1} matrices do not simplify much by retaining only $\mathcal{O}(e)$ terms in the non-J_2 terms and $\mathcal{O}(e^0)$ terms in the terms multiplied by J_2. Therefore, they are not changed. If it is desired to change them substitute

$$\mu = \mu$$
$$\eta = 1$$
$$r = a \left(1 - q_1 \cos \theta - q_2 \sin \theta\right)$$
$$r^{-1} = a^{-1} \left(1 + q_1 \cos \theta + q_2 \sin \theta\right)$$
$$v_r = \sqrt{\frac{\mu}{a}} \left(q_1 \sin \theta - q_2 \cos \theta\right) \tag{G.1}$$
$$v_t = \sqrt{\frac{\mu}{a}} \left(1 + q_1 \cos \theta + q_2 \sin \theta\right)$$
$$v_{rt} = 0$$

in the non-J_2 terms and

$$p = a$$
$$\eta = 1$$
$$r = a$$
$$v_r = 0 \tag{G.2}$$
$$v_t = an$$
$$v_{rt} = \left(q_1 \sin \theta - q_2 \cos \theta\right)$$

in the terms multiplied by J_2. In addition, as shown below in Eq. (G.12), the long-periodic variations of the elements are multiplied by e, so they are zero. In addition,

$$\Delta \omega = \dot{\omega}^{(s)} \left(t - t_0\right) \tag{G.3}$$
$$\dot{\omega}^{(s)} = 0.75 \left(\frac{R_e}{a}\right)^2 n \left(5 \cos^2 i - 1\right)$$

$\bar{B}$ Matrix

The non-zero terms in the $\bar{B}(t)$ matrix are

$$\bar{B}_{24} = -\left(\frac{n}{4a}\right) \left(5 \cos^2 i - 1\right) \sin \theta$$
$$\bar{B}_{25} = \left(\frac{n}{4a}\right) \left(5 \cos^2 i - 1\right) \cos \theta$$
$$\bar{B}_{41} = \left(\frac{7}{4} \frac{n}{a^2}\right) \cos^2 i$$
$$\bar{B}_{43} = \left(\frac{n}{2a}\right) \sin i \cos i \tag{G.4}$$
$$\bar{B}_{61} = -\left(\frac{7}{4} \frac{n}{a^2}\right) \cos \theta \sin i \cos i$$
$$\bar{B}_{63} = -\left(\frac{n}{2a}\right) \cos \theta \sin^2 i$$

Mean Element State Transition Matrix $\bar{\phi}_{\overline{\infty}}$

$$G_\theta = \frac{rn}{v_t} \approx (1 - 2q_1 \cos\theta - 2q_2 \sin\theta), \quad G_{\theta_0} = -G_\theta(t_0)$$
$$G_{q_1} = -2\sin\theta + 0.5(q_2 + 3q_1 \sin 2\theta - 3q_2 \cos 2\theta), \quad G_{q_{10}} = -G_{q_1}(t_0) \quad \text{(G.5)}$$
$$G_{q_2} = 2\cos\theta + 0.5(q_1 - 3q_1 \cos 2\theta - 3q_2 \sin 2\theta), \quad G_{q_{20}} = -G_{q_2}(t_0)$$

The non-zero elements of $\bar{\phi}_{\overline{\infty}}$ are

$$\bar{\phi}_{\overline{\infty}11} = 1 \tag{G.6}$$

$$\bar{\phi}_{\overline{\infty}21} = -\frac{3n(t - t_0)}{2aG_\theta}\left[1 + \left(\frac{7A_2}{6a^2}\right)\left(4\cos^2 i - 1\right)\right]$$

$$\bar{\phi}_{\overline{\infty}22} = -\frac{G_{\theta_0}}{G_\theta}$$

$$\bar{\phi}_{\overline{\infty}23} = -\left(\frac{2A_2}{a^2}\right)\frac{(t - t_0)}{G_\theta}\sin 2i \tag{G.7}$$

$$\bar{\phi}_{\overline{\infty}24} = -\frac{1}{G_\theta}\left[G_{q_{10}} + G_{q_1}\cos(\Delta\omega) + G_{q_2}\sin(\Delta\omega)\right]$$

$$\bar{\phi}_{\overline{\infty}25} = -\frac{1}{G_\theta}\left[G_{q_{20}} - G_{q_1}\sin(\Delta\omega) + G_{q_2}\cos(\Delta\omega)\right]$$

$$\bar{\phi}_{\overline{\infty}33} = 1 \tag{G.8}$$

$$\bar{\phi}_{\overline{\infty}44} = \cos(\Delta\omega)$$
$$\bar{\phi}_{\overline{\infty}45} = -\sin(\Delta\omega) \tag{G.9}$$

$$\bar{\phi}_{\overline{\infty}54} = \sin(\Delta\omega)$$
$$\bar{\phi}_{\overline{\infty}55} = \cos(\Delta\omega) \tag{G.10}$$

$$\bar{\phi}_{\overline{\infty}61} = \left(\frac{7}{4}\frac{A_2}{p^2}\right)\left(\frac{n\cos i}{a}\right)(t - t_0)$$

$$\bar{\phi}_{\overline{\infty}63} = \left(\frac{A_2}{2p^2}\right)(n\sin i)(t - t_0) \tag{G.11}$$

$$\bar{\phi}_{\overline{\infty}66} = 1$$

Mean-to-Osculating Transformation

$$a^{(lp)} = \theta^{(lp)} = i^{(lp)} = q_1^{(lp)} = q_2^{(lp)} = \Omega^{(lp)} = 0 \tag{G.12}$$

$$a^{(sp1)} = \theta^{(sp1)} = i^{(sp1)} = \Omega^{(sp1)} = 0$$

$$q_1^{(sp1)} = \frac{3\left(1 - 3\cos^2 i\right)}{4a^2}\cos\theta \tag{G.13}$$

$$q_2^{(sp1)} = \frac{3\left(1 - 3\cos^2 i\right)}{4a^2}\sin\theta$$

$$a^{(sp2)} = -\left(\frac{3\sin^2 i}{2a}\right)\cos 2\theta$$

$$\lambda^{(sp2)} = \frac{3\left(3 - 5\cos^2 i\right)}{8a^2}\sin 2\theta$$

$$\theta^{(sp2)} = \lambda^{(sp2)} + \left(\frac{\sin^2 i}{a^2}\right)\sin 2\theta$$

$$i^{(sp2)} = -\left(\frac{3\sin 2i}{8a^2}\right)\cos 2\theta \tag{G.14}$$

$$q_1^{(sp2)} = -\left(\frac{\sin^2 i}{8a^2}\right)\left(3\cos\theta + 7\cos 3\theta\right)$$

$$q_2^{(sp2)} = \left(\frac{\sin^2 i}{8a^2}\right)\left(3\sin\theta - 7\sin 3\theta\right)$$

$$\Omega^{(sp2)} = -\left(\frac{3\cos i}{4a^2}\right)\sin 2\theta$$

Mean-to-Osculating Jacobian D

The non-zero elements of D are:

$$D_{24}^{(lp)} = -\left(\frac{\sin^2 i}{8a^2}\right)\left(1 - 10\Theta\cos^2 i\right)\sin\theta$$

$$D_{25}^{(lp)} = -\left(\frac{\sin^2 i}{8a^2}\right)\left(1 - 10\Theta\cos^2 i\right)\cos\theta$$

$$D_{44}^{(lp)} = -\left(\frac{\sin^2 i}{16a^2}\right)\left(1 - 10\Theta\cos^2 i\right) \tag{G.15}$$

$$D_{55}^{(lp)} = \left(\frac{\sin^2 i}{16a^2}\right)\left(1 - 10\Theta\cos^2 i\right)$$

$$\Theta = \left(1 - 5\cos^2 i\right)^{-1}$$

$$D_{14}^{(sp1)} = \frac{3\left(1 - 3\cos^2 i\right)}{2a}\cos\theta$$

$$D_{15}^{(sp1)} = \frac{3\left(1 - 3\cos^2 i\right)}{2a} \sin\theta \tag{G.16}$$

$$D_{24}^{(sp1)} = \frac{9\left(1 - 5\cos^2 i\right)}{4a^2} \sin\theta$$

$$D_{25}^{(sp1)} = -\frac{9\left(1 - 5\cos^2 i\right)}{4a^2} \cos\theta \tag{G.17}$$

$$D_{41}^{(sp1)} = -\frac{3\left(1 - 3\cos^2 i\right)}{2a^3} \cos\theta$$

$$D_{42}^{(sp1)} = -\frac{3\left(1 - 3\cos^2 i\right)}{4a^2} \sin\theta$$

$$D_{43}^{(sp1)} = \left(\frac{9\sin 2i}{4a^2}\right) \cos\theta \tag{G.18}$$

$$D_{44}^{(sp1)} = \frac{3\left(1 - 3\cos^2 i\right)}{8a^2} (2 + \cos 2\theta)$$

$$D_{45}^{(sp1)} = \frac{3\left(1 - 3\cos^2 i\right)}{8a^2} \sin 2\theta$$

$$D_{51}^{(sp1)} = -\frac{3\left(1 - 3\cos^2 i\right)}{2a^3} \sin\theta$$

$$D_{52}^{(sp1)} = \frac{3\left(1 - 3\cos^2 i\right)}{4a^2} \cos\theta$$

$$D_{53}^{(sp1)} = \left(\frac{9\sin 2i}{4a^2}\right) \sin\theta \tag{G.19}$$

$$D_{54}^{(sp1)} = \frac{3\left(1 - 3\cos^2 i\right)}{8a^2} \sin 2\theta$$

$$D_{55}^{(sp1)} = \frac{3\left(1 - 3\cos^2 i\right)}{8a^2} (2 - \cos 2\theta)$$

$$D_{64}^{(sp1)} = \left(\frac{9\cos i}{4a^2}\right) \sin\theta$$

$$D_{65}^{(sp1)} = -\left(\frac{9\cos i}{4a^2}\right) \cos\theta \tag{G.20}$$

$$D_{11}^{(sp2)} = \left(\frac{3\sin^2 i}{2a^2}\right) \cos 2\theta$$

$$D_{12}^{(sp2)} = \left(\frac{3\sin^2 i}{a}\right) \sin 2\theta \tag{G.21}$$

$$D_{13}^{(sp2)} = -\left(\frac{3\sin 2i}{2a}\right)\cos 2\theta$$

$$D_{14}^{(sp2)} = -\left(\frac{9\sin^2 i}{4a}\right)(\cos\theta + \cos 3\theta)$$

$$D_{15}^{(sp2)} = \left(\frac{9\sin^2 i}{4a}\right)(\sin\theta - \sin 3\theta)$$

$$D_{21}^{(sp2)} = -\left(\frac{6 - 7\sin^2 i}{4a^3}\right)\sin 2\theta$$

$$D_{22}^{(sp2)} = \left(\frac{6 - 7\sin^2 i}{4a^2}\right)\cos 2\theta$$

$$D_{23}^{(sp2)} = -\left(\frac{7\sin 2i}{8a^2}\right)\sin 2\theta \tag{G.22}$$

$$D_{24}^{(sp2)} = \left(\frac{24 - 47\sin^2 i}{32a^2}\right)\sin\theta + \left(\frac{\cos^2 i}{4a^2}\right)\sin 3\theta$$

$$D_{25}^{(sp2)} = \left(\frac{24 - 47\sin^2 i}{32a^2}\right)\cos\theta - \left(\frac{\cos^2 i}{4a^2}\right)\cos 3\theta$$

$$D_{31}^{(sp2)} = \left(\frac{3\sin 2i}{4a^3}\right)\cos 2\theta$$

$$D_{32}^{(sp2)} = \left(\frac{3\sin 2i}{4a^2}\right)\sin 2\theta$$

$$D_{33}^{(sp2)} = -\left(\frac{3\cos 2i}{4a^2}\right)\cos 2\theta \tag{G.23}$$

$$D_{34}^{(sp2)} = -\left(\frac{\sin 2i}{8a^2}\right)(3\cos\theta + \cos 3\theta)$$

$$D_{35}^{(sp2)} = \left(\frac{\sin 2i}{8a^2}\right)(3\sin\theta - \sin 3\theta)$$

$$D_{41}^{(sp2)} = \left(\frac{\sin^2 i}{4a^3}\right)(3\cos\theta + 7\cos 3\theta)$$

$$D_{42}^{(sp2)} = \left(\frac{3\sin^2 i}{8a^2}\right)(\sin\theta + 7\sin 3\theta)$$

$$D_{43}^{(sp2)} = -\left(\frac{\sin 2i}{8a^2}\right)(3\cos\theta + 7\cos 3\theta) \tag{G.24}$$

$$D_{44}^{(sp2)} = -\left(\frac{3\sin^2 i}{16a^2}\right)(3 + 10\cos 2\theta + 3\cos 4\theta)$$

$$D_{45}^{(sp2)} = \frac{3(3 - 5\cos^2 i)}{8a^2}\sin 2\theta - \left(\frac{9\sin^2 i}{16a^2}\right)\sin 4\theta$$

$$D_{51}^{(sp2)} = -\left(\frac{\sin^2 i}{4a^3}\right)(3\sin\theta - 7\sin 3\theta)$$

$$D_{52}^{(sp2)} = \left(\frac{3\sin^2 i}{8a^2}\right)(\cos\theta - 7\cos 3\theta)$$

$$D_{53}^{(sp2)} = \left(\frac{\sin 2i}{8a^2}\right)(3\sin\theta - 7\sin 3\theta) \qquad \text{(G.25)}$$

$$D_{54}^{(sp2)} = -\frac{3(3 - 5\cos^2 i)}{8a^2}\sin 2\theta - \left(\frac{9\sin^2 i}{16a^2}\right)\sin 4\theta$$

$$D_{55}^{(sp2)} = \left(\frac{3\sin^2 i}{16a^2}\right)(3 - 10\cos 2\theta + 3\cos 4\theta)$$

$$D_{61}^{(sp2)} = \left(\frac{3\cos i}{2a^3}\right)\sin 2\theta$$

$$D_{62}^{(sp2)} = -\left(\frac{3\cos i}{2a^2}\right)\cos 2\theta$$

$$D_{63}^{(sp2)} = \left(\frac{3\sin i}{4a^2}\right)\sin 2\theta \qquad \text{(G.26)}$$

$$D_{64}^{(sp2)} = -\left(\frac{\cos i}{4a^2}\right)(3\sin\theta + \sin 3\theta)$$

$$D_{65}^{(sp2)} = -\left(\frac{\cos i}{4a^2}\right)(3\cos\theta - \cos 3\theta)$$

H YAN-ALFRIEND NONLINEAR THEORY COEFFICIENTS

This appendix contains the coefficients used in the Yan-Alfriend nonlinear theory used in Section 8.2.

$$a_{10} = -\frac{3}{L^4}$$

$$a_{11} = -\left(\frac{21}{4L^8\eta^4}\right)\left[(1 + \eta) - (5 + 3\eta)\cos^2 i\right]$$

$$a_{12} = -\left(\frac{33}{64L^{12}\eta^8}\right)\left[\left(-35 + 9\eta + 41\eta^2 + 25\eta^3\right)\right. \qquad \text{(H.1)}$$
$$+\left(90 - 162\eta - 222\eta^2 - 90\eta^3\right)\cos^2 i$$
$$\left.+\left(385 + 465\eta + 189\eta^2 + 25\eta^3\right)\cos^4 i\right]$$

$$a_{21} = -\left(\frac{3}{4L^7\eta^5}\right)\left[(4+3\eta)-(20+9\eta)\cos^2 i\right]$$

$$a_{22} = \left(\frac{3}{64L^{11}\eta^5}\right)\left[\left(480 - 63\eta - 746\eta^2 - 175\eta^3\right)\right.$$
$$+\left(-720+1134\eta+1332\eta^2+450\eta^3\right)\cos^2 i$$
$$\left.-\left(3080+3255\eta+1134\eta^2+125\eta^3\right)\cos^4 i\right]$$

(H.2)

$$a_{31} = \left(\frac{3\sin 2i}{4L^7\eta^4}\right)(5+3\eta)$$

$$a_{32} = -\left(\frac{3\sin 2i}{32L^{11}\eta^8}\right)\left[\left(45-81\eta-111\eta^2-45\eta^3\right)\right.$$
$$\left.+\left(385+465\eta+189\eta^2+25\eta^3\right)\cos^2 i\right]$$

(H.3)

$$a_{40} = \frac{12}{L^5}$$

$$a_{41} = -\frac{8}{L}a_{11} = \left(\frac{42}{L^9\eta^4}\right)\left[(1+\eta)-(5+3\eta)\cos^2 i\right]$$

$$a_{42} = -\frac{12}{L}a_{12} = \left(\frac{99}{16L^{13}\eta^8}\right)\left[\left(-35+9\eta+41\eta^2+25\eta^3\right)\right.$$
$$+\left(90-162\eta-222\eta^2-90\eta^3\right)\cos^2 i$$
$$\left.+\left(385+465\eta+189\eta^2+25\eta^3\right)\cos^4 i\right]$$

(H.4)

$$a_{51} = \left(\frac{3}{L^7\eta^6}\right)\left[(5+3\eta)-(25+9\eta)\cos^2 i\right]$$

$$a_{52} = -\left(\frac{3}{32L^{11}\eta^{10}}\right)\left[\left(1260-252\eta-861\eta^2-375\eta^3\right)\right.$$
$$-\left(3240-4536\eta-4662\eta^2-1350\eta^3\right)\cos^2 i$$
$$\left.-\left(13860+13020\eta+3969\eta^2+375\eta^3\right)\cos^4 i\right]$$

(H.5)

$$a_{61} = -\left(\frac{3\cos 2i}{2L^7\eta^4}\right)(5+4\eta)$$

$$a_{62} = -\left(\frac{3}{32L^{11}\eta^9}\right)\left[\left(280-63\eta-246\eta^2-125\eta^3\right)\right.$$
$$\left(-720+1134\eta+1332\eta^2+450\eta^3\right)\cos^2 i$$
$$\left.-\left(3080+3255\eta+1134\eta^2+125\eta^3\right)\cos^4 i\right]$$

(H.6)

$$a_{71} = -\frac{7}{L}a_{21} = \left(\frac{21}{4L^8\eta^5}\right)\left[(4+3\eta) - (20+9\eta)\cos^2 i\right]$$

$$a_{72} = -\frac{11}{L}a_{22} = -\left(\frac{33}{64L^{12}\eta^9}\right)\left[\left(280 - 63\eta - 246\eta^2 - 125\eta^3\right)\right.$$
$$-\left(720 - 1134\eta - 1332\eta^2 - 450\eta^3\right)\cos^2 i$$
$$\left.-\left(3080 + 3255\eta + 1134\eta^2 + 125\eta^3\right)\cos^4 i\right]$$

(H.7)

$$a_{81} = -\frac{7}{L}a_{31} = -\left(\frac{21\sin 2i}{4L^8\eta^4}\right)(5+3\eta)$$

$$a_{82} = -\frac{11}{L}a_{32} = \left(\frac{33\sin 2i}{32L^{12}\eta^8}\right)\left[\left(45 - 81\eta - 111\eta^2 - 45\eta^3\right)\right.$$
$$\left.+\left(385 + 465\eta + 189\eta^2 + 25\eta^3\right)\cos^2 i\right]$$

(H.8)

$$a_{91} = -\left(\frac{3}{4L^7\eta^5}\right)(20+9\eta)\sin 2i$$

$$a_{92} = -\left(\frac{3\sin 2i}{32L^{12}\eta^9}\right)\left[\left(360 + 567\eta + 666\eta^2 + 225\eta^3\right)\right.$$
$$\left.+\left(3080 + 3255\eta + 1134\eta^2 + 125\eta^3\right)\cos^2 i\right]$$

(H.9)

$$b_{11} = -\left(\frac{21}{4L^8\eta^4}\right)\left(1 - 5\cos^2 i\right)$$

(H.10)

$$b_{12} = \left(\frac{33}{64L^{12}\eta^8}\right)\left[\left(35 - 24\eta - 25\eta^2\right) + \left(-90 + 192\eta + 126\eta^2\right)\right.$$
$$\left.\cos^2 i - \left(385 + 360\eta + 45\eta^2\right)\cos^4 i\right]$$

$$b_{21} = -\left(\frac{3}{L^7\eta^5}\right)\left(1 - 5\cos^2 i\right)$$

(H.11)

$$b_{22} = \left(\frac{3}{32L^{11}\eta^9}\right)\left[\left(140 - 84\eta - 75\eta^2\right) + \left(-360 + 672\eta + 378\eta^2\right)\cos^2 i\right.$$
$$\left.-\left(1544 + 1260\eta + 135\eta^2\right)\cos^4 i\right]$$

$$b_{31} = \left(\frac{15}{4L^7\eta^4}\right)\sin 2i$$

(H.12)

$$b_{32} = \left(\frac{3\sin 2i}{32L^{11}\eta^8}\right)\left[\left(-45 + 96\eta + 63\eta^2\right) - \left(385 + 360\eta + 45\eta^2\right)\cos^2 i\right]$$

$$b_{41} = -\frac{8}{L}b_{11} = \left(\frac{42}{L^9\eta^4}\right)\left(1 - 5\cos^2 i\right)$$

(H.13)

$$b_{42} = -\frac{12}{L}c_{12} = -\left(\frac{99}{16L^{13}\eta^8}\right)\left[\left(35 - 24\eta - 25\eta^2\right)\right.$$
$$\left. + \left(90 + 192\eta + 128\eta^2\right)\cos^2 i - \left(147 + 360\eta + 15\eta^2\right)\cos^4 i\right]$$

$$b_{51} = \left(\frac{15}{L^7\eta^6}\right)\left(1 - 5\cos^2 i\right) \tag{H.14}$$

$$b_{52} = \left(\frac{9}{32L^{11}\eta^{10}}\right)\left[\left(420 - 224\eta - 175\eta^2\right)\right.$$
$$+ \left(-1080 + 1792\eta + 882\eta^2\right)\cos^2 i$$
$$\left. - \left(4632 + 3360\eta + 315\eta^2\right)\cos^4 i\right]$$

$$b_{61} = \left(\frac{15}{2L^7\eta^4}\right)\cos 2i \tag{H.15}$$

$$b_{62} = \left(\frac{3}{16L^{11}\eta^8}\right)\left[\left(45 - 96\eta - 63\eta^2\right) + 4\left(555 + 588\eta + 99\eta^2\right)\cos^2 i\right.$$
$$\left. + 8\left(385 + 360\eta + 45\eta^2\right)\cos^4 i\right]$$

$$b_{71} = -\frac{7}{L}b_{21} = \left(\frac{21}{L^8\eta^5}\right)\left(1 - 5\cos^2 i\right) \tag{H.16}$$

$$b_{72} = -\frac{11}{L}b_{22} = -\left(\frac{33}{32L^{12}\eta^9}\right)\left[\left(140 - 84\eta - 75\eta^2\right)\right.$$
$$\left. + \left(-360 + 672\eta + 378\eta^2\right)\cos^2 i - \left(1544 + 1260\eta + 135\eta^2\right)\cos^4 i\right]$$

$$b_{81} = -\frac{7}{L}b_{31} = -\left(\frac{105}{4L^7\eta^4}\right)\sin 2i \tag{H.17}$$

$$b_{82} = -\frac{11}{L}b_{32} = -\left(\frac{33\sin 2i}{32L^{11}\eta^8}\right)\left[\left(-45 + 96\eta + 63\eta^2\right)\right.$$
$$\left. - \left(385 + 360\eta + 45\eta^2\right)\cos^2 i\right]$$

$$b_{91} = -\left(\frac{15}{L^7\eta^5}\right)\sin 2i \tag{H.18}$$

$$b_{92} = -\left(\frac{3\sin 2i}{16L^{11}\eta^8}\right)\left[\left(-180 + 336\eta + 189\eta^2\right)\right.$$
$$\left. - \left(1540 + 1260\eta + 135\eta^2\right)\cos^2 i\right]$$

$$c_{11} = -\left(\frac{21}{2L^8\eta^4}\right)\cos i \tag{H.19}$$

$$c_{12} = \left(\frac{33\cos i}{16L^{12}\eta^8}\right)\left[\left(5 - 12\eta - 9\eta^2\right) + \left(35 + 36\eta + 5\eta^2\right)\cos^2 i\right]$$

$$c_{21} = -\left(\frac{6}{L^7\eta^5}\right)\cos i \qquad\qquad\qquad\text{(H.20)}$$

$$c_{22} = \left(\frac{3\cos i}{8L^{11}\eta^9}\right)\left[\left(2042\eta - 27\eta^2\right) + \left(140 + 126\eta + 15\eta^2\right)\cos^2 i\right]$$

$$c_{31} = -\left(\frac{3}{2L^7\eta^4}\right)\sin i \qquad\qquad\qquad\text{(H.21)}$$

$$c_{32} = \left(\frac{3\sin i}{16L^{11}\eta^8}\right)\left[\left(5 - 12\eta - 9\eta^2\right) + 3\left(35 + 36\eta + 5\eta^2\right)\cos^2 i\right]$$

$$c_{41} = -\frac{8}{L}c_{11} = \left(\frac{84}{L^9\eta^4}\right)\cos i \qquad\qquad\text{(H.22)}$$

$$c_{42} = -\frac{12}{L}c_{12}$$
$$= -\left(\frac{99\cos i}{4L^{13}\eta^8}\right)\left[\left(5 - 12\eta - 9\eta^2\right) + \left(35 + 36\eta + 5\eta^2\right)\cos^2 i\right]$$

$$c_{51} = \left(\frac{30}{L^7\eta^6}\right)\cos i \qquad\qquad\qquad\text{(H.23)}$$

$$c_{52} = -\left(\frac{9\cos i}{8L^{11}\eta^{10}}\right)\left[\left(60 - 112\eta - 63\eta^2\right)\right.$$
$$\left. + \left(420 + 336\eta + 35\eta^2\right)\cos^2 i\right]$$

$$c_{61} = -\left(\frac{3}{2L^7\eta^4}\right)\cos i \qquad\qquad\qquad\text{(H.24)}$$

$$c_{62} = \left(\frac{3\cos i}{16L^{11}\eta^8}\right)\left[-\left(205 + 228\eta + 39\eta^2\right)\right.$$
$$\left. +9\left(35 + 36\eta + 5\eta^2\right)\cos^2 i\right]$$

$$c_{71} = \left(\frac{42}{L^5\eta^6}\right)\cos i \qquad\qquad\qquad\text{(H.25)}$$

$$c_{72} = -\left(\frac{33\cos i}{8L^{12}\eta^9}\right)\left[\left(20 - 42\eta - 27\eta^2\right) + \left(140 + 126\eta + 15\eta^2\right)\cos^2 i\right]$$

$$c_{81} = -\frac{7}{L}c_{31} = -\left(\frac{21}{2L^8\eta^4}\right)\sin i \qquad\qquad\text{(H.26)}$$

$$c_{82} = -\frac{11}{L}c_{32}$$
$$c_{82} = -\left(\frac{33\sin i}{16L^{12}\eta^8}\right)\left[\left(5 - 12\eta - 9\eta^2\right) + 3\left(35 + 36\eta + 5\eta^2\right)\cos^2 i\right]$$

$$c_{91} = \left(\frac{6}{L^7\eta^5}\right)\sin i \tag{H.27}$$

$$c_{92} = -\left(\frac{3\sin i}{8L^{11}\eta^9}\right)\left[\left(20 - 42\eta - 27\eta^2\right) + 3\left(140 + 126\eta + 15\eta^2\right)\cos^2 i\right]$$

References

[1] Sengupta, P. and Vadali, S. R., "Relative Motion and the Geometry of Formations in Keplerian Elliptic Orbits", *Journal of Guidance, Control, and Dynamics*, Vol. 30, No. 4, 2007, pp. 953–964.

[2] Sengupta, P., Vadali, S. R., and Alfriend, K. T., "Second-order State Transition for Relative Motion near Perturbed, Elliptic Orbits", *Celestial Mechanics and Dynamical Astronomy*, Vol. 97, February 2007, pp. 101–129.

[3] Vaddi, S. S., Alfriend, K. T., Vadali, S. R., and Sengupta, P., "Formation Establishment and Reconfiguration Using Impulsive Control", *Journal of Guidance, Control, and Dynamics*, Vol. 28, No. 2, March–April 2005, pp. 262–268.

[4] Vadali, S. R., Yan, H., and Alfriend, K. T., "Formation Maintenance and Reconfiguration Using Impulsive Contol", *AIAA/AAS Astrodynamics Specialist Meeting,* August 2008, Honolulu, Hawaii, Paper AIAA-2008-7359.

[5] Breger, L. S., Inalhan, G., Tillerson, M., and How, J. P., "Cooperative Spacecraft Formation Flying: Model Predictive Control With Open- And Closed-Loop Robustness", in Gurfil, P. (Ed.) *Modern Astrodynamics*, 2006.

[6] Ren, W. and Beard, R., "Virtual Structure Based Spacecraft Formation Control with Formation Feedback", *Proceedings of the AIAA Guidance, Navigation, and Control Conference*, August 2002.

[7] Tillerson, M., Breger, L. S., and How, J. P., "Distributed Coordination and Control of Formation Flying Spacecraft", *Proceedings of the American Control Conference*, June 2003.

[8] Mccamish, S. B., Romano, M., and Yun, X., "Autonomous Distributed Control Algorithm for Multiple Spacecraft in Close Proximity Operations", *Proceedings of the AIAA Guidance, Navigation, and Control Conference*, August 2007, Paper 2007–6857.

[9] Szu, H., Xi, N., Liou, W. W., and Ro, K., "Nanotechnology Applied to Aerospace and Aeronautics: Swarming", *Proceedings of AIAA Infotech@Aerospace, September 2005, Paper 2005 6933.*

[10] Carter, T. and Humi, M., "Clohessy–Wiltshire Equations Modified to Include Quadratic Drag", *Journal of Guidance, Control, and Dynamics*, Vol. 25, No. 6, November–December 2002, pp. 1058–1063.

[11] Humi, M. and Carter, T., "Closed-Form Solutions for Near-Circular Arcs with Quadratic Drag", *Journal of Guidance, Control, and Dynamics*, Vol. 29, No. 3, May–June 2006, pp. 513–518.

[12] Vallado, D. A., *Fundamentals of Astrodynamics and Applications*, Microcosm, Second Edition, 2001.

[13] Beichman, C., "The Terrestrial Planet Finder – The Search for Life-Bearing Planets around Other Stars", *SPIE Proceedings*, Vol. 3350, No. 3, 1998, pp. 719–723.

[14] Bauer, F., Bristow, J., Folta, D., Hartman, K., Quinn, D., and How, J. P., "Satellite Formation Flying Using an Innovative Autonomous Control System (AutoCon) Environment", *Proceedings of the AIAA Guidance, Navigation, and Control Conference*, 1997, pp. 657–666.

[15] Scharf, D., Ploen, S., and Hadaegh, F., "A Survey of Spacecraft Formation Flying Guidance and Control (Part I): Guidance", *Proceedings of the American Control Conference*, June 2003.

[16] Martin, M. and Stallard, M., "Distributed Satellite Missions and Technologies – The TechSat 21 Program", *Proceedings of the AIAA Space Technology Conference and Exposition*, Albuquerque, NM, September 1999, AIAA Paper 99–4479.

[17] Shah, N. H., *"Automated Station-Keeping for Satellite Constellations"*, S. M. Thesis, Massachusetts Institute of Technology, Cambridge, Massachusetts, 1997.

[18] Smith, J. E., *"Application of Optimization Techniques to the Design and Maintenance of Satellite Constellations"*, S. M. Thesis, Massachusetts Institute of Technology, Cambridge, Massachusetts, 1999.

[19] Scharf, D., Ploen, S., and Hadaegh, F., "A Survey of Spacecraft Formation Flying Guidance and Control (Part II): Control", *American Control Conference*, June 2004, pp. 2976–2985.

[20] Bristow, J., Folta, D., and Hartman, K., "A Formation Flying Technology Vision", *Proceedings of AIAA Space 2000 Conference and Exposition, Long Beach, CA*, September 2000, AIAA Paper 2000–5194.

[21] Sabol, C., Burns, R., and Mclaughlin, C. A., "Satellite Formation Flying Design and Evolution", *Journal of Spacecraft and Rockets*, Vol. 38, No. 2, March–April 2001, pp. 270–278.

[22] Carpenter, J. R., Leitner, J. A., Folta, D., and Burns, R., "Benchmark Problems For Spacecraft Formation Flying Missions", *Proceedings of the AIAA Guidance, Navigation, and Control Conference*, August 2003, AIAA Paper 2003–5364.

[23] Vassar, R. H. and Sherwood, R. B., "Formationkeeping for a Pair of Satellites in a Circular Orbit", *Journal of Guidance, Control, and Dynamics*, Vol. 8, No. 2, March–April 1985, pp. 235–242.

[24] Kapila, V., Sparks, A. G., Buffington, J. M., and Yan, Q., "Spacecraft Formation Flying: Dynamics and Control", *Proceedings of the American Control Conference, San Diego, CA*, June 1999, pp. 4137–4141.

[25] Vadali, S. R., Vaddi, S. S., and Alfriend, K. T., "An Intelligent Control Concept for Formation Flying Satellites", *International Journal of Robust and Nonlinear Control*, Vol. 12, No. 2–3, February–March 2002, pp. 97–115.

[26] Tillerson, M., Inalhan, G., and How, J. P., "Coordination and Control of Distributed Spacecraft Systems Using Convex Optimization Techniques", *International Journal of Robust and Nonlinear Control*, Vol. 12, No. 2–3, February–March 2002, pp. 207–242.

[27] Mishne, D., "Formation Control of LEO Satellites Subject to Drag Variations and J_2 Perturbations", *Proceedings of the Astrodynamics Specialist Conference, Monterey, California*, August 2002.

[28] Gurfil, P., "Control-Theoretic Analysis of Low-Thrust Orbital Transfer Using Orbital Elements", *Journal of Guidance, Control, and Dynamics*, Vol. 26, No. 6, November–December 2003, pp. 979–983.

[29] Schaub, H. and Junkins, J. L., *Analytical Mechanics of Space Systems, AIAA Education Series*, AIAA, Reston, VA, 2003.

[30] Naasz, B., "*Classical Element Feedback Control for Spacecraft Orbital Maneuvers*", S. M. Thesis, Deptartment of Aerospace Engineering, Virginia Polytechnic Institute and State University, 2002.

[31] Sengupta, P. and Vadali, S. R., "Satellite Orbit Transfer and Formation Reconfiguration via an Attitude Control Analogy", *Journal of Guidance, Control, and Dynamics*, Vol. 28, No. 6, November–December 2005, pp. 1200–1209.

[32] Schaub, H. and Alfriend, K. T., "J_2 Invariant Relative Orbits for Formation Flying", *Celestial Mechanics and Dynamical Astronomy*, Vol. 79, No. 2, February 2001, pp. 77–95.

[33] Battin, R. H., *An Introduction to the Mathematics and Methods of Astrodynamics*, AIAA, Reston, VA, 1999.

[34] Schaub, H. and Alfriend, K. T., "Impulse Feedback Control to Establish Specific Mean Orbit Elements of Spacecraft Formations", *Journal of*

Guidance, Control, and Dynamics, Vol. 24, No. 4, July–August 2001, pp. 739–745.

[35] Ilgen, M. R., "Low Thrust OTV Guidance using Lyapunov Optimal Feedback Control Techniques", *Advances in the Astronautical Sciences*, Vol. 85, No. 2, August 1993, pp. 527–1545, Also Paper AAS 93-680 of the AAS/AIAA Astrodynamics Specialists Conference.

[36] Gurfil, P., Idan, M., and Kasdin, N. J., "Neural Adaptive Control for Deep-Space Formation Flying", *Journal of Guidance, Control, and Dynamics*, Vol. 26, No. 3, March–April 2003, pp. 491–501.

[37] Gurfil, P. and Mishne, D., "Cyclic Spacecraft Formations: Relative Motion Control Using Line-of-Sight Measurements Only", *Journal of Guidance, Control, and Dynamics*, Vol. 30, No. 1, 2007, pp. 214–225.

[38] Tillerson, M. and How, J. P., "Analysis of the Impact of Sensor Noise on Formation Flying Control", *Proceedings of the 2001 American Control Conference*, Arlington, VA, June 25–27, 2001.

[39] Tillerson, M. and How, J. P., "Formation Flying Control in Eccentric Orbits", *Proceedings of the AIAA Guidance, Navigation, and Control Conference*, Montreal, Canada, August 2001.

[40] Tillerson, M. and How, J. P., "Advanced Guidance Algorithms for Spacecraft Formation Flying", *Proceedings of the American Control Conference*, May 2002, pp. 2830–2835.

[41] Hofmann-Wellenhof, B., Lichtenegger, H., and Collins, J., *Global Positioning System Theory and Practice*, Springer-Verlag, 1994.

[42] Parkinson, B. W., Spilker, J. J., and Enge, P., *Global Positioning System: Theory and Applications, Volumes 1 and 2*, American Institute of Aeronautics and Astronautics, 1996.

[43] Busse, F., *"Precise Formation-State Estimation in Low Earth Orbit Using Carrier Differential GPS"*, PhD Dissertation, Stanford University, Deptartment of Aeronautics and Astronautics, 2003.

[44] Bate, R. R., Mueller, D. D., and White, J. E., *Fundamentals of Astrodynamics*, Dover, 1971.

[45] Hintz, G., "Survey of Orbit Element Sets", *Journal of Guidance, Control, and Dynamics*, Vol. 31, No. 3, May–June 2008, pp. 785–790.

[46] Deprit, A. and Rom, A., "The Main Problem of Artificial Satellite Theory for Small and Moderate Eccentricities", *Celestial Mechanics*, Vol. 2, 1970, pp. 166–206.

[47] Broucke, R. A., "Solution of the Elliptic Rendzevous Problem with the Time as Independent Variable", *Journal of Guidance, Control, and Dynamics*, Vol. 26, No. 4, 2003, pp. 615–621.

[48] Cohen, C. J. and Hubbard, E. C., "A Nonsingular Set of Orbital Elements", *The Astronomical Journal*, Vol. 67, No. 1, February 1962, pp. 10–15.

[49] Walker, M. J. H., Ireland, B., and Owens, J., "A Set of Modified Equinoctial Orbit Elements", *Celestial Mechanics*, Vol. 36, 1985, pp. 409–419.

[50] Gurfil, P., "Euler Parameters as Natural Nonsingular Orbital Elements in Near-Equatorial Orbits", *Journal of Guidance, Control, and Dynamics*, Vol. 28, No. 5, September–October 2005, pp. 1079–1084.

[51] Junkins, J. L. and Turner, J. D., "On the Analogy Between Orbital Dynamics and Rigid Body Dynamics", *Journal of the Astronautical Sciences*, Vol. 27, No. 4, October–December 1979, pp. 345–358.

[52] Kane, T. R., Likins, P. W., and Levinson, D. A., *Spacecraft Dynamics*, McGraw Hill, New York, 1983.

[53] Kustaanheimo, P. and Stiefel, E., "Perturbation Theory of Kepler Motion based on Spinor Regularization", *J. Reine Angew. Math*, Vol. 218, 1965, p. 204.

[54] Stiefel, E. and Scheifele, G., *Linear and Regular Celestial Mechanics*, Springer, New York, 1971.

[55] Arakida, H. and Fukushima, T., "Long-Term Integration Error of Kustaanheimo-Stiefel Regularized Orbital Motion", *Astronomical Journal*, Vol. 120, 2000, pp. 3333–3339.

[56] Euler, L., *Opera Mechanica et Astronomica*, Birkhauser-Verlag, Switzerland, 1999.

[57] Lagrange, J. L., "Sur la Théorie Générale de la Variation des Constants Arbitraries dans tous les Problèmes de la Méchanique", *Lu*, Vol. le 13 Mars 1809 à l'Institut de France, 1809.

[58] Newman, W. and Efroimsky, M., "The Method of Variation of Constants and Multiple Time Scales in Orbital Mechanics", *Chaos*, Vol. 13, 2003, pp. 476–485.

[59] Brouwer, D. and Clemence, G. M., *Methods of Celestial Mechanics*, Academic Press, New York, 1961.

[60] King-Hele, D. G., "The Effect of the Earth's Oblateness on the Orbit of a Near Satellite", *Proceedings of the Royal Society of London. Series A, Mathematical and Physical Scinces*, Vol. 247, No. 1248, September 1958, pp. 49–72.

[61] Cook, A. H., "The Contribution of Observations of Satellites to the Determination of the Earth's Gravitational Potential", *Space Science Reviews*, Vol. 2, 1963, pp. 355–437.

[62] Efroimsky, M., "Equations for the Orbital Elements. Hidden Symmetry", *Preprint No. 1844 of the Institute of Mathematics and its Applications, University of Minnesota, USA,* http://www.ima.umn.edu/preprints/feb02/1844.pdf, 2002.

[63] Efroimsky, M. and Goldreich, P., "Gauge Symmetry of the N-body Problem in the Hamilton–Jacobi Approach", *Journal of Mathematical Physics*, Vol. 44, 2003, pp. 5958–5977.

[64] Efroimsky, M. and Goldreich, P., "Gauge Freedom in the N-body Problem of Celestial Mechanics", *Astronomy & Astrophysics*, Vol. 415, 2004, pp. 1187–1199.

[65] Vinti, J. P., "Theory of the Orbit of an Artificial Satellite with Use of Spheroidal Coordinates", *Astronomical Journal*, Vol. 65, 1960, pp. 353–354.

[66] Brouwer, D., "Solution of the Problem of Artificial Satellite Theory without Drag", *The Astronomical Journal*, Vol. 64, November 1959, pp. 378–397.

[67] Kozai, Y., "The Motion of a Close Earth Satellite", *Astronomical Journal*, Vol. 64, 1959, pp. 367–377.

[68] Kozai, Y., "Second-Order Solution of Artificial Satellite Theory without Air Drag", *Astronomical Journal*, Vol. 67, No. 7, 1962, pp. 446–461.

[69] Brumberg, V. A., *Essential Relativistic Celestial Mechanics*, Taylor and Francis, Florida, USA, 1991.

[70] Sanders, J. A. and Verhulst, F., *Averaging Methods in Nonlinear Dynamical Systems*, Springer-Verlag, New York, 1985.

[71] Marsden, J. E. and Ratiu, T. S., *Introduction to Mechanics and Symmetry*, Springer, Second Edition, 1999.

[72] Hori, G., "Theory of General Perturbations with Unspecified Canonical Variables", *Publications of the Astronomical Society of Japan*, Vol. 18, 1966, pp. 287–295.

[73] Nayfey, A. H., *Perturbation Methods*, Wiley, 1973.

[74] Coffey, S. and Deprit, A., "Third-Order Solution to the Main Problem in Satellite Theory", *Journal of Guidance, Control, and Dynamics*, Vol. 5, No. 4, October–November 1980, pp. 366–371.

[75] Gim, D.-W. and Alfriend, K. T., "State Transition Matrix of Relative Motion for the Perturbed Noncircular Reference", *Journal of Guidance, Control, and Dynamics*, Vol. 26, No. 6, November–December 2003, pp. 956–971.

[76] Lyddane, R. H., "Small Eccentricities or Inclinations in the Brouwer Theory of the Artificial Satellite", *Astronomical Journal*, Vol. 68, No. 8, 1963, pp. 555–558.

[77] Haykin, S., *Neural Networks: A Comprehensive Foundation*, Prentice Hall, Third Edition, 2008.

[78] Khalil, H. K., *Nonlinear Systems*, Prentice-Hall, NJ, 1996.

[79] Lewis, F. L. and Syrmos, V. L., *Optimal Control*, 2nd ed., 1995.

[80] Brown, R. and Hwang, P., *Introduction to Random Signals and Applied Kalman Filtering*, Wiley, 1997.

[81] Gelb, A., *Applied Optimal Estimation*, MIT Press, 1974.

[82] Bucy, R. and Joseph, P., *Filtering for Stochastic Processes with Applications to Guidance*, Interscience Publishers, New York, 1968.

[83] Julier, S. and Uhlmann, J., "Unscented Filtering and Nonlinear Estimation", *Proceedings of the IEEE*, Vol. 92, No. 3, March 2004, pp. 401–422.

[84] Julier, S. and Uhlmann, J., "A New Method for the Nonlinear Transformation on Means and Covariances in Filters and Estimators", *IEEE Transactions on Automatic Control*, Vol. 45, No. 3, March 2000, pp. 477–482.

[85] Lefebvre, T., Bruyninckx, H., and Schutter, J. D., "Comment on 'A New Method for the Nonlinear Transformation of Means and Covariances in Filters and Estimators'", *IEEE Transactions on Automatic Control*, Vol. 47, No. 8, August 2002, pp. 1406–1408.

[86] Julier, S. and Uhlmann, J., "Authors' Reply to Comment on 'A New Method for the Nonlinear Transformation of Means and Covariances in Filters and Estimators'", *IEEE Transactions on Automatic Control*, Vol. 47, No. 8, August 2002, pp. 1408–1409.

[87] Julier, S. and Uhlmann, J., "A New Extension of the Kalman Filter to Nonlinear Systems", *SPIE AeroSense Symposium*, April 1997.

[88] Haykin, S., *Kalman Filtering and Neural Networks*, Wiley, 2001.

[89] van der Merwe, R. and Wan, E., The Square-Root Unscented Kalman Filter for State and Parameter Estimation, *International Conference on Acoustics, Speech, and Signal Processing*, May 2001.

[90] Gurfil, P., "Relative Motion Between Elliptic Orbits: Generalized Boundedness Conditions and Optimal Formationkeeping", *Journal of Guidance, Control, and Dynamics*, Vol. 28, No. 4, July–August 2005, pp. 761–767.

[91] Leitner, J., "A Hardware-in-the-loop Testbed for Spacecraft Formation Flying Applications", *Proceedings of the 2001 IEEE Aerospace Conference*, March 2001, Big Sky, MT, Vol. 2, pp. 615–620.

[92] Szebehely, V. and Giacaglia, G. E. O., "On the Elliptic Restricted Problem of Three Bodies", *Astronomical Journal*, Vol. 69, 1964, pp. 230–235.

[93] Gurfil, P. and Kasdin, N. J., "Canonical Modelling of Coorbital Motion in Hill's Problem using Epicyclic Orbital Elements", *Astronomy & Astrophysics*, Vol. 409, 2003, pp. 1135–1140.

[94] Kasdin, N. J., Gurfil, P., and Kolemen, E., "Canonical Modelling of Relative Spacecraft Motion via Epicyclic Orbital Elements", *Celestial Mechanics and Dynamical Astronomy*, Vol. 92, No. 4, August 2005, pp. 337–370.

[95] Palmer, P. and İmre, E., "Relative Motion Between Satellites on Neighboring Keplarian Orbits", *Journal of Guidance, Control, and Dynamics*, Vol. 30, No. 2, March–April 2007, pp. 521–528.

[96] Kaplan, M. H., *Modern Spacecraft Dynamics and Control*, John Wiley, New York, 1976.

[97] Irigoyen, M. and Simo, C., "Non Integrability of the J_2 Problem", *Celestial Mechanics and Dynamical Astronomy*, Vol. 55, 1993, pp. 281–287.

[98] Celletti, A. and Negrini, P., "Non-integrability of the Problem of Motion around an Oblate Planet", *Celestial Mechanics and Dynamical Astronomy*, Vol. 61, 1995, pp. 253–260.

[99] Vadali, S. R., Schaub, H., and Alfriend, K. T., "Initial Conditions and Fuel-Optimal Control for Formation Flying of Satellites", *Proceedings of the AIAA GNC Conference*, AIAA, Portland, OR, August 1999, AIAA 99–4265.

[100] Gim, D.-W. and Alfriend, K. T., "Satellite Relative Motion using Differential Equinoctial Elements", *Celestial Mechanics and Dynamical Astronomy*, Vol. 92, No. 4, August 2005, pp. 295–336.

[101] Clohessy, W. and Wiltshire, R., "Terminal Guidance System for Satellite Rendezvous", *Journal of the Astronautical Sciences*, Vol. 27, No. 9, 1960, pp. 653–678.

[102] Tschauner, J. and Hempel, P., "Optimale Beschleunigeungsprogramme fur das Rendezvous-Manover", *Astronautica Acta*, Vol. 10, 1964, pp. 296–307.

[103] Ross, I. M., "Space Trajectory Optimization and L^1-Optimal Control Problems", in Gurfil P., (Ed.) *Modern Astrodynamics*, 2006, pp. 155–186.

[104] Bond, V. R., "A New Solution for the Rendezvous Problem", *Advances in the Astronautical Sciences*, Vol. 102, 1999, pp. 1115–1144.

[105] Mitchell, J. W. and Richardson, D. L., "Invariant Manifold Tracking for First-Order Nonlinear Hill's Equations", *Journal of Guidance, Control, and Dynamics*, Vol. 26, No. 4, 2003, pp. 622–627.

[106] Vaddi, S. S., Vadali, S. R., and Alfriend, K. T., "Formation Flying: Accommodating Nonlinearity and Eccentricity Perturbations", *Journal of Guidance, Control, and Dynamics*, Vol. 26, No. 2, 2003, pp. 214–223.

[107] Sengupta, P., Sharma, R., and Vadali, S. R., "Periodic Relative Motion Near a Keplerian Elliptic Orbit with Nonlinear Differential Gravity", *Journal of Guidance, Control, and Dynamics*, Vol. 29, No. 5, 2006, pp. 1110–1121.

[108] Jiang, F., Li, J., and Baoyin, H., "Approximate Analysis for Relative Motion of Satellite Formation Flying in Elliptical Orbits", *Celestial Mechanics and Dynamical Astronomy*, Vol. 98, No. 1, 2007, pp. 31–66.

[109] Melton, R. G., "Time Explicit Representation of Relative Motion Between Elliptical Orbits", *Journal of Guidance, Control, and Dynamics*, Vol. 23, No. 4, July–August 2000, pp. 604–610.

[110] Lawden, D. F., *Optimal Trajectories for Space Navigation*, Butterworths, London, 1963.

[111] de Vries, J. P., "Elliptic Elements in Terms of Small Increments of Position and Velocity Components", *AIAA Journal*, Vol. 1, No. 11, November 1963, pp. 2626–2629.

[112] Carter, T. E., "State Transition Matrices for Terminal Rendezvous Studies: Brief Survey and New Examples", *Journal of Guidance, Control, and Dynamics*, Vol. 21, No. 1, January–February 1998, pp. 148–155.

[113] Yamanaka, K. and Ankersen, F., "New State Transition Matrix for Relative Motion on an Arbitrary Elliptical Orbit", *Journal of Guidance, Control, and Dynamics*, Vol. 25, No. 1, January–February 2002, pp. 60–66.

[114] Broucke, R. A., "Solution of the Elliptic Rendezvous Problem with the Time as Independent Variable", *Journal of Guidance, Control, and Dynamics*, Vol. 26, No. 4, July–August 2003, pp. 615–621.

[115] Lee, D., Cochran, J. E., and Lo, J. H., "Solutions to the Variational Equations for Relative Motion of Satellites", *Journal of Guidance, Control, and Dynamics*, Vol. 30, No. 3, 2007, pp. 669–678.

[116] Nazarenko, A. I., "State Transition Matrix of Relative Motion for the Noncircular Orbit: Relation with Partial-derivative Matrix in the Satellite Coordinate System", *International Astronatical Congress*, No. IAC-06-C1.6.08, Valencia, Spain, October 2006.

[117] Inalhan, G., Tillerson, M. J., and How, J., "Relative Dynamics and Control of Spacecraft Formations in Eccentric Orbits", *Journal of Guidance, Control, and Dynamics*, Vol. 25, No. 1, 2002, pp. 48–59.

[118] Hill, G. W., "Researches in the Lunar Theory", *American Journal of Mathematics*, Vol. 1, 1878, pp. 5–26.

[119] Schaub, H., "Relative Orbit Geometry Through Classical Orbit Element Differences", *Journal of Guidance, Control, and Dynamics*, Vol. 27, No. 1, September–October 2004, pp. 839–848.

[120] Karlgaard, C. D. and Lutze, F. H., "Second-Order Relative Motion Equations", *Journal of Guidance, Control, and Dynamics*, Vol. 26, No. 1, January–February 2003, pp. 41–49.

[121] Gurfil, P. and Kasdin, N. J., "Nonlinear Modeling of Spacecraft Relative Motion in the Configuration Space", *Journal of Guidance, Control, and Dynamics*, Vol. 27, No. 1, January–February 2004, pp. 154–157.

[122] Gurfil, P. and Kholshevnikov, K. V., "Manifolds and Metrics in the Relative Spacecraft Motion Problem", *Journal of Guidance, Control, and Dynamics*, Vol. 29, No. 4, July–August 2006, pp. 1004–1010.

[123] Yan, H., Sengupta, P., Vadali, S. R., and Alfriend, K. T., "Development of a State Transition Matrix for Relative Motion Using the Unit Sphere Approach", *AAS/AIAA Space Flight Mechanics Conference*, 2004, Paper No. AAS 04-163.

[124] Vadali, S. R., "An Analytical Solution for Relative Motion of Satellites", *Dynamics and Control of Systems and Structures in Space 2002*, Cranfield University, Cranfield, UK, July 2002, pp. 309–316.

[125] Sengupta, P., Vadali, S. R., and Alfriend, K. T., "Averaged Relative Motion and Applications to Formation Flight near Perturbed Orbits", *Journal of Guidance, Control, and Dynamics*, Vol. 31, No. 2, 2008, pp. 258–272.

[126] Hill, K., Sabol, C., Luu, K., Murai, M., and Mclaughlin, C., "Relative Orbit Trajectories of Geosynchronous Satellites using the COWPOKE Equations", *Proceedings of the 6th US Russian Space Surveillance Workshop*, St. Petersburg, Russia, August 2005, pp. 274–285.

[127] Sabol, C., Mclaughlin, C., and Luu, K., "Meet the Cluster Orbits With Perturbations Of Keplerian Elements (COWPOKE) Equations", *AAS/AIAA Spaceflight Mechanics Conference*, Ponce, Puerto Rico, February 2003, Paper No. AAS-03-138.

[128] Balaji, S. K. and Tatnall, A., "Relative Trajectory Analysis of Dissimilar Formation Flying Spacecraft", *AAS/AIAA Spaceflight Mechanics Conference*, Ponce, Puerto Rico, February 2003, Paper No. AAS-03-134.

[129] Junkins, J. L. and Turner, J., *Optimal Spacecraft Rotational Maneuvers*, Elsevier, New York, 1986.

[130] Garrison, J. L., Gardner, T. G., and Axelrad, P., "Relative Motion in Highly Elliptical Orbits", *Advances in the Astronautical Sciences*, Vol. 89, 1995, pp. 1359–1376.

[131] Gim, D.-W., "*A Precise Analytic Solution to the Relative Motion of Formation Flying Satellites*", PhD Thesis, Texas A&M University, December 2002.

[132] Gim, D.-W. and Alfriend, K. T., "State Transition Matrix of Relative Motion for the Perturbed Noncircular Reference Orbit", *Journal of Guidance, Control, and Dynamics*, Vol. 26, No. 6, November–December 2003, pp. 956–971.

[133] Kechichian, J. A., "Motion in General Elliptic Orbits with Respect to a Dragging and Precessing Coordinate Frame", *Journal of the Astronautical Sciences*, Vol. 46, No. 1, 1998, pp. 25–46.

[134] Hamel, J. and de Lafontaine, J., "Linearized Dynamics of Formation Flying Spacecraft on a J_2-perturbed Elliptical Orbit", *Journal of Guidance, Control, and Dynamics*, Vol. 30, No. 6, November–December 2007, pp. 1649–1658.

[135] Born, G. H., Goldstein, D. B., and Thompson, B., "An Analytical Theory for Orbit Determination", *Journal of the Astronautical Sciences*, Vol. 49, No. 2, April–June 2001, pp. 345–361.

[136] Vadali, S. R., "A Model for J_2-perturbed Linear Relative Motion about Mean Circular Orbits", *Journal of Guidance, Control, and Dynamics*, Vol. 32, No. 5, September–October 2009, pp. 1687–1691.

[137] Izzo, D., Sabatini, M., and Valente, C., "A New Linear Model Describing Formation Flying Dynamics under J2 Effects", *Proceedings of the 17th AIDAA National Congress*, Vol. 1, 2003, pp. 493–500.

[138] Sabatini, M. and Palmerini, G. B., "Linearized Formation-Flying Dynamics in a Perturbed Orbital Environment", *Proceedings of the IEEE Aerospcae Conference*, March 2008.

[139] Vadali, S. R., Sengupta, P., Yan, H., and Alfriend, K. T., "Fundamental Frequencies of Satellite Relative Motion and Control of Formations", *Journal of Guidance, Control, and Dynamics*, Vol. 31, No. 5, 2008, pp. 1239–1248.

[140] Wnuk, E. and Golebiewska, J., "The Relative Motion of Earth Orbiting Satellites", *Celestial Mechanics and Dynamical Astronomy*, Vol. 91, No. 3–4, March 2005, pp. 373–389.

[141] İmre, E. and Palmer, P., "High-Precision Symplectic Numerical, Relative Orbit Propagation", *Journal of Guidance, Control, and Dynamics*, Vol. 30, No. 4, July–August 2007, pp. 965–973.

[142] Wiesel, W. E., "Relative Satellite Motion About an Oblate Planet", *Journal of Guidance, Control, and Dynamics*, Vol. 25, No. 4, July–August 2002, pp. 776–785.

[143] Vadali, S. R., Alfriend, K. T., and Vaddi, S. S., "Hill's Equations Mean Orbital Elements and Formation Flying of Satellites", *Advances in the Astronautical Sciences*, Vol. 106, March 2000, pp. 187–201, Also Paper AAS 00-258 of the Richard H. Battin Astrodynamics Symposium.

[144] Sengupta, P., Vadali, S. R., and Alfriend, K. T., "Orbit Design and Maintenance for Satellites Accessing a Terrestrial Target", *Journal of Spacecraft and Rockets*, to appear, 2009, doi:10.2514/1.44120.

[145] London, H. S., "Second Approximation to the Solution of the Rendezvous Equations", *AIAA Journal*, Vol. 1, No. 7, July 1963, pp. 1691–1693.

[146] Mitchell, J. W. and Richardson, D. L., "A Third Order Analytical Solution for Relative Motion with a Circular Reference Orbit", *Journal of the Astronautical Sciences*, Vol. 51, No. 1, January–March 2003, pp. 1–12.

[147] Vaddi, S. S., Vadali, S. R., and Alfriend, K. T., "Formation Flying: Accommodating Nonlinearity and Eccentricity Perturbations", *Journal of Guidance, Control, and Dynamics*, Vol. 26, No. 2, March–April 2003, pp. 214–223.

[148] Karlgaard, C. D. and Lutze, F. H., "Second-Order Relative Motion Equations", *Journal of Guidance, Control, and Dynamics*, Vol. 26, No. 1, January–February 2003, pp. 41–49.

[149] Junkins, J. L., Akella, M. L., and Alfriend, K. T., "Non-Gaussian Error Propagation in Orbital Mechanics", *Journal of the Astronautical Sciences*, Vol. 44, No. 4, October–December 1996, pp. 541–563.

[150] Yan, H., *"Dynamics and Real-Time Optimal Control of Satellite and Satellite Formation Systems"*, PhD Thesis, Texas A&M University, August 2006.

[151] Junkins, J. L. and Singla, P., "How Nonlinear Is It? A Tutorial on Nonlinearity of Orbit and Attitude Dynamics", *Journal of the Astronautical Sciences*, Vol. 52, No. 1–2, January–June 2004, pp. 7–60.

[152] Vadali, S. R., Vaddi, S. S., and Alfriend, K. T., "A New Concept for Controlling Formation Flying Satellite Constellations", *Advances in the Astronautical Sciences*, Vol. 108, No. 2, February 2001, pp. 1631–1648. Also Paper AAS 01-218 of the AAS/AIAA Space Flight Mechanics Meeting.

[153] Vadali, S. R., Vaddi, S. S., Naik, K., and Alfriend, K. T., "Control of Satellite Formations", *Proceedings of the AIAA Guidance, Navigation, and Control Conference*, August 2001, Paper AIAA-2001-4028.

[154] Sabatini, M., Izzo, D., and Bevilacqua, R., "Special Inclinations Allowing Minimal Drift Orbits for Formation Flying Satellites", *Journal of*

Guidance, Control, and Dynamics, Vol. 31, No. 1, January–February 2008, pp. 94–100.

[155] Izzo, D. and Sabatini, M., "Magic (Special) Inclinations for Formation Flying", *Proceedings of the 3rd International Symposium on Formation Flying, Missions and Technologies*, ESTEC, 2009.

[156] Vadali, S. R., Vaddi, S. S., and Alfriend, K. T., "An Intelligent Control Concept for Formation Flying Satellite Constellations", *International Journal of Nonlinear and Robust Control*, Vol. 12, 2002, pp. 97–115.

[157] Williams, T. and Wang, Z.-S., "Uses of Solar Radiation Pressure for Satellite Formation Flight", *International Journal of Robust and Nonlinear Control*, Vol. 12, No. 2–3, February–March 2002, pp. 163–183.

[158] Pan, H. and Kapila, V., "Adaptive Nonlinear Control for Spacecraft Formation Flying with Coupled Translational and Attitude Dynamics", *Proceedings of the 40th IEEE Conference on Decision and Control*, Orlando, Florida, USA, December 2001, pp. 2057–2062.

[159] Kim, S., Crassidis, J. L., Cheng, Y., Fosbury, A. M., and Junkins, J. L., "Kalman Filtering for Relative Spacecraft Attitude and Position Estimation", *Journal of Guidance, Control, and Dynamics*, Vol. 30, No. 1, January–February 2007, pp. 133–143.

[160] Ploen, S. R., Hadaegh, F. Y., and Scharf, D. P., "Rigid Body Equations of Motion for Modeling and Control of Spacecraft Formations – Part 1: Absolute Equations of Motion," *Proceedings of the 2004 American Control Conference*, Boston, Massachusetts, June 2004, pp. 3646–3653.

[161] Beard, R. W., Lawton, J., and Hadaegh, F. Y., "A Coordination Architecture for Spacecraft Formation Control", *IEEE Transactions on Control Systems Technology*, Vol. 9, No. 6, 2001, pp. 777–790.

[162] Landau, L. D. and Lifshitz, E. M., *Mechanics*, Pergamon Press, 1960.

[163] Segal, S. and Gurfil, P., "Effect of Kinematic Rotation-Translation Coupling on Relative Spacecraft Translational Dynamics", *Journal of Guidance, Control and Dynamics*, Vol. 32, No. 3, May–June 2009, pp. 1045–1050.

[164] Wertz, J. R., *Spacecraft Attitude Determination and Control*, Kluwer, 1997.

[165] de Queiroz, M. S., Kapila, V., and Yan, Q., "Adaptive Nonlinear Control of Multiple Spacecraft Formation Flying", *Journal of Guidance, Control, and Dynamics*, Vol. 23, No. 3, May–June 2000, pp. 385–390.

[166] Vaddi, S. S. and Vadali, S. R., "Linear and Nonlinear Control Laws for Formation Flying", *Proceedings of the AAS/AIAA Space Flight Mechanics Conference*, February 2003.

[167] Starin, S. R., Yedavalli, R. K., and Sparks, A. G., "Spacecraft Formation Flying Maneuvers Using Linear Quadratic Regulation with No Radial Axis Inputs", *AIAA Guidance, Navigation and Control Conference and Exhibit*, No. AIAA-2001-4029, AIAA, Montreal, Canada, August 2001.

[168] Vassar, R. H. and Sherwood, R. B., "Formationkeeping for a Pair of Satellites in a Circular Orbit", *Journal of Guidance, Control, and Dynamics*, Vol. 8, No. 2, March–April 1985, pp. 235–242.

[169] Ulybyshev, Y., "Long-Term Formation Keeping of Satellite Constellation Using Linear Quadratic Controller", *Journal of Guidance, Control, and Dynamics*, Vol. 21, No. 1, January–February 1998, pp. 109–115.

[170] Kumar, R. and Seywald, H., "Fuel-optimal Station Keeping via Differential Inclusions", *Journal of Guidance, Control, and Dynamics*, Vol. 18, No. 5, September–October 1995, pp. 1156–1162.

[171] Gill, P. E., Murray, W., and Saunders, M. A., *User's Guide to SNOPT Version 7: Software for Large-Scale Nonlinear Programming*, Department of Mathematics, University of California, San Diego, La Jolla, CA 92093-0112, June 2008.

[172] Breger, L. S. and How, J. P., "J_2-Modified GVE-Based MPC for Formation Flying in Space", *Proceedings of the AIAA Guidance, Navigation, and Control Conference*, August 2005.

[173] Tillerson, M., "*Coordination and Control of Multiple Spacecraft using Convex Optimization Techniques*", S. M. Thesis, Massachusetts Institute of Technology, Deptartment of Aeronautics and Astronautics, 2002.

[174] Busse, F. D. and How, J. P., "Real-Time Experimental Demonstration of Precise Decentralized Relative Navigation for Formation-Flying Spacecraft", *Proceedings of the AIAA Guidance, Navigation, and Control Conference*, August 2002, AIAA Paper 2002–5003.

[175] Leung, S. and Montenbruck, O., "Real-Time Navigation of Formation-Flying Spacecraft Using Global-Positioning-System Measurements", *Journal of Guidance, Control, and Dynamics*, Vol. 28, No. 2, March–April 2005, pp. 226–235.

[176] Gill, E., D'amico, S., and Montenbruck, O., "Autonomous Formation Flying for the PRISMA Mission", *Journal of Spacecraft and Rockets*, Vol. 44, No. 3, May–June 2007, pp. 671–681.

[177] Lam, D. C., Friberg, K. H., Brown, J. M., Sarani, S., and Lee, A. Y., "An Energy Burn Algorithm for Cassini Saturn Orbit Insertion", *Proceedings of the AIAA Guidance, Navigation, and Control Conference*, August 2005, AIAA Paper 2005–5994.

[178] Frauenholz, R. B., Bhat, R. S., Chesley, S. R., Mastrodemos, Jr. N., and Ryne, W. M. O., "Deep Impact Navigation System Performance",

Journal of Spacecraft and Rockets, Vol. 45, No. 1, January–February 2008, pp. 39–56.

[179] Stengel, R. F., *Optimal Control and Estimation*, Dover Publications, 1994.

[180] Tolson, R. H., Keating, G. M., Zurek, R. W., Bougher, S. W., Justus, C. G., and Fritts, D. C., "Application of Accelerometer Data to Atmospheric Modeling During Mars Aerobraking Operations", *Journal of Spacecraft and Rockets*, Vol. 44, No. 6, November–December 2007, pp. 1172–1179.

[181] Franklin, G. F., Powell, J. D., and Workman, M., *Digital Control of Dynamic Systems*, Addison-Wesley, 3rd Edition, 1998.

[182] Jeffrey, M. M., *"Closed-loop Control of Spacecraft Formations with Applications on SPHERES"*, S. M. Thesis, Massachusetts Institute of Technology, Deptartment of Aeronautics and Astronautics, 2008.

[183] Rudel, M. P. and Gurfil, P., "Precise Spacecraft Relative Positioning Using Single-Frequency Pseudorange Measurements", *Journal of Navigation*, Vol. 62, No. 1, January 2009, pp. 119–134.

[184] Misra, P. and Enge, P., *Global Positioning System – Signals, Measurements, and Performance*, Ganga–Jamuna Press, Lincoln, MA, 2001.

[185] Kaplan, E., *Understanding GPS Principles and Applications*, Artech House Publishers, Boston, MA, 1996.

[186] A.I. Solutions, *FreeFlyer User's Guide*, Version 4.0, 1999.

[187] Feess, W. A. and Stephens, S., "Evaluations of GPS Ionospheric Models", *IEEE Transactions on Aerospace and Electronic Systems*, Vol. AES-23, No. 3, 1987, pp. 332–338.

[188] "Spacecraft Parameters for Research on GNC Architecture for Future Spacecraft Formation Flight", *Mitsubishi Electric Corporation Internal Document*, February 2005.

[189] Dvorak, D., Rasmussen, R. D., Reeves, G., and Sacks, A., "Software Architecture Themes in JPL's Mission Data System", *Proceedings of the AIAA Space Technology Conference & Exposition*, September 1999, AIAA Paper 1999–4553.

[190] Ingham, M. D., "Robust Model-based Execution of Critical Spacecraft Sequences", *Proceedings of the AIAA 1st Intelligent Systems Technical Conference*, September 2004, AIAA Paper 2004–6352.

[191] Rasmussen, R. D., Ingham, M. D., and Dvorak, D., "Achieving Control and Interoperability through Unified Model-based Systems and Software Engineering", *Proceedings of AIAA Infotech@Aerospace*, September 2005, AIAA Paper 2005–6918.

[192] Ingham, M. D., Rasmussen, R. D., Bennett, M. B., and Moncada, A. C., "Engineering Complex Embedded Systems with State Analysis and the Mission Data System", *Journal of Aerospace Computing, Information, and Communication*, Vol. 2, No. 12, December 2005, pp. 507–536.

Index

Printed in the United States
By Bookmasters